AF411403

Comprehensive Research in Earthquake Forecasting and Seismic Hazard Assessment

Comprehensive Research in Earthquake Forecasting and Seismic Hazard Assessment

Editors

Alexey Dm. Zavyalov
Eleftheria E. Papadimitriou

Basel • Beijing • Wuhan • Barcelona • Belgrade • Novi Sad • Cluj • Manchester

Editors
Alexey Dm. Zavyalov
O.Yu. Schmidt Institute of
Physics of the Earth,
Russian Academy of Sciences,
Moscow, Russia

Eleftheria E. Papadimitriou
Aristotle University of Thessaloniki
Thessaloniki, Greece

Editorial Office
MDPI
St. Alban-Anlage 66
4052 Basel, Switzerland

This is a reprint of articles from the Special Issue published online in the open access journal *Applied Sciences* (ISSN 2076-3417) (available at: https://www.mdpi.com/journal/applsci/special_issues/earthquake_forecasting_seismic_hazard_reduction).

For citation purposes, cite each article independently as indicated on the article page online and as indicated below:

Lastname, A.A.; Lastname, B.B. Article Title. *Journal Name* **Year**, *Volume Number*, Page Range.

ISBN 978-3-0365-9480-4 (Hbk)
ISBN 978-3-0365-9481-1 (PDF)
doi.org/10.3390/books978-3-0365-9481-1

Cover image courtesy of Alexey Zavyalov

Contents

About the Editors

Alexey Zavyalov

Alexey Zavyalov graduated with honours from the Moscow Mining Institute in 1972 as a mining engineer-physicist and was sent to the O.Yu. Schmidt Institute of Physics of the Earth of the USSR Academy of Sciences, where he has been involved in expeditions and research since 1967. He has been PhD of Physical and Mathematical Sciences since 1985 and defended his thesis for the degree of Doctor of Physical and Mathematical Sciences in 2003.

Since 1981, his main fields of scientific interest have been earthquake physics, physics of seismic process and earthquake precursors, and development of algorithms for earthquake forecast by complex physically based features. In 1998, together with his colleagues A. Ponomarev and V. Smirnov, he was awarded the E.F. Savarensky Prize for the series of papers on "Structural Properties of Seismicity: Field Observations and Laboratory Simulation". A. Zavyalov is one of the authors involved in the discovery of the Round-the-World seismic echo effect.

He is the author of more than 200 scientific publications, including monographs, articles, and reports.

A. Zavyalov is a member of the editorial boards of a number of Russian and foreign journals, a member of the Bureau of the National Geophysical Committee of the Russian Academy of Sciences. From 2006 to 2013, he was the Chair of the Commission "Earthquake Source: Modelling and Monitoring for Forecasting" of the International Association for Seismology and Physics of the Earth's Interior (IASPEI). From 2011 to 2015, he was a member of the Executive Committee of IASPEI. In 2010, he was elected Vice-President of the European Seismological Commission (ESC). From 2012 to 2014, he was President of the ESC. He is currently the Titular Member of Russia in the ESC.

A. Zavyalov has led and implemented a number of Russian and international projects supported by the Russian Foundation for Basic Research, the Russian Ministry of Education and Science, and the governments of Greece, China and India.

Eleftheria E. Papadimitriou

Eleftheria E. Papadimitriou is a geologist and seismologist with a deep commitment to advancing the field of geophysics. With a Bachelor's degree in Geology and a Ph.D. specializing in Seismology, her career has been dedicated to furthering our understanding of earthquakes and seismic activity. Fluent in Greek, English, French, and Russian, she has fostered international collaborations and established connections with approximately 55 foreign institutions.

She has actively engaged in teaching seismology to undergraduate and postgraduate students, nurturing their passion for geophysics, mentoring and guiding the theses of undergraduate, master's, and Ph.D. students.

She is also a member of eight scientific societies and has actively contributed to 14 special committees and councils. In recognition of her dedication to advancing scientific knowledge, she currently serves as the Editor-in-Chief of Acta Geophysica, a prestigious publication managed by the Polish Academy of Sciences and Springer. Additionally, she holds the position of Associate Editor in five scientific peer-reviewed journals.

She also actively participated in more than 97 scientific meetings and made significant contributions to over 65 research projects, assuming scientific responsibility for 18 of them. With over 240 publications and more than 2200 citations (excluding self and co-authors), her research has made a lasting impact on the field of geophysics.

She also holds leadership positions, including serving on boards of directors and as the Head of the Department and Laboratory of Geophysics at the School of Geology, Aristotle University of Thessaloniki.

Her commitment to the awareness of seismic activity is evident in her continuous efforts to inform both the public and the state. She recognizes the importance of communicating information about earthquakes and works tirelessly to raise awareness.

 *applied sciences*

Editorial

Special Issue on Comprehensive Research in Earthquake Forecasting and Seismic Hazard Assessment

Alexey Zavyalov [1],* and Eleftheria Papadimitriou [2],*

[1] Schmidt Institute of Physics of the Earth, Russian Academy of Sciences, Bol'Shaya Gruzinskaya Str., 10, Bldg. 1, Moscow 123242, Russia

[2] Geophysics Department, Aristotle University of Thessaloniki, 54124 Thessaloniki, Greece

* Correspondence: zavyalov@ifz.ru (A.Z.); ritsa@geo.auth.gr (E.P.)

Citation: Zavyalov, A.; Papadimitriou, E. Special Issue on Comprehensive Research in Earthquake Forecasting and Seismic Hazard Assessment. *Appl. Sci.* **2023**, *13*, 11564. https://doi.org/10.3390/app132011564

Received: 18 October 2023

Accepted: 19 October 2023

Published: 23 October 2023

Dear Colleagues,

Despite some success, the issue of earthquake forecasting has yet to be resolved. There are occasional discussions within the scientific community about the principal feasibility of earthquake forecasting, particularly in the short-term aspect. However, the bulk of these discussions were set in the Resolution of the General Assembly of the International Association of Seismology and Physics of the Earth's Interior (IASPEI) in 2009 in Cape Town: "Resolution 4: Earthquake Forecasting and Predictability Studies—IASPEI RECOGNIZING the opportunities provided by recent developments in earthquake science and technology RECOMMENDS that research on forecasting and predictability of earthquakes, and the validation and comparative testing of prediction methods be supported".

However, it is not sufficient to precisely predict a future strong earthquake. It is necessary to make a correct, scientifically based assessment of the level of seismic hazard and the intensity of seismic shocks to be expected in a particular region, city and settlement. What should the administration of a megapolis do when it receives information about the likelihood of a strong earthquake? The problems of earthquake forecasting and seismic hazard assessment are, therefore, closely related to the problems of high-quality anti-seismic constructions.

More than 13 years have passed since the adoption of the IASPEI Resolution. New earthquakes have occurred. Their study increased our knowledge regarding the physics of the seismic process, the physics of earthquake preparation processes and the search for earthquake precursors. The new data obtained became the basis for the development of new models of the behaviour of the ground under the influence of seismic waves and provided initial information for the development and parameterization of earthquake occurrence zone models and ground motion prediction equations.

More than one and a half years have passed since the announcement of the Special Issue "Comprehensive Research in Earthquake Forecasting and Seismic Hazard Assessment" in the MDPI Journal of *Applied Sciences*. We invited representatives of the seismological community to present their results on these topics, to show the current view of the state of the problem, what has been achieved in the field of earthquake forecasting and seismic hazard assessment, what needs to be done next and in which direction to move forward. We expected to discuss the results and directions of further research on the physics of the seismic process—from experiments under laboratory conditions to rock bursts in mines and earthquakes in seismically active regions at the stage of preparation for strong earthquakes.

As a result, 14 articles were published in the Special Issue, with authors representing different thematic areas and working in different institutions and organisations in Russia, Greece, Italy, Colombia, New Zealand, China, Argentina and Japan. The total number of authors was around 50. Thus, we managed to attract a sufficiently wide range of representatives of the scientific geophysical community to participate in this Special Issue. In this sense, our hopes and assumptions were fulfilled. In addition, this Special Issue is fully in line with Resolution 3 "Sharing Geophysical Data across Borders" adopted at the 28th IUGG General Assembly in Berlin on 18 July 2023.

All published articles can be roughly divided into three unequal groups in terms of the number of articles presented. The first group includes theoretical and methodological articles [1,2]. The second group includes articles confirming one or another model of seismicity behaviour in anticipation of a strong earthquake [3–5]. Finally, the third and most numerous group of articles consists of those analysing the results of long-term observations of the behaviour of various geophysical fields (seismic noise [6], seismicity [7–9], magneto-telluric field [10,11], deformation field [12], infrared radiation [13], vertical electric field in the atmosphere [14]) before strong earthquakes. We are confident that each of these articles will find an interested reader, and the whole collection will deserve the attention of representatives of the scientific community dealing with the problem of earthquake forecasting and the search for their precursors.

Acknowledgments: We are grateful to all the contributors who made this Special Issue, "Comprehensive Research in Earthquake Forecasting and Seismic Hazard Assessment", a success. We thank and congratulate all the authors for submitting their papers. Our sincere gratitude is also extended to all the reviewers for their efforts and time spent in helping the authors to improve their papers. We would like to express our gratitude to the Editorial Team of *Applied Sciences* for their effective and tireless editorial support for the success of this Special Issue.

Conflicts of Interest: The authors declare no conflict of interest.

References

1. Zavyalov, A.; Zotov, O.; Guglielmi, A.; Klain, B. On the Omori Law in the Physics of Earthquakes. *Appl. Sci.* **2022**, *12*, 9965. [CrossRef]
2. Rastin, S.J.; Rhoades, D.A.; Rollins, C.; Gerstenberger, M.C. How Useful Are Strain Rates for Estimating the Long-Term Spatial Distribution of Earthquakes? *Appl. Sci.* **2022**, *12*, 6804. [CrossRef]
3. Kaviris, G.; Zymvragakis, A.; Bonatis, P.; Kapetanidis, V.; Voulgaris, N. Probabilistic and Scenario-Based Seismic Hazard Assessment on the Western Gulf of Corinth (Central Greece). *Appl. Sci.* **2022**, *12*, 11152. [CrossRef]
4. Avgerinou, S.-E.; Anyfadi, E.-A.; Michas, G.; Vallianatos, F. A Non-Extensive Statistical Physics View of the Temporal Properties of the Recent Aftershock Sequences of Strong Earthquakes in Greece. *Appl. Sci.* **2023**, *13*, 1995. [CrossRef]
5. Petrillo, G.; Lippiello, E. Incorporating Foreshocks in an Epidemic-like Description of Seismic Occurrence in Italy. *Appl. Sci.* **2023**, *13*, 4891. [CrossRef]
6. Lyubushin, A. Spatial Correlations of Global Seismic Noise Properties. *Appl. Sci.* **2023**, *13*, 6958. [CrossRef]
7. Console, R.; Vannoli, P.; Carluccio, R. The 2022 Seismic Sequence in the Northern Adriatic Sea and Its Long-Term Simulation. *Appl. Sci.* **2023**, *13*, 3746. [CrossRef]
8. Kopylova, G.; Kasimova, V.; Lyubushin, A.; Boldina, S. Variability in the Statistical Properties of Continuous Seismic Records on a Network of Stations and Strong Earthquakes: A Case Study from the Kamchatka Peninsula, 2011–2021. *Appl. Sci.* **2022**, *12*, 8658. [CrossRef]
9. Kourouklas, C.; Tsaklidis, G.; Papadimitriou, E.; Karakostas, V. Analyzing the Correlations and the Statistical Distribution of Moderate to Large Earthquakes Interevent Times in Greece. *Appl. Sci.* **2022**, *12*, 7041. [CrossRef]
10. Vargas, C.A.; Caneva, A.; Solano, J.M.; Gulisano, A.M.; Villalobos, J. Evidencing Fluid Migration of the Crust during the Seismic Swarm by Using 1D Magnetotelluric Monitoring. *Appl. Sci.* **2023**, *13*, 2683. [CrossRef]
11. Vargas, C.A.; Gomez, J.S.; Gomez, J.J.; Solano, J.M.; Caneva, A. Space–Time Variations of the Apparent Resistivity Associated with Seismic Activity by Using 1D-Magnetotelluric (MT) Data in the Central Part of Colombia (South America). *Appl. Sci.* **2023**, *13*, 1737. [CrossRef]
12. Gapeev, M.; Marapulets, Y. Modeling Locations with Enhanced Earth's Crust Deformation during Earthquake Preparation near the Kamchatka Peninsula. *Appl. Sci.* **2023**, *13*, 290. [CrossRef]
13. Yue, Y.; Chen, F.; Chen, G. Statistical and Comparative Analysis of Multi-Channel Infrared Anomalies before Earthquakes in China and the Surrounding Area. *Appl. Sci.* **2022**, *12*, 7958. [CrossRef]
14. Han, R.; Cai, M.; Chen, T.; Yang, T.; Xu, L.; Xia, Q.; Jia, X.; Han, J. Preliminary Study on the Generating Mechanism of the Atmospheric Vertical Electric Field before Earthquakes. *Appl. Sci.* **2022**, *12*, 6896. [CrossRef]

Article

On the Omori Law in the Physics of Earthquakes

Alexey Zavyalov [1,*], Oleg Zotov [1,2], Anatol Guglielmi [1] and Boris Klain [2]

[1] Schmidt Institute of Physics of the Earth, Russian Academy of Sciences, Bol'shaya Gruzinskaya Str., 10, bld. 1, 123242 Moscow, Russia
[2] Borok Geophysical Observatory of Schmidt Institute of Physics of the Earth, Russian Academy of Sciences, Nekouzsky District, Yaroslavl Region, 152742 Borok, Russia
* Correspondence: zavyalov@ifz.ru

Abstract: This paper proposes phenomenological equations that describe various aspects of aftershock evolution: elementary master equation, logistic equation, stochastic equation, and nonlinear diffusion equation. The elementary master equation is a first-order differential equation with a quadratic term. It is completely equivalent to Omori's law. The equation allows us to introduce the idea of proper time of earthquake source "cooling down" after the main shock. Using the elementary master equation, one can pose and solve an inverse problem, the purpose of which is to measure the deactivation coefficient of an earthquake source. It has been found for the first time that the deactivation coefficient decreases with increasing magnitude of the main shock. The logistic equation is used to construct a phase portrait of a dynamical system simulating the evolution of aftershocks. The stochastic equation can be used to model fluctuation phenomena, and the nonlinear diffusion equation provides a framework for understanding the spatiotemporal distribution of aftershocks. Earthquake triads, which are a natural trinity of foreshocks, main shock, and aftershocks, are considered. Examples of the classical triad, the mirror triad, the symmetrical triad, as well as the Grande Terremoto Solitario, which can be considered as an anomalous symmetrical triad, are given. Prospects for further development of the phenomenology of earthquakes are outlined.

Keywords: earthquake; source deactivation; logistic equation; nonlinear diffusion equation; Omori epoch; round-the-world echo; mirror triad

Citation: Zavyalov, A.; Zotov, O.; Guglielmi, A.; Klain, B. On the Omori Law in the Physics of Earthquakes. *Appl. Sci.* **2022**, *12*, 9965. https://doi.org/10.3390/app12199965

Academic Editor: Nicholas Vassiliou Sarlis

Received: 14 August 2022
Accepted: 30 September 2022
Published: 4 October 2022

Publisher's Note: MDPI stays neutral with regard to jurisdictional claims in published maps and institutional affiliations.

1. Introduction

Omori Law [1] describes the evolution of the aftershocks of a strong earthquake. Established at the end of the century before last, the law is characterized by the beauty of its form, quite definite clarity, as a result of which it still attracts considerable attention from the geophysical community (e.g., see [2–4]).

Initially, Omori law, which can be called hyperbolic, was formulated as follows:

$$n(t) = k/(c + t). \tag{1}$$

Here n is the frequency of aftershocks [1]. Formula (1) is one-parameter since the parameter c is free and is completely determined by an arbitrary choice of the time origin. In contrast to the aspirations of Hirano [5] and Utsu [6], who introduced a two-parameter modification of Formula (1) into widespread use, we came to the conclusion that it is reasonable to put the differential equation of evolution into the basis of the phenomenological theory of aftershocks [7–9]. Guided by this consideration, in recent years we have accumulated considerable experience in the study of the evolution of aftershocks.

The purpose of this paper is to summarize our results. The main attention is focused on the phenomenological theory of aftershocks. The paper indicates successful examples of the use of theory in the analysis of experimental material. We also paid some attention to the presentation of our position on controversial issues of a methodological nature. For example, in the literature there are erroneous statements about the deep physical content of the parameter c in Formula (1).

Another misconception is associated with the idea that the two-parameter Hirano–Utsu formula is preferable to the one-parameter Omori formula. As an argument, it is argued that the presence of two parameters facilitates the approximation of observation data. On the contrary, we consider the presence of only one physical parameter in Formula (1) as an important sign of the fundamental nature of the Omori law.

2. Elementary Master Equation

The differential approach to modeling aftershocks opens up a wide scope for searches. From the richest set of differential equations available here, we take the simplest implementation of our idea, namely the truncated Bernoulli equation:

$$dn/dt + \sigma n^2 = 0, \tag{2}$$

Here σ is the so-called deactivation coefficient of the earthquake source, "cooling down" after the main shock [7–9]. Elementary master Equation (2) is useful in making the law of evolution simpler and easier to understand. It expresses the essence of the Omori hyperbolic law (1) that everyone understands. Moreover, it will serve as an initial basis for interesting generalizations (see below Sections 3–5).

Equation (2) contains only one phenomenological parameter σ. It is easy to make sure that both formulations of the law, (1) and (2), are completely equivalent to each other for $\sigma = const$. However, firstly, in contrast to (1), Formula (2) makes it possible to take into account the nonstationarity of the geological medium in the source, which undergoes a complex relaxation process after the discontinuity has formed during the main shock. The second advantage of Formula (2) is no less important. We can seek and find natural generalizations of the differential law of evolution of aftershocks, which opens up new, sometimes unexpected, approaches to processing and analyzing experimental data.

Concluding this section of the paper, let us show how easy it is to take into account nonstationarity when formulating the Omori law in the form of the evolution Equation (2). For this, it is sufficient to assume that the deactivation factor depends on time. Let us rewrite Omori's law in the most compact form:

$$dn/d\tau + n^2 = 0, \tag{3}$$

where $\tau = \int_0^t \sigma(t')dt'$. The general solution to Equation (3) is

$$n(\tau) = n_0/(1 + n_0\tau), \tag{4}$$

It is seen that solution (4) retains the hyperbolic structure of the law, which was originally established thanks to Omori's discernment. The difference between (4) and (1) is only that time in the source, figuratively speaking, flows unevenly. For $\sigma = const$, (4) coincides with (1) up to notation. Thus, Equation (2) and its solution (4), in a certain sense, complete Omori's plan, as well as what Hirano and Utsu were striving for in their attempt to improve the law using a two-parameter modification of Formula (1).

3. Logistic Equation

Faraoni [10] considered the possibility of representing the Omori law, written in the form (2), as the Euler–Lagrange equation. The Lagrangian formulation of Omori law is interesting in many ways. In particular, it provides a basis for searching for possible generalizations of the law [11]. But the search for a suitable Lagrangian is not the only way to derive the evolution equation. One can proceed, for example, from a fairly general integro-differential equation:

$$\frac{dn}{dt} = \int_0^{\infty} K(t - t') \cdot F[n(t')] \cdot dt'. \tag{5}$$

If we put $F(n) = -\sigma n^2$, then, when choosing the trivial kernel $K(t - t') = \delta(t - t')$ from (5) follows the Omori law in the form (2).

Derivation of (2) from (5) is useful in the sense that a natural generalization of the Omori law is suggested to us. The need for generalization is dictated by the following consideration. It follows from (2) that $\lim n(t) = 0$ for $t \to \infty$. Meanwhile, experience shows that the flow of aftershocks ends with a transition to a certain background seismicity of the source. It is desirable for us to take this circumstance into account by using minimal changes in the form of the classical Omori law. It turns out that for this it is enough to take into account the linear term in the formula $F(n)$: $F(n) = \gamma n - \sigma n^2$. Here γ is the second phenomenological parameter of our theory. As a result, we get the master equation in the following form:

$$\frac{dn}{dt} = n(\gamma - \sigma n).\tag{6}$$

This is the logistic equation of Verhulst [12], well known in biology, chemistry, and sociology. It turns out to be useful in the physics of earthquakes [9,11].

We divide the family of solutions to logistic Equation (6) into two classes. The first class includes growing, and the second, falling functions of time. The separation principle is easiest to show in the phase portrait shown in Figure 1, where we have used the dimensionless quantities

$$X = \frac{n}{n_{max}}, \quad P = \frac{n_\infty}{\gamma n_{max}} \cdot \frac{dX}{dT}, \quad T = \gamma t\tag{7}$$

instead of the original quantities. Here $n_\infty = \gamma/\sigma$.

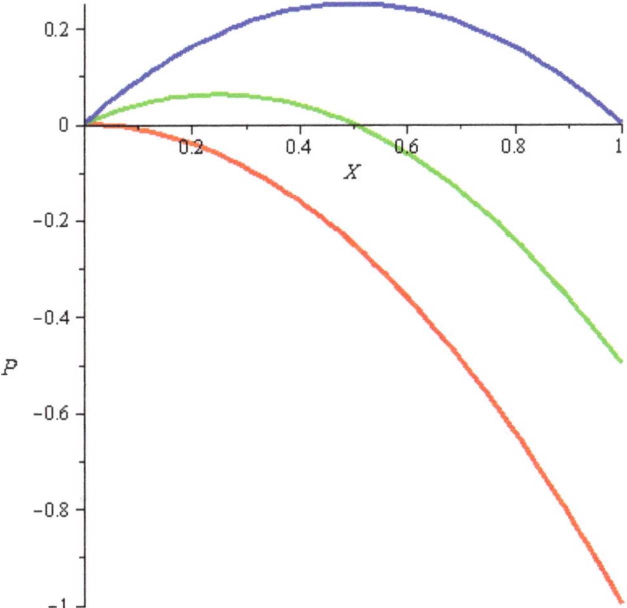

Figure 1. Phase portrait on the phase plane of Equation (6). The red, green, and blue phase trajectories are plotted at $X_\infty = 0$, 0.5 and 1, respectively (see text).

Faraoni [10] proposed the introduction of the phase plane of the dynamic system simulating the evolution of aftershocks according to the Omori law (2). The corresponding phase portrait is shown in Figure 1 with the red line. The point (0,0) corresponds to the equilibrium state. The representative point moves from bottom to top along the phase trajectory with deceleration. The portrait consists of one phase trajectory that starts at point $(1,-1)$ and ends at point $(0,0)$.

However, we are interested in the family of phase trajectories for Equation (6), constructed for different values of the parameter $X_\infty = n_\infty / n_{max}$. The red, green, and blue trajectories in Figure 1 are plotted with X_∞ values of 0, 0.5, and 1, respectively. Equilibrium point $(0,0)$ is stable at $X_\infty = 0$ and unstable at $X_\infty > 0$. Equilibrium point $(X_\infty, 0)$ is stable at any values of parameter $X_\infty > 0$. It can be shown that the velocity of motion of the imaging point along the phase trajectory asymptotically tends to zero with approaching $(X_\infty, 0)$. The segment of the trajectory located above the horizontal axis corresponds to the Verhulst logistic curve, widely known in biology, chemistry, sociology, and other sciences. The segment located below the horizontal axis corresponds to the evolution of aftershocks (Figure 2).

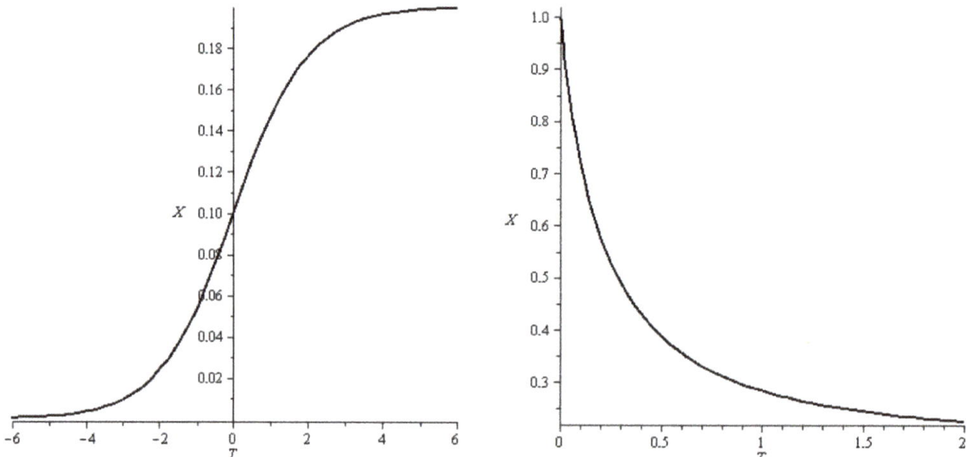

Figure 2. Logistic curve (on left) and aftershocks curve (on right) (second branch of logistic equation) at $X_\infty = 0.2$. Dimensionless time $T = \gamma t$ is plotted along the horizontal axis.

The choice between the logistic and aftershock branches is made when setting the Cauchy problem for Equation (6). Evolution proceeds along the aftershock branch if the initial condition satisfies the inequality $n(0) = n_0 > n_\infty$, where $n_\infty = \gamma / \sigma$. Thus, in the physics of aftershocks, when setting the Cauchy problem, one should set the initial conditions under the additional constraint $n_0 > n_\infty$. Moreover, it is reasonable to use the strong inequality $n_0 \gg n_\infty = \gamma / \sigma$. Indeed, for $t \to \infty$, the frequency of aftershocks asymptotically approaches from above to the background (equilibrium) value n_∞. Experience shows that as a rule $n_0 \gg n_\infty$ after a strong earthquake. The analysis of Equation (6) under the condition $n_0 \gg n_\infty$ indicates that at the first stages of evolution, the frequency of aftershocks decreases with time in accordance with the classical Omori Formula (1). Let us take a closer look at this important circumstance, since the existence of the aftershock branch is not so widely known.

The aftershock branch is entirely located above the saturation level n_∞ and is a monotonically decreasing function of time. When $t \to +\infty$ it tends asymptotically from above the saturation level (see the right panel in Figure 2). When setting the Cauchy problem in the physics of aftershocks, the initial condition should be asked the restriction $n_0 \gg n_\infty$.

Let us show that the decrease in the frequency of aftershocks with time at the first stage of evolution occurs according to the Omori hyperbola (1). It is natural to call this stage of evolution the *Omori epoch*.

Let us introduce the notation

$$t_\infty = \frac{1}{\gamma} \ln(1 - \frac{n_\infty}{n_0}), \tag{8}$$

and write the solution of evolution Equation (6) in the following form:

$$n(t) = n_\infty \{1 - \exp[\gamma(t_\infty - t)]\}^{-1}. \tag{9}$$

In the Omori epoch $t_\infty < t \ll 1/\gamma$ and, respectively,

$$n(t) = 1/\sigma(t - t_\infty). \tag{10}$$

Formula (10) coincides with the classical Omori Formula (1) up to notation.

Observational experience indirectly testifies to the plausibility of our logistic model. It is known, for example, that over time the frequency n tends not to zero, as follows from the Omori law, but to some equilibrium value n_∞. Further, some combination of the logistic and aftershock branches makes it possible to propose a scenario for the occurrence of an earthquake swarm (see details in [11]).

4. Stochastic Equation

Changing variables in a differential equation is often a powerful tool for finding solutions to it. We already know the solutions for the Omori Equation (2) and the logistic Equation (6). Nevertheless, we will still change the variable $n(t)$ in order to linearize both of these equations. This will make it easier for us to search for a stochastic generalization of the equation for the evolution of aftershocks.

The following replacement will help us transform nonlinear Equations (2) and (6) into linear ones [3]:

$$n(t) \to g(t) = 1/n(t). \tag{11}$$

Omori Equation (2) takes on an extremely simple form:

$$\dot{g} = \sigma. \tag{12}$$

Here the dot above the symbol means time differentiation. Logistic Equation (6) becomes a first-order linear differential equation:

$$\dot{g} + \gamma g = \sigma. \tag{13}$$

Generally speaking, this circumstance is quite interesting in itself, but we use it precisely to make the stochastic generalization of the evolution equation the simplest possible way. Namely, let us imagine that the deactivation coefficient experiences small fluctuations. This assumption is formalized as follows: $\sigma \to \sigma + \xi(t)$, where $\xi(t)$ is a random function of time, and $max|\xi| \ll \sigma$. As a result, we have

$$\dot{g} + \gamma g = \sigma + \xi(t). \tag{14}$$

Let us formally solve the Equation (14):

$$g(t) = g_\infty + (g_0 - g_\infty) \exp(-\gamma t) + \int_0^t \xi(t_1) \exp[\gamma(t_1 - t)]dt_1. \tag{15}$$

Here $g_\infty = \sigma/\gamma$, $g_0 = g(0)$.

Our second assumption is that $\xi(t)$ is the Langevin source, i.e., delta-correlated random function with zero mean

$$\overline{\xi(t)} = 0, \qquad \overline{\xi(t)\xi(t_1)} = N\delta(t - t_1), \tag{16}$$

where the line at the top means averaging.

Now Equation (14) should be considered as the Langevin stochastic equation (e.g., see [13], where the Langevin equation is studied in detail). The phenomenological parameter N is determined by the intensity of noise affecting our dynamic system.

5. Nonlinear Diffusion Equation

If we ask ourselves how to describe the evolution of aftershocks not only in time, but also in space–time, then this immediately puts us in a difficult position. On the one hand, a number of methods are known for modeling space–time distributions, but on the other hand, in our case, there is a strong limitation, which consists in the fact that when averaging over the epicentral zone, we want to obtain the Omori law in the form (2), or in the form (6). Fortunately for us, it turns out here that we can use the well-known Kolmogorov–Petrovsky–Piskunov equation (abbreviated KPP), which describes nonlinear diffusion [14]. It is convenient for us to represent it in the following form:

$$\frac{\partial n}{\partial t} = n(\gamma - \sigma n) + D\frac{\partial^2 n}{\partial x^2},\qquad(17)$$

where $n(x,t)$ is the spatio-temporal distribution of aftershocks, the x axis is directed along the earth's surface (for simplicity, we limited ourselves to a one-dimensional model), and D is a new phenomenological parameter (diffusion coefficient). At $D = 0$, (17) turns into the logistic equation of the evolution of aftershocks (6), and under the additional condition $\gamma = 0$ into Omori law (2).

It is useful to derive (17) from the integro-differential equation

$$\frac{\partial n}{\partial t} = \Phi(n) + \int\limits_{-\infty}^{\infty} K(x - y)\cdot n(y,t)\cdot dy.\qquad(18)$$

Here $\Phi(n)$ is some kind of functional. This will make it possible to express the parameters γ and D in terms of the kernel $K(x - y)$. Indeed, suppose that $K(x - y) = K(y - x)$, i.e., the core is symmetrical. If $K \to 0$ for $|x - y| \to \infty$, then expanding $n(x - z, t)$ into a Taylor series in powers of x, we obtain

$$\frac{\partial n}{\partial t} = \gamma n + \Phi(n) + D\frac{\partial^2 n}{\partial x^2} + \dots\qquad(19)$$

where

$$\gamma = \int_{-\infty}^{\infty} K(z)dz, \; D = \frac{1}{2}\int_{-\infty}^{\infty} z^2 K(z)dz,\qquad(20)$$

and $z = x - y$ (e.g., see [11,15,16]). Let us $\Phi(n) = -\sigma n^2$, and confine the first two terms in the series. In this approximation we obtain Equation (17) from which after phenomenological reduction follows the Omori law in the form (2).

When constructing a phenomenological theory of aftershocks, there is one most important condition, only one, but absolutely necessary: the phenomenological coefficients, whatever meaning we put into them, we must be able to measure experimentally, since they cannot be calculated on the basis of a more fundamental theory—because we simply do not have such a theory. In Section 6.1 we show how the deactivation factor we introduced in Section 2 is estimated. In the current section, we have significantly complicated the theory and introduced the diffusion coefficient D. We want to briefly describe the result of observing aftershocks, which actually led us to master equation (17), and then indicate how, at least in principle, the parameter D can be estimated experimentally. (In this regard, it is appropriate to mention the recent interesting work [17]. It provides additional arguments in favor of the idea of the applicability of the KPP equation for modeling aftershocks).

The study of aftershocks in space and time led us to the idea of using the KPP equation as the master equation. The main step forward in the study of the space–time distribution was the discovery that, apparently, at least some of the aftershocks tend to propagate like waves with a speed much lower than the speed of seismic waves [8,18]. The rate of propagation varies widely from case to case. It is roughly a few kilometers per hour. This value is three orders of magnitude less than the velocities of elastic waves in the crust, which suggests the propagation of a nonlinear diffusion wave excited by the main shock.

Equation (17) has self-similar solutions in the form of a traveling wave $n(x,t) = n(x \pm Ut)$ [14,19]. It is this circumstance that played a role in our choice of the KPP equation as the master equation. The estimation of the wave propagation velocity can be performed by analyzing the dimensions of the coefficients of the master equation: $U \sim \sqrt{\gamma D}$. Knowing the propagation velocity U, and estimating the parameter γ according to the formula $\gamma = n_\infty \sigma$, we can give an oriented estimate of the diffusion coefficient $D = U^2 / \gamma$.

6. Discussion

We have outlined a phenomenological basis, united by the general idea of a differential approach, to describing and understanding the dynamics of aftershocks. Starting with an elementary nonlinear differential Equation (2), completely equivalent to the Omori law in its classical expression (1) at $\sigma = const$, we tried to use minimal modifications in order to go first to the logistic equation (6), then to the stochastic Equation (14), and, finally, to the nonlinear diffusion Equation (17). Perhaps it would be useful to note that we explicitly used the methodological principle of Descartes, the essence of which is that one should go from simple to complex, using clear and precise modifications of the theoretical description of the problem under study.

Taken together, four phenomenological master equations, united by a common idea, make it possible to comprehend a fairly wide range of properties and patterns of aftershocks found experimentally. Moreover, phenomenological theory allows certain predictions to be made that can be verified experimentally. Let us illustrate what has been said with a number of examples.

6.1. Inverse Problem

The inverse problem of the physics of earthquake source is to determine the phenomenological coefficients from the observation data of aftershocks. Omori law in the form (2) makes it possible to formulate and solve the problem of determining the deactivation coefficient σ from the observation data of the frequency of aftershocks n. The auxiliary value g, which we introduced in Section 3, is conveniently written in the form of $g = (n_0 - n)/nn_0$. It is easy to see that

$$\sigma = \frac{d}{dt}\langle g \rangle. \tag{21}$$

Here, the angle brackets denote the regularization of the function $g(t)$ calculated from observation $n(t)$. Regularization is reduced in this case to the smoothing procedure.

Our experience indicates that the deactivation coefficient σ undergoes complex variations over time [8,20]. However (and this seems to us extremely important) at the first stages of evolution $\sigma = const$. The time interval during which $\sigma = const$, we called the Omori epoch. In the Omori epoch, the classical Omori law is fulfilled (1), according to which the frequency of aftershocks hyperbolically decreases over time. In our experience, the duration of the Omori epoch varies from case to case from a few days to two or three months. We noticed a tendency for the duration of the Omori epoch to increase with the increase in the magnitude of the main shock.

An interesting prediction follows from the existence of the Omori epoch, which has been reliably confirmed by experience [21]. Namely, the question of dependence of the deactivation factor on the magnitude of main shock is analyzed theoretically and experimentally. A monotonic decrease in the deactivation factor with an increase in the magnitude of the main shock, M_0, has been reliably established. Figure 3 shows the result of measurements of σ at different values of M_0. To measure the deactivation factor, we used the USGS/NEIC earthquake catalog and the technique developed during the compilation of the Atlas of Aftershocks [20]. We see that, on average, σ decreases monotonically with the increase in M_0. The dependence $\sigma(M_0)$ is approximated by the formula

$$\sigma = A - BM_0, \tag{22}$$

where $A = 0.64$, $B = 0.07$ with a sufficiently high coefficient of determination $R^2 = 0.82$. Thus, the theoretical inequality $d\sigma/dM_0$ is reliably confirmed by direct measurements. We have a wonderful harmony between theory and experiment.

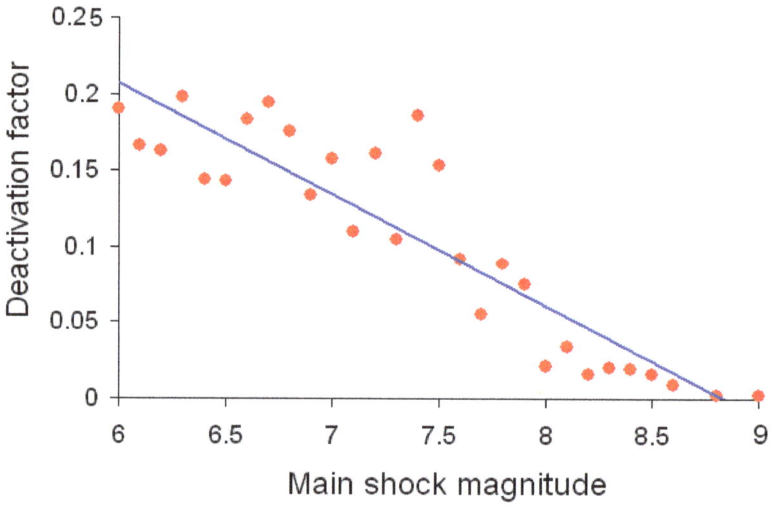

Figure 3. Dependence of the deactivation factor of the earthquake source on the magnitude of the main shock. Other explanations see in the text below.

It is quite clear that the question of how best to formulate the Omori law, in the form (1) or (2), could only be solved by observation and experience. Equation (2) turned out to be more effective, since it made it possible to introduce a simple and useful concept of deactivation of the source, to pose the inverse problem of the source, and to reveal the existence of the Omori epoch. In addition to this, we have shown that with the help of (2) one can make a meaningful statement about the deactivation coefficient and, moreover, check this statement experimentally.

Finally, the question of whether it is not better to use the one-parameter Formula (2) for modeling aftershocks than the two-parameter Hirano–Utsu formula $n = k/(c + t)^p$ [2] deserves discussion. We give preference to Formula (2), since the inverse problem solved on its basis indicates the existence of the Omori epoch [8,9]. The Hirano–Utsu formula is unacceptable, since it contradicts the existence of the Omori epoch at $p \neq 1$, and at $p = 1$ it coincides with the Omori Formula (1).

Perhaps it would be appropriate to draw a distant historical analogy here. According to the law of gravitation, the interaction potential $\varphi \propto 1/r$ leads to the ellipticity of the planetary orbit. The deviation from ellipticity, for example, of the orbit of Mercury, could in no way serve as a reason for choosing the interaction, say, in the form $\propto 1/r^p$. Another understanding of the deviation of orbit from strictly elliptical had to be looked for, and it was found within the framework of general relativity. Perhaps, finding themselves in a similar situation, Hirano and Utsu should not have immediately abandoned the excellent Omori law, but should have looked for other explanations for the deviation of the real flow of aftershocks from strict hyperbolicity.

6.2. Triggers

Deviations from the classical Omori law (1) are caused not only by the nonstationarity of the parameters of the geological environment in the source, which we have expressed in the form of a possible dependence of the deactivation factor on time. Deviations can occur under the influence of so-called triggers, i.e., relatively small disturbances of geophysical fields of endogenous or exogenous origin.

We point here to two endogenous triggers that we have recently discovered. Both are aroused on the main shock. One of them has the form of a round-the-world seismic echo, and the second represents free elastic oscillations of the Earth as a whole, excited by the main shock. We have described both triggers in detail in a number of papers, so we will restrict ourselves here to references [22–26].

Let us dwell on exogenous triggers in more detail. For a long time, the cosmic effects on seismicity have been widely discussed, but there is still no agreement in the geophysical community on the effectiveness of such impacts. The controversy about the influence of geomagnetic storms on the global activity of earthquakes arises especially often (see, for example, [27–30]). The question is really difficult. On the one hand, observations indicate a correlation between seismicity and geomagnetic storms and a complex of electromagnetic phenomena associated with them (see papers [31–37] and the literature cited therein). On the other hand, the mechanism of the impact of geomagnetic storms on rocks, leading to modulation of seismicity, is not entirely clear. In this regard, the idea of the magnetoplasticity of rocks [32,33] seems to us very encouraging, but a discussion of this deep idea would lead us far astray.

A wide class of exogenous triggers of anthropogenic origin is known. We will restrict ourselves here to an indication of the weekend effect discovered in [38], and the so-called Big Ben effect, or the effect of hour markers [39,40]. Both effects pose a difficult question for the researcher about the global impact of the industrial activity of mankind on the lithosphere.

6.3. Triads

Apparently, the idea of a peculiar trinity of foreshocks, main shock, and aftershocks in a sequence of tectonic earthquakes [41–43] was formed in seismology not without the influence of mathematics, in which a binary relation between elements of a set can give rise to a trichotomic relation. The trinity of foreshocks, main shock, and aftershocks was proposed to be called the classical triad [44]. The magnitude of the main shock M_0 is always greater than the maximum magnitudes of foreshocks and aftershocks. The classical triad satisfies the inequalities

$$M_- < M_+, \tag{23}$$

and

$$N_- < N_+. \tag{24}$$

Here $M_-(M_+)$ and $N_-(N_+)$ are the maximum magnitude and the number of foreshocks (aftershocks), respectively.

Quite often $N_- = 0$, i.e., foreshocks are absent even before rather strong earthquakes. Figure 4 illustrates this situation (the database and the plotting method will be described in detail below). With regard to aftershocks, there is a stable opinion that after a sufficiently strong earthquake, repeated tremors are always observed, i.e., $N_+ \neq 0$.

In this section, we want to present rare but extremely interesting types of anomalous triad, for which inequalities that are directly opposite to inequalities (23) and (24) hold [45]. These are the so-called mirror triads, for which $M_- > M_+$, $N_- > N_+$ and symmetric triads, for which $N_- = N_+$. Moreover, $N_+ = 0$ in a significant part of the mirror triads.

Mirror triads. Extensive literature is devoted to the experimental study of the classical triads. We point here to work [21], since in the study of anomalous triads we used a database and general methods of analysis similar to those used here in the study of classical triads (see also [9]).

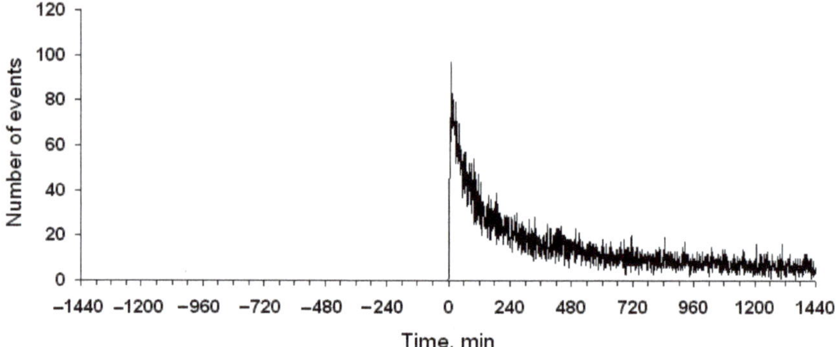

Figure 4. Generalized picture of a shortened classical triad. Zero moment of time corresponds to the moment of the main shock.

We used data on earthquakes that occurred on Earth from 1973 to 2019 and were registered in the world USGS/NEIC catalog of earthquakes (https://earthquake.usgs.gov; last accessed on 30 April 2022). There were found $N_0 = 2508$ main shocks with a magnitude of $M_0 \geq 6$ and a hypocenter depth not exceeding 250 km. For each main shock, a circular epicentral zone was determined by the formula $lgL = 0.43M_0 - 1.27$, where the radius of the zone L is expressed in kilometers [46]. According to our definition, the classical triad is formed by earthquakes, which occurred in the epicentral zone in the interval of ± 24 h relative to the moment of the main shock, provided that the inequalities (23), (24) are satisfied. The total number of earthquakes was distributed among the members of the triad in the following way: $N_- = 1105$, $N_0 = 2398$, $N_+ = 31865$. Note that here and below N_0 is the number of main shocks. Figure 4 is based on truncated triads, in which there are no foreshocks. The graph was constructed by the method of overlapping epochs, and the main shock of the earthquake was used as a benchmark. For truncated triads, the distribution looks like this: $N_- = 0$, $N_0 = 2066$, $N_+ = 21422$. The distribution at $N_- \neq 0$: $N_- = 1105$, $N_0 = 332$, $N_+ = 10443$. We see that in the presence of foreshocks, the activity of aftershocks is higher than in the absence of foreshocks and the number of truncated triads significantly exceeds the number of complete ones.

In the course of studying classical triads, the idea arose to make a selection of observational data by replacing inequalities (23), (24) with the opposite ones. As a result, it was possible to find the mirror triads. Figure 5 shows the truncated mirror triads: $N_- = 237$, $N_0 = 156$, $N_+ = 0$. If $N_+ \neq 0$, then $N_- = 1375$, $N_0 = 104$, $N_+ = 755$.

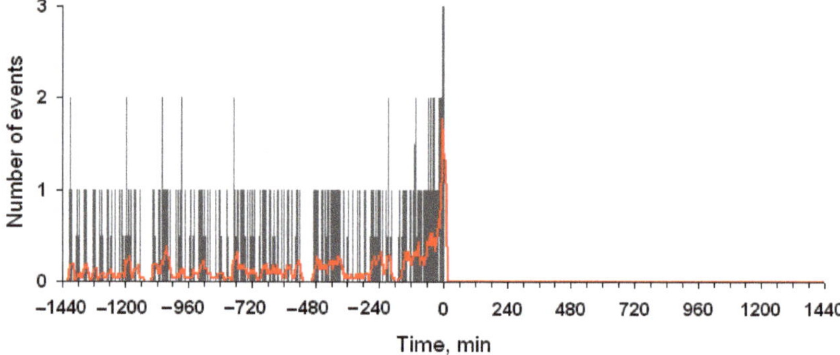

Figure 5. Generalized view of truncated mirror triads. Zero moment of time corresponds to the moment of the main shock. The red line is obtained by averaging over a sliding window of 20 min, with the step 1 min.

We see that mirror triads are relatively rare phenomenon. They appear about an order of magnitude less frequently than classical triads. To make the picture of mirror triads more visual, we will show Figure 6. It shows mirror triads in the range of main shocks magnitude $5 \leq M_0 < 6$. Here $N_- = 4189$, $N_0 = 2430$, $N_+ = 201$.

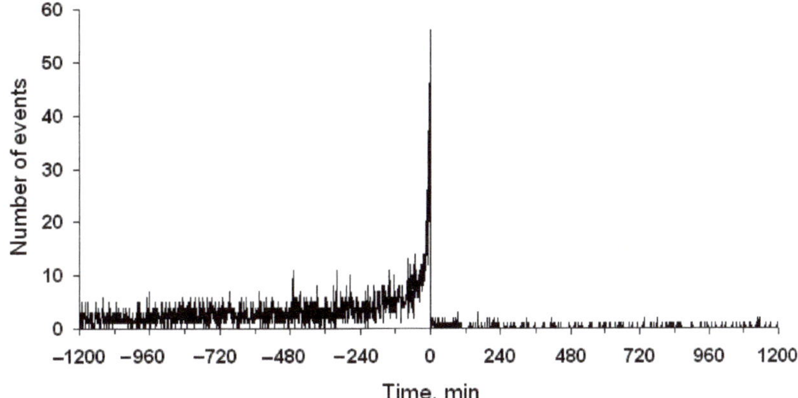

Figure 6. Time distribution of foreshocks and aftershocks of mirror triads in the range of magnitudes of the main shocks $5 \leq M_0 < 6$.

GTS. So, we found that there is a rare but rather interesting subclass of tectonic earthquakes, in which the number of aftershocks in the interval of 24 h after the main shock is significantly less than the number of foreshocks in the same interval before the main shock. In many cases, there are no aftershocks at all. We asked the question: Are there earthquakes with magnitudes $M_0 \geq 6$, neither before nor after which there are neither foreshocks nor aftershocks? The search result was amazing. We have discovered a wide variety of this kind of earthquake and named it *Grande terremoto solitario* (Italian), or GTS for short [47]. In Figure 7, we see that the number of GTS (2460) is approximately equal to the number of classical triads (2398).

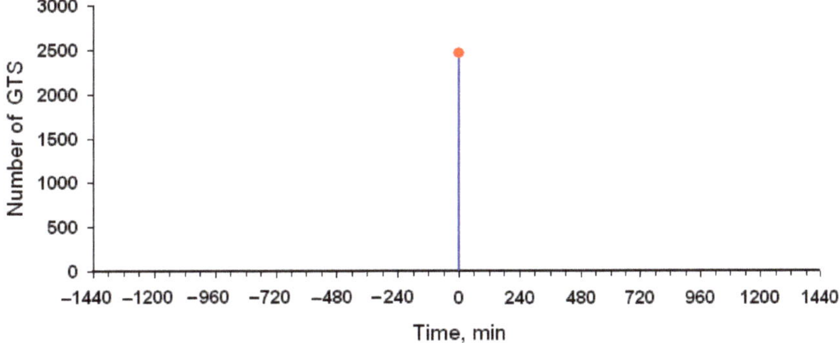

Figure 7. Solitary earthquakes with magnitudes $M_0 \geq 6$.

GTS arise spontaneously under very calm seismic conditions and are not accompanied by aftershocks. This suggests an analogy between the GTS and the so-called "Rogue waves" (or "Freak waves")—isolated giant waves that occasionally emerge on a relatively quiet ocean surface (e.g., see [48]). This analogy may prove to be quite profound, since the spontaneous occurrence of pulses with anomalously high amplitudes is a common property of the nonlinear evolution of dynamic systems [49].

For completeness, we also present the data for the symmetric triads in which $M_0 \geq 6$ and $N_+ = N_-$: $N_- = 186$, $N_0 = 121$, $N_+ = 186$. It is interesting to note that, formally, GTS can be related to a variety of symmetric triads, since for them $N_+ = N_- = 0$.

Activation factor. Figures 5 and 6 shows that foreshocks in the mirror triad appear to have a temporal distribution similar to the Omori distribution for aftershocks in the classical triad. Let us dwell on this in more detail. We represent the classical Omori law [1] in the simplest differential form

$$\dot{g} = \sigma_+. \tag{25}$$

Here $g = 1/n$, $n(t)$ is the frequency of aftershocks, $t > 0$, the dot above the symbol means time differentiation, σ_+ is the so-called deactivation factor of the earthquake source, "cooling down" after the main shock (see [3,8,21]). Suppose that for foreshocks of the mirror triad, the evolution law (25) is fulfilled with the replacement of σ_+ by σ_-. It is natural to call the σ_- value the activation factor.

In this paper, we will limit ourselves to presenting the interesting Figure 8. It shows the generalized evolution of foreshocks and aftershocks in symmetric triads satisfying the condition $5 \leq M_0 < 6$. Here $N_- = 1050$, $N_0 = 742$, $N_+ = 1050$. The top panel shows an amazing mirror image. In the bottom panel, we have shown the variations of the σ_- and σ_+ functions as the first step towards studying the activation and deactivation coefficients of an earthquake source in mirror triads. (For the procedure for calculating $\sigma_\pm$, see [8]).

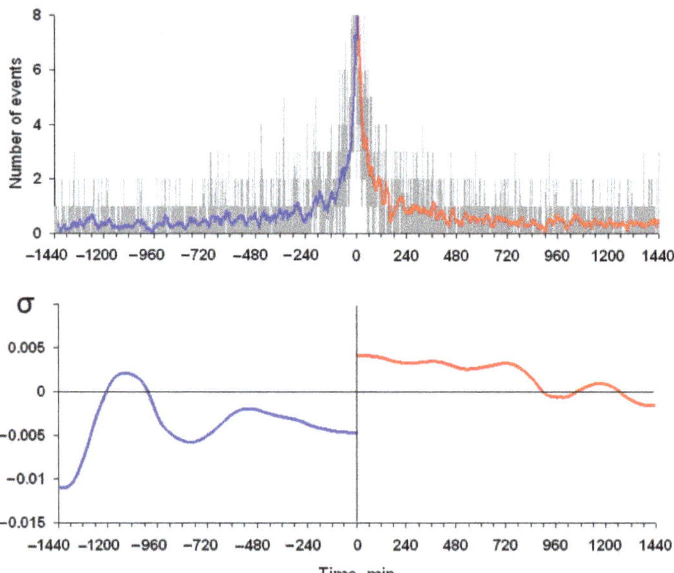

Figure 8. Time dependence of the properties of symmetric triads. From top to bottom: earthquake frequency, activation (blue), and deactivation (red) factors.

Origin of mirror triads. In conclusion of this section, we would like, with all the necessary reservations, to express a careful judgment on the question of the origin of mirror triads. Let us assume that a system of faults in a certain volume of rocks is under the influence of a slowly growing total shear stress τ. Threshold tension τ_* at which destruction occurs, i.e., the sides of the fault shift and a rupture occurs, generally speaking, is the lower, the larger the linear dimensions of the fault l:

$$\tau_* = Cl^{-m}. \tag{26}$$

Here C is the dimensional coefficient of proportionality, depending on the properties of rocks in the selected volume, $m > 0$. Then, the largest fault reaches the threshold first. Its destruction is manifested in the form of the main shock of an earthquake with a magnitude M_0.

If the C parameter is uniformly distributed over the source volume, then foreshocks do not arise. Aftershocks appear due to the fact that after the main shock, the general external stress is partially removed, and the remaining stress is redistributed in a complex way throughout the volume. The local over-tensions arise, in such a way that smaller faults than the one that generated the main shock can be activated and give repeated tremors. This is how we can imagine the emergence of a shortened classical triad.

In some cases, the specific distribution of faults in terms of l and the distribution of local stresses may turn out to be such that not a single aftershock occurs. It is possible that such a situation occurs when the GTS is excited.

The appearance of the mirror triad can be understood if we assume that the parameter C is not uniformly distributed over the volume, or rather, that there is a strong scatter in the $C(l)$ values. Then a situation is possible when, before the largest fault is destroyed, smaller faults are activated, and foreshocks appear. A triad of tectonic earthquakes will appear. Whether it will be classical or mirror-like depends on the distribution of faults by the value of l, on the dispersion of the $C(l)$ coefficient, and on the mosaic of local stresses that arose after the main shock.

As a summary, we point out that classical (normal) triads make up approximately 85% of all triads. Anomalous triads account for 15%, with mirrored, 10% and symmetrical, 5%. In this calculation, we excluded the GTS, which seem to form a special set of earthquakes. In the class of tectonic earthquakes, we found a subclass of the so-called mirror triads. A specific property of mirror triads is that, in contrast to classical triads, in which the number of aftershocks is greater than the number of foreshocks, in mirror triads the number of aftershocks is less than the number of foreshocks in the interval 24 h before and 24 h after the main shock. In many cases, there are no aftershocks at all. In addition to this, strong solitary earthquakes were discovered, which are not preceded by foreshocks, and after which there are no aftershocks.

The mirror triads, these ghosts of the classical triads, are not only curious in themselves, but can most decisively influence our understanding of the alternative possibilities of the dynamics of lithosphere, leading to catastrophic earthquakes. In particular, we face a fundamental problem, the essence of which is to find the physical and geotectonic reasons for the apparent predominance of truncated classical triads in the seismic activity of the Earth.

Concluding the discussion, we want to make a judgment, perhaps controversial, that Omori's law, like no other, gives us the opportunity to realize the uncertainty, incompleteness, and, in a certain sense, immaturity of the physics of earthquakes. So far, it is still possible and, in reality, there is a wide range of opinions on the unresolved issues of the Earth's seismicity. So far, everyone, in accordance with their idea of the nature of cognition, can see in the Omori law either an empirical formula convenient for approximating observations, or an indication of the deep physical meaning of hyperbolicity in the frequency of aftershocks after a strong earthquake.

7. Conclusions

We will deviate from tradition and, instead of simply listing the results of the work (which is still far from complete), we will briefly present some prospects of the research that we propose to carry out in the near future. The study will be devoted to the geometro-dynamics of a tectonic earthquake source. We proceed from the fact that in the study of earthquakes some geometrically visual representations and considerations are necessary, and that analytics alone is insufficient. Representing the source as the interior of the convex hull of a point manifold of aftershocks, we have outlined a program whose purpose is to reduce the mosaic of very complex, intricate realities of the evolving source to geometric objects. An interesting object is the space–time trajectory of the shell's center of gravity. The curvature of the envelope surface of the epicenters in dynamics and other geometrically visual images can also turn out to be quite interesting. As conceived, the geometrodynamics will become a source of new ideas for the development of the phenomenological theory of earthquakes.

However, let us return to the work presented by us and try to make some summary of all the results. The general conclusion is as follows: the methodological approach based on differential equations of evolution opens up new possibilities for the analysis of experimental material. The phenomenological equations of evolution proposed by us allows for the posing of inverse problems of the source physics and makes it possible to formulate unexpected questions regarding the dynamics of earthquakes. The phenomenological theory, a sketch of which we have given here, not only enriches the system of ideas about the source, but, we hope, indicates the possibility of searching for approaches to solving problems of a fundamental nature.

We are fully aware of the fact that neither the totality of facts, even if it is represented by a set of empirical formulas, nor a logically consistent phenomenological theory, by itself, lead us to a deep penetration into the essence of earthquakes. A deep understanding would be much more facilitated by a theory based on the fundamental laws of physics, taking into account the characteristics of the geological environment. Theoretical constructions of this kind are known, but for a completely understandable reason they refer only to individual aspects of the phenomenon, and not to the phenomenon as a whole. Under these conditions, it is reasonable and natural to consistently continue the development of the phenomenological theory of earthquakes, the foundations of which were laid by Fusakichi Omori.

Author Contributions: Conceptualization, A.G., O.Z., A.Z. and B.K.; data curation, O.Z. and B.K.; formal analysis, A.Z., O.Z., A.G. and B.K.; investigation, O.Z., A.G. and B.K.; methodology, O.Z., A.G. and B.K.; Validation, A.Z., O.Z., A.G. and B.K.; writing—original draft, A.G.; writing—review and editing, A.Z., O.Z., A.G. and B.K.; project administration, A.Z. and A.G. All authors have read and agreed to the published version of the manuscript.

Funding: This work was supported by the Ministry of Science and Higher Education of the Russian Federation as part of the state assignment of the Schmidt Institute of Physics of the Earth of the Russian Academy of Sciences # 075-00693-22-00.

Institutional Review Board Statement: Not applicable.

Informed Consent Statement: Not applicable.

Data Availability Statement: The USGS earthquakes catalog from 1973 to 2019 is accessible from https://earthquake.usgs.gov/earthquakes/search/ (last accessed on 30 April 2022).

Acknowledgments: We express our deep gratitude to A.L. Buchachenko, F.Z. Feygin, N.A. Kurazhkovskaya, and A.S. Potapov for their interest in this work. The authors thank the staff U.S. Geological Survey for providing the catalogues of earthquakes.

Conflicts of Interest: The authors declare no conflict of interest.

References

1. Omori, F. On the aftershocks of earthquake. *J. Coll. Sci. Imp. Univ. Tokyo* **1894**, *7*, 111–200.
2. Utsu, T.; Ogata, Y.; Matsu'ura, R.S. The centenary of the Omori formula for a decay law of aftershock activity. *J. Phys. Earth* **1995**, *43*, 1–33. [CrossRef]
3. Guglielmi, A.V. Omori's law: A note on the history of geophysics. *Phys.-Uspekhi* **2017**, *60*, 319–324. [CrossRef]
4. Guglielmi, A.V.; Zavyalov, A.D. The 150th anniversary of Fusakichi Omori. *J. Volcanol. Seismol.* **2018**, *12*, 353–358. [CrossRef]
5. Hirano, R. Investigation of aftershocks of the great Kanto earthquake at Kumagaya. *Kishoshushi* **1924**, *2*, 77–83. (In Japanese)
6. Utsu, T. A statistical study on the occurrence of aftershocks. *Geophys. Mag.* **1961**, *30*, 521–605.
7. Guglielmi, A.V. Interpretation of the Omori Law. *Izv. Phys. Solid Earth* **2016**, *52*, 785–786. [CrossRef]
8. Zavyalov, A.D.; Guglielmi, A.V.; Zotov, O.D. Three problems in aftershock physics. *J. Volcanol. Seismol.* **2020**, *14*, 341–352. [CrossRef]

9. Guglielmi, A.V.; Klain, B.I.; Zavyalov, A.D.; Zotov, O.D. A Phenomenological theory of aftershocks following a large earthquake. *J. Volcanol. Seismol.* **2021**, *15*, 373–378. [CrossRef]
10. Faraoni, V. Lagrangian formulation of Omori's law and analogy with the cosmic Big Rip. *Eur. Phys. J. C* **2020**, *80*, 445. [CrossRef]
11. Guglielmi, A.V.; Klain, B.I. The phenomenology of aftershocks. *arXiv* **2020**, arXiv:2009.10999.
12. Verhulst, P.F. Notice sur la loi que la population poursuit dans son accroissement. *Corresp. Math. Phys.* **1838**, *10*, 113–121.
13. Klimontovich, Y. *Statistical Physics*; CRC Press: Boca Raton, FL, USA, 1986; 734p.
14. Kolmogorov, A.N.; Petrovsky, I.G.; Piskunov, N.S. A study of the diffusion equatio related to increase of material, and its application to a biological problem. *Bull. MGU Ser. A Mat. Mekhanika* **1937**, *1*, 1–26.
15. Murray, J.D. *Mathematical Biology. V.1. An Introduction*; Springer: Berlin/Heidelberg, Germany, 2002; 551p.
16. Murray, J.D. *Mathematical Biology. V.2. Spatial Model and Biomedical Applications*; Springer: Berlin/Heidelberg, Germany, 2003; 839p.
17. Rodrigo, M.R. A spatio-temporal analogue of the Omori-Utsu law of aftershock sequences. *arXiv* **2021**, arXiv:2111.02955.
18. Zotov, O.D.; Zavyalov, A.D.; Klain, B.I. On the spatial-temporal structure of aftershock sequences. In *Problems of Geocosmos–2018*; Springer Proceedings in Earth and Environmental Sciences; Yanovskaya, T., Kosterov, A., Bobrov, N., Divin, A., Saraev, A., Zolotova, N., Eds.; Springer: Cham, Switzerland, 2019. [CrossRef]
19. Polyanin, A.D.; Zaitsev, V.F. *Handbook of Nonlinear Partial Differential Equations*, 2nd ed.; Chapman & Hall/CRC Press: Boca Raton, FL, USA, 2012.
20. Guglielmi, A.V.; Zotov, O.D.; Zavyalov, A.D. A project for an Atlas of aftershocks following large earthquakes. *J. Volcanol. Seismol.* **2019**, *13*, 415–419. [CrossRef]
21. Guglielmi, A.V.; Zotov, O.D. Dependence of the source deactivation factor on the earthquake magnitude. *arXiv* **2021**, arXiv:2108.02438.
22. Guglielmi, A.V.; Zotov, O.D. Impact of the Earth's oscillations on the earthquakes. *arXiv* **2012**, arXiv:1207.0365.
23. Guglielmi, A.V.; Zotov, O.D.; Zavyalov, A.D. The aftershock dynamics of the Sumatra–Andaman earthquake. *Izv. Phys. Solid Earth* **2014**, *50*, 64–72. [CrossRef]
24. Zotov, O.D.; Zavyalov, A.D.; Guglielmi, A.V.; Lavrov, I.P. On the possible effect of round-the-world surface seismic waves in the dynamics of repeated shocks after strong earthquakes. *Izv. Phys. Solid Earth* **2018**, *54*, 178–191. [CrossRef]
25. Zavyalov, A.D.; Guglielmi, A.V.; Zotov, O.D. Deviation from the Omori law as the result of the trigger impact of round-the-world surface seismic waves on the source of strong earthquakes. *Acta Geol. Sin.* **2019**, *93* (Suppl. S1), 271. [CrossRef]
26. Guglielmi, A.V. On self-excited oscillations of the Earth. *Izv. Phys. Solid Earth* **2015**, *51*, 920–923. [CrossRef]
27. Zotov, O.D.; Guglielmi, A.V.; Sobisevich, A.L. On magnetic precursors of earthquakes. *Izv. Phys. Solid Earth* **2013**, *49*, 882–889. [CrossRef]
28. Guglielmi, A.V. On the relationship between earthquakes and geomagnetic disturbances. *Geophys. Res.* **2020**, *21*, 78–83. (In Russian) [CrossRef]
29. Guglielmi, A.V.; Zotov, O.D. On geoseismic noise and helioseismic oscillations. *Izv. Phys. Solid Earth* **2021**, *57*, 567–572. [CrossRef]
30. Guglielmi, A.V.; Klain, B.I.; Kurazhkovskaya, N.A. On the correlation of Earthquakes with geomagnetic storms. *Izv. Phys. Solid Earth* **2021**, *57*, 994–998. [CrossRef]
31. Sobolev, G.A.; Zakrzhevskaya, N.A.; Migunov, I.N.; Sobolev, D.G.; Boiko, A.N. Effect of magnetic storms on the low-frequency seismic noise. *Izv. Phys. Solid Earth* **2020**, *56*, 291–315. [CrossRef]
32. Buchachenko, A.L. Magneto-plasticity and the physics of earthquakes. Can a catastrophe be prevented? *Phys.-Uspekhi* **2014**, *57*, 92–98. [CrossRef]
33. Buchachenko, A.L. Magnetic control of the earthquakes. *Open J. Earthq. Res.* **2021**, *10*, 138–152. [CrossRef]
34. Guglielmi, A.V.; Lavrov, I.P.; Sobisevich, A.L. Storm sudden commencements and earthquakes. *Sol.-Terr. Phys.* **2015**, *1*, 98–103. [CrossRef]
35. Guglielmi, A.V.; Klain, B.I. The solar influence on terrestrial seismicity. *Sol.-Terr. Phys.* **2020**, *6*, 111–115. [CrossRef]
36. Guglielmi, A.V.; Klain, B.I.; Kurazhkovskaya, N.A. Earthquakes and geomagnetic disturbances. *Sol.-Terr. Phys.* **2020**, *6*, 80–83. [CrossRef]
37. Zotov, O.D.; Guglielmi, A.V.; Silina, A.S. On possible relation of earthquakes with the sign change of the interplanetary magnetic field radial component. *Sol.-Terr. Phys.* **2021**, *7*, 59–66. [CrossRef]
38. Zotov, O.D. Weekend effect in seismic activity. *Izv. Phys. Solid Earth* **2007**, *43*, 1005–1011. [CrossRef]
39. Zotov, O.D.; Guglielmi, A.V. Problems of synchronism of electromagnetic and seismic events in the magnetosphere-technosphere-lithosphere dynamic system. *Solnechno-Zemn. Fiz.* **2010**, *16*, 19–25.
40. Guglielmi, A.V.; Zotov, O.D. The phenomenon of synchronism in the magnetosphere-technosphere-lithosphere dynamical system. *Izv. Phys. Solid Earth* **2012**, *48*, 486–495. [CrossRef]
41. Kasahara, K. *Earthquake Mechanics*; Cambridge University Press: Cambridge, UK, 1981; 284p.
42. Mogi, K. *Earthquake Prediction*; Academic Press Inc.: Tokyo, Japan; Orlando, FL, USA, 1985; 355p.
43. Aki, K.; Richards, P.G. *Quantitative Seismology*; University Science Books: Sausalito, CA, USA, 2002; 700p.
44. Guglielmi, A.V. Foreshocks and aftershocks of strong earthquakes in the light of catastrophe theory. *Phys.-Uspekhi* **2015**, *58*, 384–397. [CrossRef]
45. Zotov, O.D.; Guglielmi, A.V. Mirror triad of tectonic earthquakes. *arXiv* **2021**, arXiv:2109.05015.

46. Zavyalov, A.D.; Zotov, O.D. A new way to determine the characteristic size of the source zone. *J. Volcanol. Seismol.* **2021**, *15*, 19–25. [CrossRef]
47. Guglielmi, A.V.; Zotov, O.D.; Zavyalov, A.D.; Klain, B.I. On the Fundamental Laws of Earthquake Physics. *J. Volcanol. Seismol.* **2022**, *16*, 143–149. [CrossRef]
48. Zakharov, V.E.; Shamin, R.V. Probability of the occurrence of freak waves. *JETP Lett.* **2010**, *91*, 62–65. [CrossRef]
49. Zakharov, V.E.; Kuznetsov, E.A. Solitons and collapses: Two evolution scenarios of nonlinear wave systems. *Phys.-Uspekhi* **2012**, *182*, 569–592. [CrossRef]

Article

How Useful Are Strain Rates for Estimating the Long-Term Spatial Distribution of Earthquakes?

Sepideh J. Rastin *, David A. Rhoades, Christopher Rollins and Matthew C. Gerstenberger

GNS Science, 1 Fairway Drive, Avalon, P.O. Box 30-368, Lower Hutt 5040, New Zealand; d.rhoades@gns.cri.nz (D.A.R.); c.rollins@gns.cri.nz (C.R.); m.gerstenberger@gns.cri.nz (M.C.G.)
* Correspondence: s.rastin@gns.cri.nz; Tel.: +64-21-2810336

Abstract: Strain rates have been included in multiplicative hybrid modelling of the long-term spatial distribution of earthquakes in New Zealand (NZ) since 2017. Previous modelling has shown a strain rate model to be the most informative input to explain earthquake locations over a fitting period from 1987 to 2006 and a testing period from 2012 to 2015. In the present study, three different shear strain rate models have been included separately as covariates in NZ multiplicative hybrid models, along with other covariates based on known fault locations, their associated slip rates, and proximity to the plate interface. Although the strain rate models differ in their details, there are similarities in their contributions to the performance of hybrid models in terms of information gain per earthquake (IGPE). The inclusion of each strain rate model improves the performance of hybrid models during the previously adopted fitting and testing periods. However, the hybrid models, including strain rates, perform poorly in a reverse testing period from 1951 to 1986. Molchan error diagrams show that the correlations of the strain rate models with earthquake locations are lower over the reverse testing period than from 1987 onwards. Smoothed scatter plots of the strain rate covariates associated with target earthquakes versus time confirm the relatively low correlations before 1987. Moreover, these analyses show that other covariates of the multiplicative models, such as proximity to the plate interface and proximity to mapped faults, were better correlated with earthquake locations prior to 1987. These results suggest that strain rate models based on only a few decades of available geodetic data from a limited network of GNSS stations may not be good indicators of where earthquakes occur over a long time frame.

Keywords: earthquake forecasting; strain rates; multiplicative hybrid models; reverse testing; spatial distribution

Citation: Rastin, S.J.; Rhoades, D.A.; Rollins, C.; Gerstenberger, M.C. How Useful Are Strain Rates for Estimating the Long-Term Spatial Distribution of Earthquakes? *Appl. Sci.* **2022**, *12*, 6804. https://doi.org/10.3390/app12136804

Academic Editors: Eleftheria E. Papadimitriou and Alexey Zavyalov

Received: 25 May 2022
Accepted: 2 July 2022
Published: 5 July 2022

Publisher's Note: MDPI stays neutral with regard to jurisdictional claims in published maps and institutional affiliations.

1. Introduction

Estimation of the long-term spatial distribution of earthquakes is an essential component of Probabilistic Seismic Hazard Analysis (PSHA). In the traditional approach to PSHA, the estimation is based mainly on the combined information from studies of the long-term slip rates and rupture histories of mapped faults and spatial smoothing of the locations of past earthquakes. However, other emerging data streams, including geodetic observations from the global navigation satellite system (GNSS) and derived strain-rate models, seem to offer promise for improving the estimation [1–5]. In a revision of the New Zealand (NZ) National Seismic Hazard Model (NZNSHM) currently being undertaken, we are interested in assessing the relative value of different available data inputs, including information from fault studies, tectonics, the earthquake catalogue, and strain rate models, for estimating the long-term spatial distribution of earthquakes. In particular, we are interested in assessing how strongly strain rate estimates based on a few decades of geodetic data should be weighted when estimating the long-term spatial distribution of earthquakes.

A multiplicative hybrid modelling framework [6] was developed to optimally combine a set of spatially gridded forecasts submitted to the five-year regional earthquake likelihood models (RELM) experiment in California [7,8]. It was found that, although one best model strongly outperformed all others individually [9], it was possible to find several hybrid models that could outperform the best individual model over the 5-year period of the experiment. Such hybrid models were formed from pairs of models, including the best model as the baseline.

Multiplicative hybrid modelling has since been used to combine gridded information from fault, earthquake, and strain rate covariates in NZ [10,11] with a simple Stationary Uniform Poisson (SUP) baseline earthquake likelihood model. When hybrids only of fault- and earthquake-based variables were considered, any hybrid model that included one particular covariate called "Proximity to Mapped Faults" (PMF) outperformed all hybrids without PMF in both a 20-year fitting period and an independent testing period [10]. The PMF covariate was derived from the locations of NZ mapped faults and their estimated slip rates. Subsequently, Rhoades et al. [11] used models for maximum shear, rotational and dilatational strain rates by Beavan [12], based on GNSS data from 1996–2011, as additional covariates. They found that any hybrid including shear strain rate (BSS) outperformed all hybrids not including BSS in both the 20-year fitting period and an independent testing period.

In the present study, we restrict our attention to (maximum) shear strain rate models. In addition to BSS, we consider the global shear strain rate model of Kreemer et al. [13] based on GNSS data over the period 1996–2013 (GSS) and an updated NZ strain rate model by Haines and Wallace [14] based on GNSS data from 1995–2013 (HWS). We analyse the performance of multiplicative hybrids during the fitting and testing periods used by Rhoades et al. [11] with GSS and HWS as alternative strain rate covariates. We then examine the stability of the multiplicative modelling results by testing them against earthquakes from 1951 to 1986. Furthermore, we evaluate the strain rate and other covariates as earthquake predictors using Molchan error diagrams. We also produce smoothed scatter plots of covariate values corresponding to targeted earthquakes against time from 1951 to 2020. These results are relevant to the question of how useful strain rate estimates are for the assessment of long-term seismic hazards.

2. Method and Data

2.1. Multiplicative Hybrid Model

We follow the method used by Rhoades et al. in [6,10,11] to produce hybrid earthquake likelihood models from a baseline model and a set of covariates. As in [10,11], the stationary uniform Poisson (SUP) model is used as the baseline model, and subsets of the available spatially varying covariates are assimilated into it to form a range of hybrid models.

The hybrid and baseline models are regional earthquake likelihood models [8] defined by a set of expected numbers of earthquakes $\{\lambda(j,k)\}$ in cells of location, indexed by j, and magnitude, indexed by k, where $j = 1, \ldots, n_j$, and $k = 1, \ldots, n_k$. The baseline model is denoted by $\{\lambda_0(j,k)\}$, and the hybrid model by $\{\lambda_H(j,k)\}$. The n_i covariates take values in the location cells only and are denoted by $\{\rho_i(j), j = 1, \ldots, n_j; i = 1, \ldots, n_i\}$. The covariates may be real-valued or categorical (e.g., binary) variables. The multiplicative hybrid model is of the form

$$\lambda_H(j,k) = \lambda_0(j,k)\exp(a)\Pi_{i=1}^{n_i}f_i[\rho_i(j)]. \tag{1}$$

Here, a is a normalising parameter that ensures the total expected numbers of earthquakes in the hybrid model match with the actual number in the catalogue to which it is fitted. The function f_i converts a cell value of the ith covariate into a multiplier to be applied to the corresponding cells of the baseline model. For a real-valued covariate, the multiplier f_i is a non-negative, monotone, non-decreasing and nonlinear function of the form

$$f_i[\rho] = \exp(b_i[\ln(1+\rho)]^{c_i}), \ (b_i \geq 0; c_i \geq 0). \tag{2}$$

It thus preserves the ordering of values in cells but not the ratio of values between cells. For a binary covariate, where $\rho(j)$ takes only the values 0 and 1, f_i is of the form

$$f_i[\rho] = \exp(d_i\rho), \tag{3}$$

where adjustable parameters d_i are allowed to take positive or negative values.

The adjustable parameters in the hybrid model are the normalising parameter a, the shape parameters b_i and c_i for each real-valued covariate, and d_i for each binary covariate. In fitting a hybrid, the parameters are chosen to optimize the log-likelihood, $\ln L$, of the target earthquakes, given by Rhoades et al. [15] as

$$\ln L = \sum_{n=1}^{N} \ln \lambda_H(j_n, k_n) - \hat{N}_H, \tag{4}$$

in which the target earthquakes occur in cells $\{(j_n, k_n), n = 1, \cdots, N\}$, and $\hat{N}_H$ denotes the total expected number of target earthquakes, i.e.,

$$\hat{N}_H = \Sigma_{j=1}^{n_j} \Sigma_{k=1}^{n_k} \lambda_H(j, k). \tag{5}$$

Following Rhoades et al. [6], the corrected information gain per earthquake (IGPEc) of a fitted hybrid model $\{\lambda_H(j,k)\}$ over the baseline model $\{\lambda_0(j,k)\}$ is calculated using

$$\text{IGPEc} = \frac{-\Delta}{2N}, \tag{6}$$

in which Δ is the change in the corrected Akaike Information Criterion (AICc) between the hybrid model and the baseline model. AICc is defined by

$$\text{AICc} = -2\ln L + 2p + \frac{p+1}{N-p-1}, \tag{7}$$

in which p is the number of fitted parameters in the hybrid model, and N is the number of target earthquakes [16].

Confidence limits on IGPEc are estimated using an adaptation of the T-test [15] for comparing one earthquake likelihood model to another based on independent data, as described by Rhoades et al. [6].

For testing of the fitted models on independent data, in which there are no adjustable parameters, the information gain per earthquake (IGPE) is simply the change in log-likelihood between the hybrid and baseline models divided by the number of target earthquakes, and the T-test given in [15] applies.

The covariates are defined on the NZ Collaboratory for the Study of Earthquake Predictability (CSEP) testing grid of 0.1-degree rectangular cells [17]. The covariates used in this study are strain rates (BSS, GSS, and HWS), Proximity to the Plate Interface (PPI), Proximity to Mapped Faults (PMF), and Fault in cell (FLT). We have provided details of strain rate covariates in Appendix A. The non-strain-rate covariates are slightly modified compared to those used by Rhoades et al. [10]. We have provided information on these modifications in Appendix B. It is not meant to include the smoothed seismicity covariates from Rhoades et al. [10] because they were fitted to the earthquakes prior to 1986 that are now used for reverse testing. Figure 1 displays the spatial distributions of the non-strain-rate covariates.

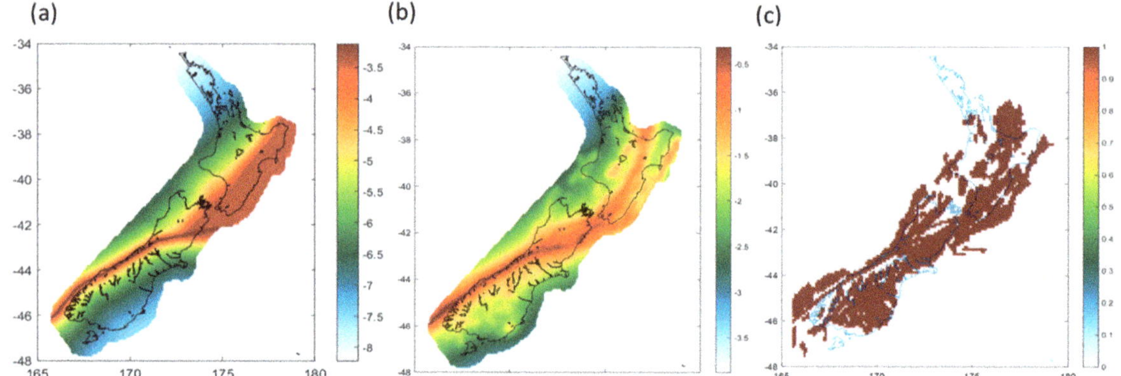

Figure 1. Spatial distributions of the modified (**a**) Proximity to plate interface (PPI) and (**b**) Proximity to mapped faults (PMF) covariates (log of grid cell values). The colour scale shows the logarithm of the relative rate. (**c**) Spatial distribution of updated Fault in cell (FLT) binary covariate using all faults from the New Zealand Community Fault Model [18].

The strain rate covariates (BSS, GSS, and HWS) are included separately in hybrid models as single covariates and with all possible combinations of the other (non-strain-rate) covariates. Hybrid models are fitted to the period 1987–2006, with forward testing over the period 2012–2015 and reverse testing over the period 1951–1986. The spatial distributions of the strain rate covariates are shown in Figure 2.

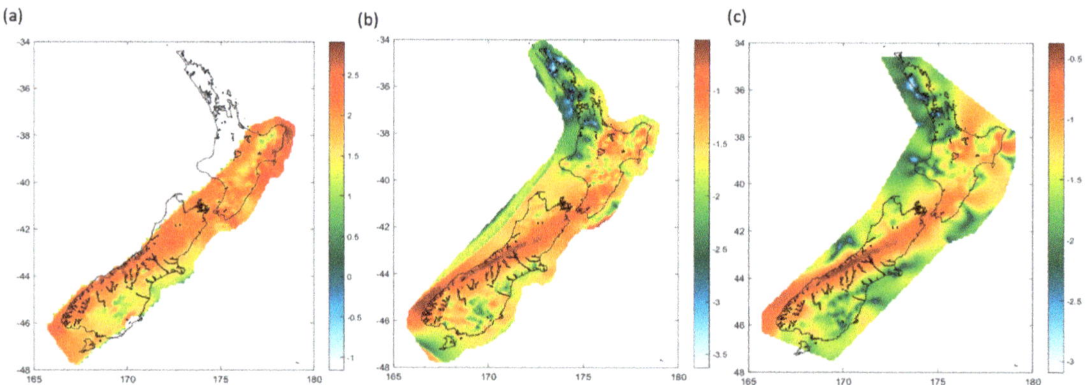

Figure 2. Spatial distribution of (**a**) GSS (1996–2013), (**b**) BSS (1996–2011), and (**c**) HWS (1995–2013) strain rate models in the New Zealand CSEP testing region. The colour scale shows the logarithm of the relative rate.

2.2. Molchan Error Diagram

We obtain Molchan error diagrams [19,20] for all covariates at different time intervals. Molchan error diagrams are equivalent to Receiver Operating Characteristic curves, which have been widely used in other contexts [21,22]. They can be used to investigate the correlation between a spatially distributed covariate and the spatial distribution of target earthquakes.

The error diagram is based on the concept of an alarm being declared wherever a covariate exceeds a certain threshold. As the threshold is decreased, the proportion of space occupied by alarms increases, and the proportion of target earthquakes outside of the alarm space (i.e., the proportion of "unpredicted earthquakes") decreases. In the error diagram, the proportion of unpredicted earthquakes is plotted against the proportion of space occupied by alarms, but the threshold for declaring an alarm is not shown. The error diagram is, therefore, a monotone decreasing curve.

For an ideal predictor, all target earthquakes are predicted by an alarm occupying almost zero area. The error diagram curve then consists of the line segments joining the point (0,1) to (0,0) and the point (0,0) to (1,0). For a totally uninformative predictor, the proportion of space occupied would be the same as the proportion of predicted earthquakes. The error diagram curve would then be the diagonal line joining (1,0) to (0,1).

The area above the error diagram curve within the unit box is called the Area Skill Score (ASS) [23]. The ASS is a rough measure of the overall potential the covariate has for predicting the locations of target earthquakes. An ideal predictor has an ASS of 1, and a totally uninformative predictor has an ASS of 0.5.

If a covariate is better correlated with the locations of target earthquakes during one time period than another, then the error diagram will vary with time accordingly. Therefore, we plot and compare error diagrams for different periods.

2.3. Scatter Plot of Covariates for the Target Earthquakes

We have investigated gradual changes over time in the correlation of a covariate with earthquake locations using a simple scatter plot. The scatter plot displays values of a covariate in cells corresponding to target earthquakes versus time. Trends in these values, whether decreasing or increasing, are revealed by a smoothed local regression fit to the data.

2.4. Data

The target earthquakes in this study are events from 1951 to 2020 with a hypocentral depth between 0 and 40 km and a magnitude greater than 4.95 (GeoNet preferred magnitude) within the CSEP testing region [17]. The events of interest were extracted from the GeoNet catalogue of NZ earthquakes (www.geonet.org.nz; last accessed on 1 July 2022). Magnitudes assigned by GeoNet are mostly local magnitudes, but moment magnitudes are preferred for larger events. The catalogue is consistent with that used by Rhoades et al. [10,11] within the fitting and testing periods of those studies.

3. Results

3.1. Hybrid Models

IGPE estimates and 95% confidence intervals of the hybrid models with BSS for the fitting period (1987–2006), forward testing period (2012–2015), and reverse testing period (1951–1986) are presented in Figure 3. Corresponding results with GSS and HWS are presented in Figures 4 and 5, respectively. The IGPEs are relative to the SUP baseline model. To compare the IGPEs of two hybrid models, one needs to obtain the differences between their log-likelihoods for the target earthquakes and perform a T-test on those differences. In this study, we have only presented T-tests for comparing each hybrid model to the baseline model.

The combination of covariates contributing to each hybrid is indicated by a model number composite. This composite combines the baseline model identifier with those of the selected covariates: "1" for SUP, "2" for PPI, "3" for PMF, "4" for FLT, and "5" for strain rate covariate (BSS, GSS, or HWS). Table 1 contains the name, acronym, and identifier for the baseline model and covariates. For example, the composite "123" represents a hybrid with SUP as baseline and PPI and PMF as covariates.

Table 1. Name, acronym, and identifier for the baseline model and covariates.

Covariate or Model Name	Acronym	Identifier
Stationary uniform Poisson baseline model	SUP	"1"
Proximity to plate interface covariate	PPI	"2"
Proximity to mapped faults covariate	PMF	"3"
Fault in cell covariate (binary)	FLT	"4"
Beavan shear strain rate covariate	BSS	"5"
Global shear strain rate covariate	GSS	"5"
Haines and Wallace shear strain rate covariate	HWS	"5"

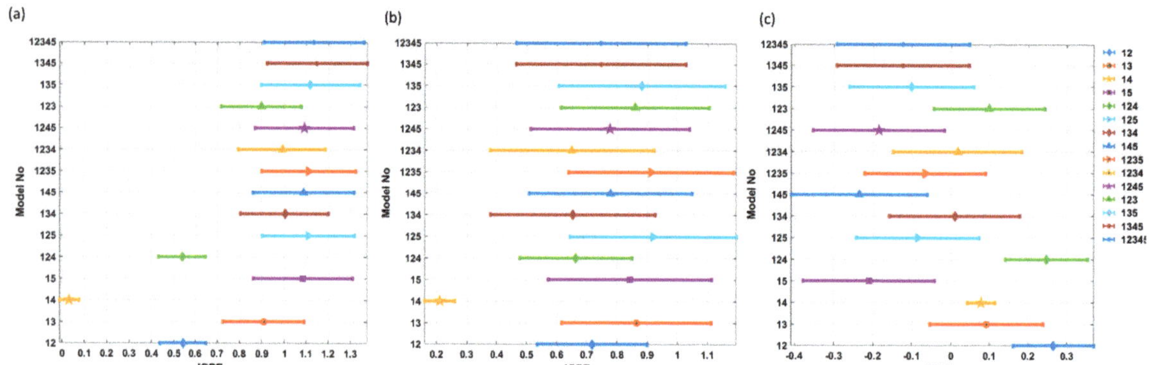

Figure 3. Information gain per earthquake (IGPE) of hybrid models relative to SUP baseline model with the Beavan shear strain rate (BSS) covariate. The combination of covariates contributing to each hybrid is indicated by the model number, which is a composite of the numbers corresponding to the baseline model and selected covariates, as follows: 1-SUP; 2-PPI; 3-PMF; 4-FLT; 5-BSS. (**a**) Fitting period (1987–2006); (**b**) Forward testing period (2012–2015); (**c**) Reverse testing period (1951–1986). Error bars are 95% confidence intervals.

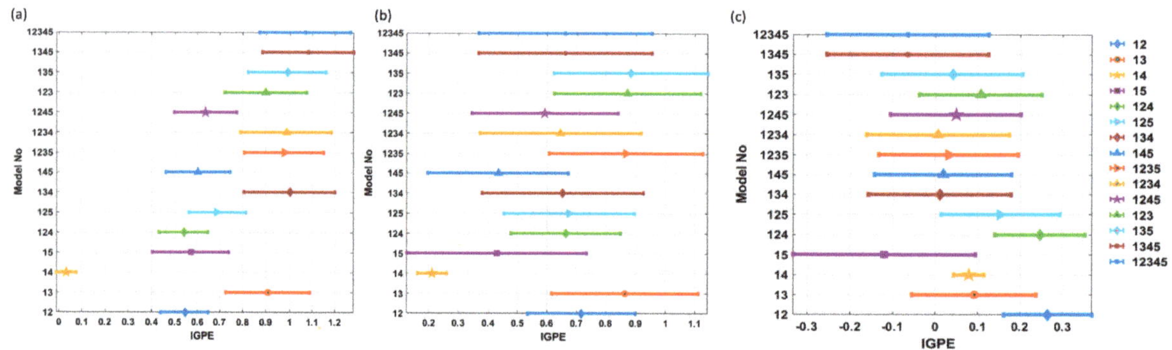

Figure 4. Information gain per earthquake (IGPE) of hybrid models relative to SUP baseline model with the GSS strain rate covariate. The combination of covariates contributing to each hybrid is indicated by the model number, which is a composite of the numbers corresponding to the baseline model and selected covariates, as follows: 1-SUP; 2-PPI; 3-PMF; 4-FLT; 5-GSS. (**a**) Fitting period (1987–2006); (**b**) Forward testing period (2012–2015); (**c**) Reverse testing period (1951–1986). Error bars are 95% confidence intervals.

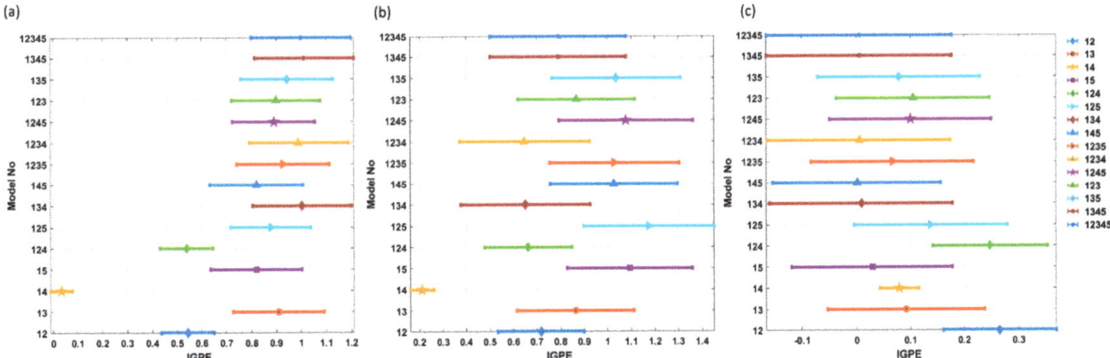

Figure 5. Information gain per earthquake of hybrid models relative to SUP baseline model with the HWS strain rate covariate. The combination of covariates contributing to each hybrid is indicated by the model number, which is a composite of the numbers corresponding to the baseline model and selected covariates, as follows: 1-SUP; 2-PPI; 3-PMF; 4-FLT; 5-HWS. (**a**) Fitting period (1987–2006); (**b**) Forward testing period (2012–2015); (**c**) Reverse testing period (1951–1986). Error bars are 95% confidence intervals.

Results for the fitting and forward testing period with the BSS covariate were previously reported by Rhoades et al. [11] and showed that the IGPE was higher for all hybrid models including BSS than for all models not including BSS. Here, with modified non-strain-rate covariates for the fitting period (Figure 3a), all hybrids including BSS (composites containing "5") have IGPE > 1.1. The largest IGPE is 1.48 for "1345". On the other hand, all hybrids excluding BSS have IGPE < 1.1. For these, the largest IGPE is 1.00 for "134".

When the BSS covariate is replaced by either GSS (Figure 4a) or HWS (Figure 5a), the results for the fitting period are different. With BSS, "15" (SUP + BSS) has the largest IGPE amongst all the simple hybrids. However, with GSS and HWS, "13" (SUP + PMF) outperforms "15". However, similar to BSS, the largest IGPE values for GSS and HWS are obtained for "1345" (1.08 and 1.01, respectively).

For the forward testing period, the hybrids again perform differently with different strain rate covariates. With BSS (Figure 3b), the hybrids including strain rates are not as dominant as in Rhoades et al. [11], due to the revision of the non-strain-rate covariates (Appendix B). Hybrid "13" performs almost as well as "15", i.e., the IGPEs of both hybrids are similar. With GSS (Figure 4b), "13" performs much better than "15". With HWS, "15" outperforms "13" and all other simple hybrids. Moreover, all hybrid models including HWS outperform most hybrids without HWS except "123".

Hybrids with HWS perform not quite as well as those with BSS in the fitting period (Figures 4a and 5a) but better than those with BSS in the forward testing period (Figures 4b and 5b). Larger IGPEs in the forward testing period may be due to a partial overlap between this period (2012–2015) and the time range of GNSS data used to derive HWS (1995–2013). Hybrids with GSS perform not as well in either the fitting or forward testing period as their counterparts with BSS and HWS (Figure 4a, Figure 5a, and Figure 6a, and Figure 4b, Figure 5b, and Figure 6b, respectively).

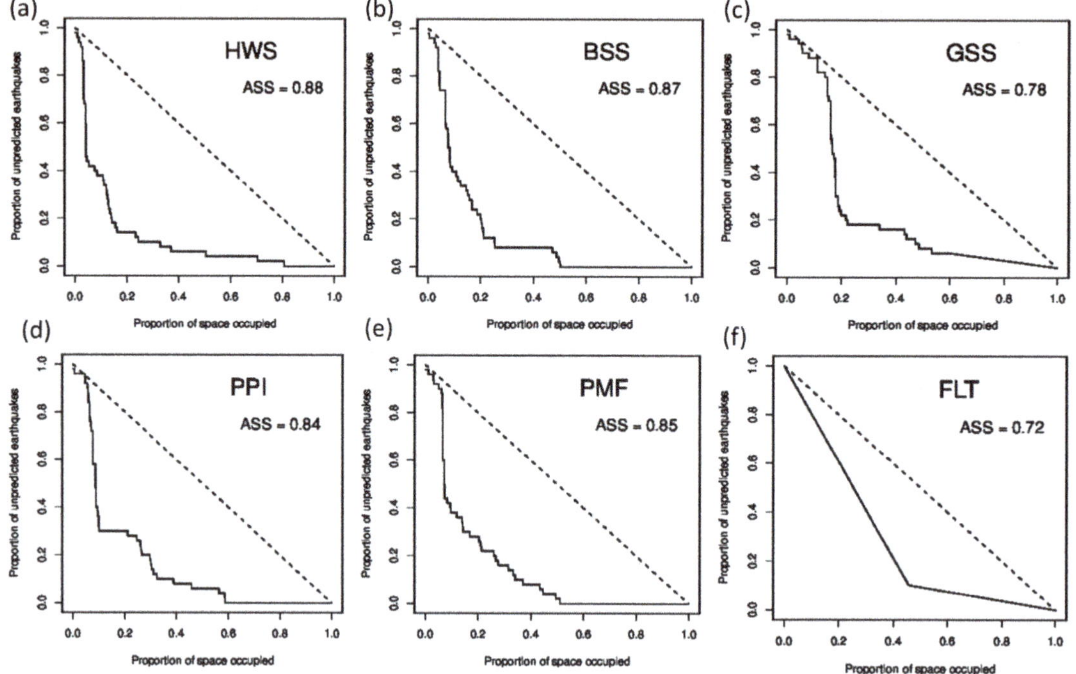

Figure 6. Molchan error diagrams for covariates (**a**) HWS, (**b**) BSS, (**c**) GSS, (**d**) PPI, (**e**) PMF, and (**f**) FLT, and target earthquakes in the forward test period 2012–2015.

Hybrids with BSS perform poorly in the reverse testing period 1951–1986 (Figure 3c) and even worse than those without BSS, including the SUP baseline (since their IGPEs are negative). For all models without BSS, IGPE is mostly lower than in the fitting and forward testing periods. The largest IGPE for the reverse testing period was 0.27, obtained by hybrid "12", and is only half of the corresponding IGPE for the fitting period.

With GSS, the IGPE of the "15" hybrid is again negative in the reverse testing period, as are the IGPEs for "12345" and "1345". With HWS, the "15" hybrid has a small positive IGPE of 0.03 for the reverse testing period, and all the hybrids have small positive IGPE values (Figure 5c). Thus, hybrid models with HWS perform slightly better than the corresponding models with BSS or GSS in the reverse testing period (c.f. Figures 3c, 4c and 5c). However, overall, the performance of hybrids including strain rates is poor in the reverse testing period.

3.2. Error Diagrams

A hybrid model can perform poorly in an independent testing period either because the covariates that contribute to it are not well correlated with earthquake locations or because the model is overfitted to a training period in which the covariates are better correlated with earthquake locations than in the testing period. Molchan error diagrams give an indication of the correlation between covariates and earthquake locations in different time periods independent of any model fitting. For the error diagram analysis, we have split the 36-year reverse testing period into three separate 12-year periods.

In the forward testing period 2012–2015, the non-strain-rate covariates PPI and PMF and strain covariates BSS and HWS are well-correlated with the locations of target earthquakes, as shown by the error diagrams deviating far below the diagonal (Figure 6). The ASS for each of these covariates is rather high, between 0.85 and 0.88. HWS has the highest ASS of 0.88. Again, this may be due to the partial overlap between the forward testing period and the time range of GNSS data used to derive HWS. In contrast, GSS and FLT have smaller ASS values of 0.78 and 0.72, respectively, which implies they are not as strongly correlated with the locations of target earthquakes. This is reflected in the error diagrams not reaching as far below the diagonal as those of the other covariates.

The correlation between the covariates and target earthquake locations varies appreciably over time (Figure 7). The strain rate covariates, especially BSS, show this variation more than the other covariates. Figure 7 compares error diagrams and ASS values for each covariate in five selected time periods—1951–1962; 1963–1974; 1975–1986; 1987–2006; and 2012–2015.

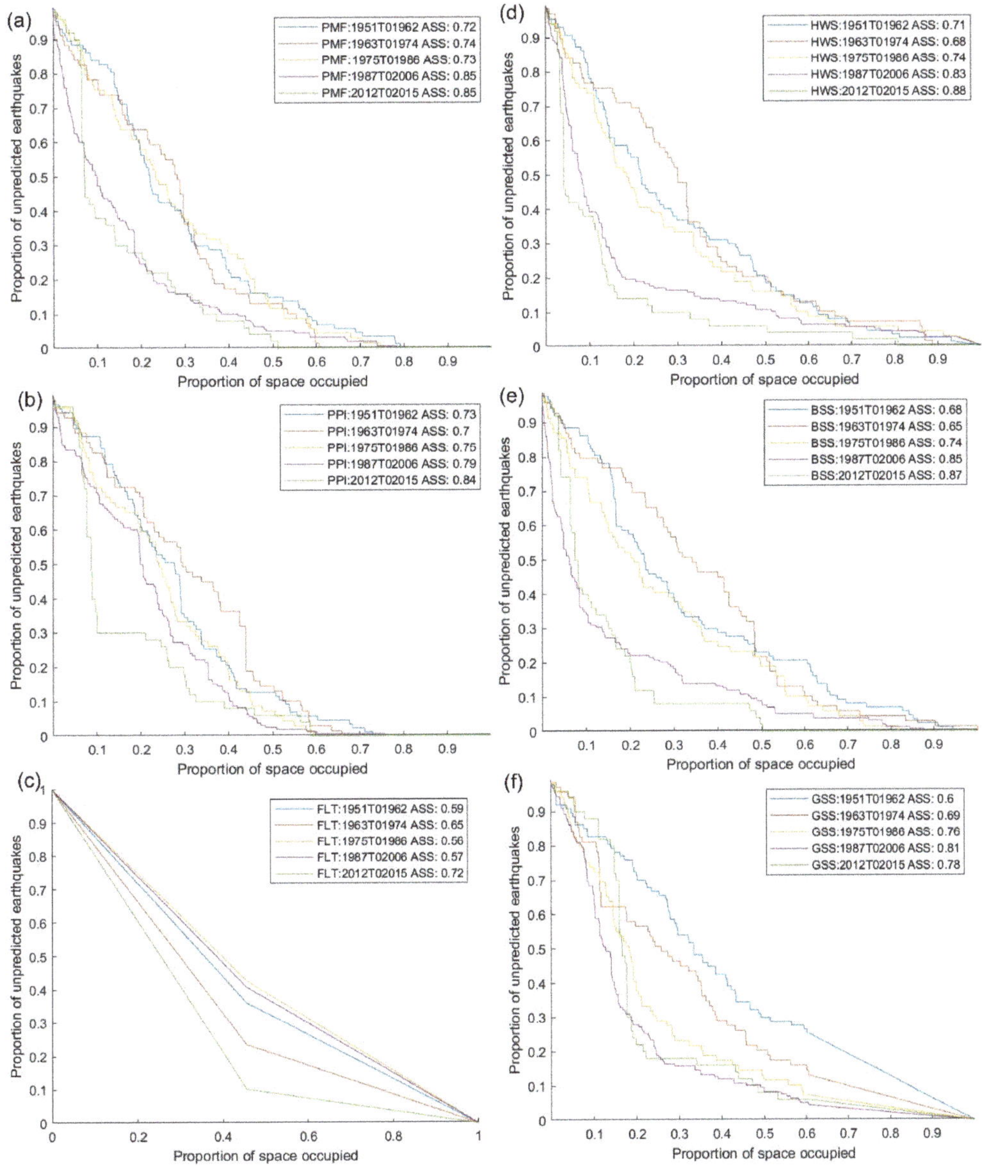

Figure 7. Molchan error diagrams for the covariates during different time intervals. (**a**) HWS; (**b**) PMF; (**c**) BSS; (**d**) PPI; (**e**) GSS; (**f**) FLT.

The range of variations in ASS values for a single covariate is an indicator of how widely the correlation varies between time periods. For the PMF covariate (Figure 7a), the correlation is equally high during fitting and forward testing periods ($ASS_{1987–2006} = ASS_{2012–2015} = 0.85$) and lowest during 1951–1962 ($ASS_{1951–1962} = 0.72$). For the PPI covariate (Figure 7b), the correlation with earthquake locations is highest during 2012–2015 ($ASS_{2012–2015} = 0.84$) and lowest during 1963–1974 ($ASS_{1963–1974} = 0.70$). For the FLT covariate (Figure 7c), the correlation is highest during the forward testing period ($ASS_{2012–2015} = 0.72$) and lowest during 1975–2006 ($ASS_{1975–1986} = 0.56$, $ASS_{1987–2006} = 0.57$). For the HWS covariate (Figure 7d), the correlation is highest during 2012–2015 ($ASS_{2012–2015} = 0.88$) and lowest during 1963–1974 ($ASS_{1963–1974} = 0.68$). For the BSS covariate (Figure 7e), the correlation is highest during 2012–2015 ($ASS_{1987–2006} = 0.87$) and lowest during 1963–1974 ($ASS_{1963–1974} = 0.65$). For the GSS covariate (Figure 7f), the correlation is highest during 1987–2006 ($ASS_{1987–2006} = 0.81$) and lowest during 1951–1962 ($ASS_{1951–1962} = 0.60$). The low ASS can be explained by the fact that many of the target events in 1951–1962 were located where GSS had a value of zero (Figure 8).

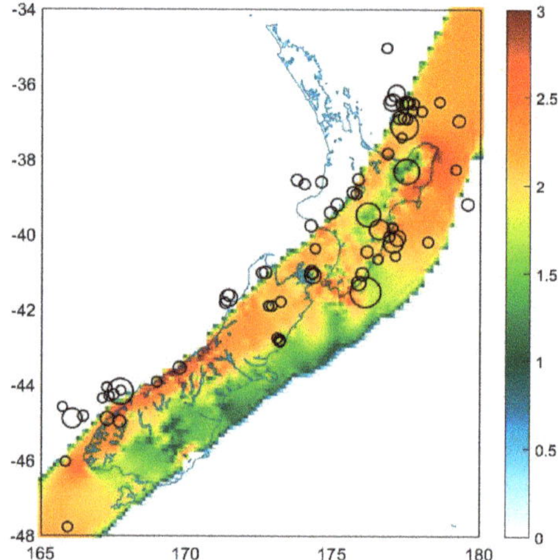

Figure 8. Spatial distribution of GSS and target earthquakes during 1951–1962.

The variation of ASS values with time period is plotted for each covariate in Figure 9. No covariate has a consistently higher ASS than the other covariates across all time periods. As can be seen, the highest ASS among the covariates in 1951–1962 is that of PPI, in 1963–1974 that of PMF, in 1975–1986 that of GSS, in 1987–2006 that of BSS, and in 2012–2015 that of HWS. At the other end of the scale, the FLT covariate has the lowest ASS value in all time periods except 1963–1974, in which it shares the lowest value with BSS. Among the strain rate covariates, the highest ASS value is that of the HWS from 1951 to 1962 and 2012 to 2015, GSS from 1963 to 1974 and 1975 to 1986, and BSS from 1987 to 2006. Figure 9 also shows that the HWS covariate has a higher ASS value than BSS in all time periods except 1987–2006—the fitting period for the multiplicative models.

3.3. Smoothed Scatter Plots

To give a better insight into changes over time in the correlation of covariates with earthquake locations, we plotted the covariate values for the cells corresponding to target earthquakes versus time over the period 1951–2020. We then fitted a trend curve using locally weighted scatter plot smoothing (lowess) based on Cleveland's method in [24]. We observed clear trends in the strain rate covariate values corresponding to target earthquake locations over the period 1951–2020, as shown by Figure 10.

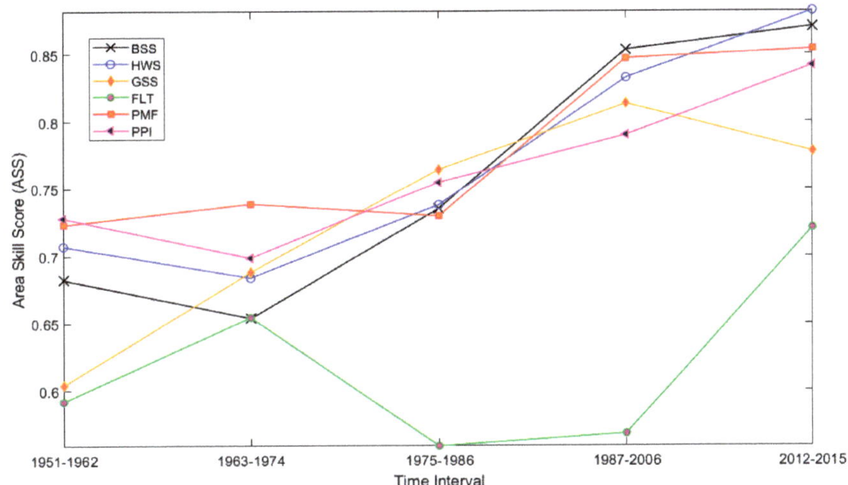

Figure 9. Area Skill Score (ASS) of Molchan error diagram for covariates over different times.

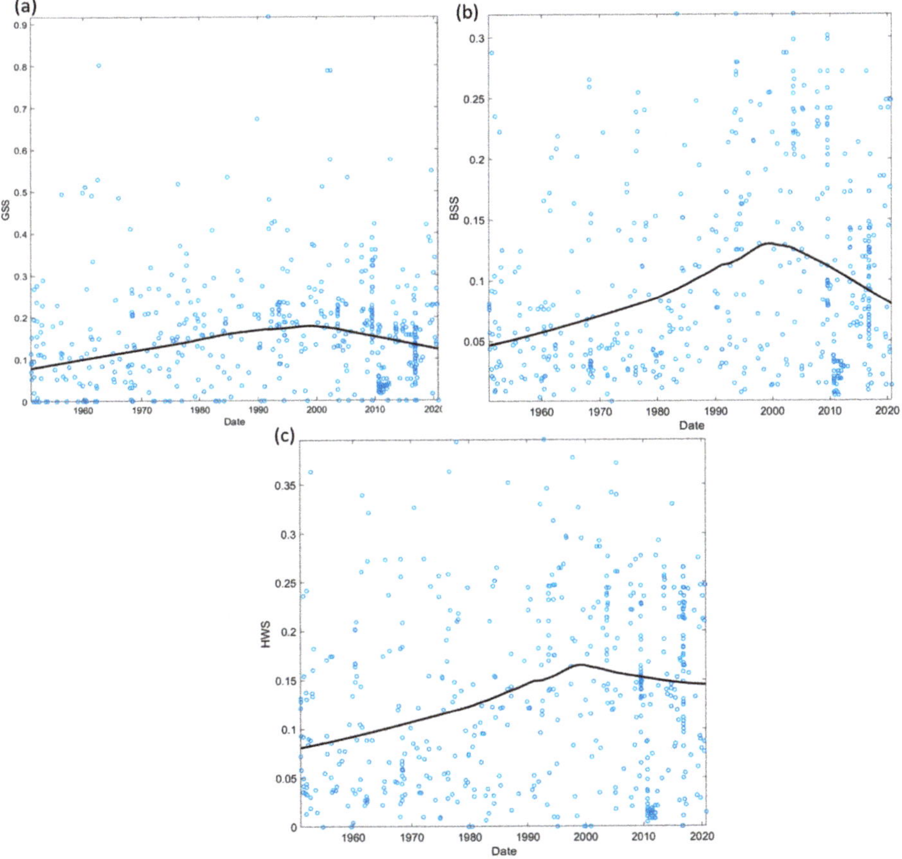

Figure 10. Smoothed scatter plots of (**a**) GSS, (**b**) BSS, and (**c**) HWS strain rate covariates for targeted earthquakes against time from 1951 to 2020.

The trend is stronger for BSS and HWS than for GSS. However, for all three covariates, the smoothed trend line starts from its lowest level in 1951, rises gradually to a peak in 1999, and then tails off somewhat through to 2020. However, the HWS curve also tails off less than the others. These results confirm and elaborate on the trends over time that was observed in Figures 8 and 10. The HWS model was extracted from the longest period of recordings (1995–2013) and a larger number of GNSS stations compared with the GSS and BSS models. In addition, unlike BSS and GSS that were both based on an older method of Haines and Holt [25–27], the HWS model was derived using the novel VDOHS technique [14]. Such differences could explain the differences observed in the smoothed scatter plots versus time.

We observed that FLT was the only covariate to show a clear trend with the magnitude of target earthquakes. Figure 11 plots the FLT values versus the magnitudes of the target earthquakes from 1951 to 2020. The locally weighted smoothed curve, shown by the solid black line, reveals an increasing trend with magnitude. This may not necessarily imply that future large earthquakes are more likely to occur on known faults than smaller earthquakes. It may instead reflect the fact that large earthquakes cause surface ruptures more frequently than small ones; consequently, new ruptures over the past 70 years have revealed the associated faults and allowed them to be mapped.

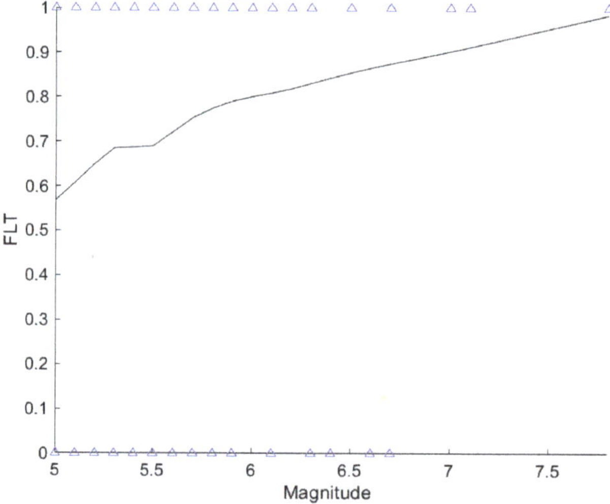

Figure 11. Fault covariate values for targeted earthquakes against magnitude, 1951–2020. The solid black line shows the smoothed scatter plot. The blue triangles depict the values of the FLT covariate for each target earthquake.

4. Discussion

The low information gain of hybrid models including strain rate covariates in reverse testing is consistent with the relatively low ASS values of the strain rate covariates and low strain rate values associated with target earthquakes from 1951 to 1986. The ASS values of the strain rate covariates in the reverse testing period are not only lower than in the fitting period but are also lower than the ASS values of the PPI and PMF covariates during the reverse testing period (Figures 7 and 8). The contrast of ASS values between the fitting and reverse testing periods is particularly strong for BSS. This explains the negative IGPEs for hybrids with BSS in the reverse testing period (Figure 3c).

Similarly, the large IGPE for hybrids with HWS in forward testing is consistent with the high ASS for HWS in 2012–2015 and with the slow tailing off between 2010 and 2020 of the trend in HWS associated with target earthquakes. Evidently, if the forward testing period were extended through to 2020, hybrids including HWS would have continued to perform well.

Differences between the three present strain rate models can be attributed to the differences in the geodetic datasets on which they were based and the evolving techniques for estimating strain rates from the geodetic data. The most recently developed of the three models is the HWS model, which was published in 2020. HWS was derived from the most complete data and up-to-date techniques and showed the best correlation with the locations of the most recent earthquakes (Figure 7). However, all three strain rate models exhibit a disappointingly low correlation with the locations of earthquakes from 1951 to 1986.

The implicit assumption behind the fitting of hybrid models to estimate long-term earthquake rates is that the correlation between the covariates and the locations of target earthquakes is stable over time. Alternatively, if not stable over time, the correlation during the fitting period should be indicative of the long-term correlation. The correlations for PPI and PMF appear to be more stable over time than those for the strain rate covariates, based on the ASS variations. PPI and PMF represent features that are expected to affect earthquake occurrence over a long time frame. The greater instability of the correlation for strain rate covariates may be attributed to the short-term non-stationarity of earthquake occurrence. The strain rate covariates are most highly correlated with earthquake occurrence within the time period of the data on which they are based or within a few years after (Figure 7).

Without a dense GNSS network and extensive historical and paleo-seismicity records, it is not possible to pinpoint temporal and spatial variations of strain rates. Recently, Iezzi et al. obtained high temporal histories of slip rates for three parallel normal faults in central Greece with progressive clusters using in situ 36Cl cosmogenic dating [28]. They also had access to associated GNSS recordings of a dense network. They compared regional decadal geodetic strain rates with strain rates derived from slip-rate pulses over a few thousand years on each fault. They found that strain builds up across a fault system over thousands of years, but not all individual faults actively slip at the same time. They discussed how the available geodetic data alone was not capable of pinpointing the switchover of slip from one fault to another over time. In comparison, New Zealand has neither a dense GNSS network nor a comparable record of fault slip histories. The density of GNSS stations has been progressively improving but is still sparse in some parts and limited to onshore regions [14]. Obviously, this affects the accuracy of the strain rate models in NZ offshore regions, where many of the earthquakes occur.

Given the low earthquake location–strain rate correlations in the reverse testing period here, it is expected that there will be future periods during which such correlations are similarly low.

In fitting hybrid models for estimating long-term seismicity rates, it seems best to use the longest possible fitting period subject to earthquake catalogue quality considerations. The NZ catalogue covers about 180 years of historical and instrumental records. With the improvement of the NZ seismographic network, the magnitude of completeness has lowered to 5.0 for the recent 70 years of the GeoNet catalogue. Such a level of magnitude completeness fulfils the magnitude threshold of 4.95 applied to the target earthquakes for this study. However, we cannot be confident that a 70-year period, or any other fixed period, is long enough to represent the long-term spatial distribution of earthquakes. The main purpose of the hybrid fitting is to realistically weight the information offered by the available covariates, none of which depend directly on the catalogue. Here we showed that it would be unwise to base this weighting on a 20-year subset of the available catalogue. Performing tests with the 180-year catalogue could potentially help us to understand how adequate a 70-year catalogue is for fitting such hybrid models.

The robustness of our results against time-varying errors in earthquake locations could be questioned. However, since the strain rate models and all other covariates were defined on a rather coarse grid of 0.1×0.1-degree cells, they naturally can accommodate location errors without having much effect on the correlation with earthquake locations. We examined the robustness of the correlations between all covariates and target earthquake locations by introducing synthetic location errors within the range reported by Bondár et al. [29] to be the most probable values for the ISC-GEM catalogue. The location errors were added to the epicentres of target events from 1987 onwards. We tested the hypothesis that events prior to 1987 had larger location errors due to the seismograph network's sparsity, and that could be the reason for low correlations between the covariates and target earthquake locations. However, this hypothesis was rejected; despite the added location errors for all the covariates, the main features of the smoothed curves were preserved.

5. Conclusions

We analysed the performance of multiplicative hybrid models in forward and reverse testing periods. To complement the performance analysis, we also obtained Molchan error diagrams and smoothed scatter plots of strain rate covariates versus time. All the analyses reveal strong variability over time in the correlation of strain rates with earthquake locations. These results suggest that strain rate models based on the available geodetic data for only a few decades may not be good indicators of where earthquakes occur over a long time frame. Therefore, the longest available high-quality earthquake catalogue should be used in constructing hybrid models to estimate long-term earthquake rates for seismic hazard analysis.

Author Contributions: Conceptualization, S.J.R., D.A.R. and M.C.G.; methodology, S.J.R. and D.A.R.; software, S.J.R., D.A.R. and C.R.; formal analysis, S.J.R. and D.A.R.; writing—original draft preparation, S.J.R. and D.A.R.; writing—review and editing, S.J.R., D.A.R., C.R. and M.C.G. All authors have read and agreed to the published version of the manuscript.

Funding: This research was funded by the Strategic Science Investment Fund (SSIF) and the National Seismic Hazard Model Revision Project of the Ministry of Business, Innovation and Employment of New Zealand.

Institutional Review Board Statement: Not applicable.

Informed Consent Statement: Not applicable.

Data Availability Statement: The New Zealand Earthquake Catalogue was obtained from GeoNet. Available online: http://www.geonet.org.nz (accessed on 30 May 2022). Data on earthquake faults and slip rates were obtained from the New Zealand Community Fault Model v1.0 (Data Set), GNS Science, https://doi.org/10.21420/NMSX-WP67 (accessed on 30 May 2022).

Acknowledgments: We acknowledge the New Zealand GeoNet project and its sponsors EQC, GNS Science, LINZ, NEMA, and MBIE for providing the data used in this study. Kiran Kumar Thingbaijam and David Burbidge provided helpful internal reviews of the manuscript.

Conflicts of Interest: The authors declare no conflict of interest. The funders had no role in the design of the study; in the collection, analyses, or interpretation of data; in the writing of the manuscript, or in the decision to publish the results.

Appendix A

Appendix A.1 Strain Rate Covariates

Beavan and Haines [30] utilized GNSS data from 9 geodetic networks across NZ, comprising 362 GNSS stations. They calculated continuous horizontal velocity and associated strain rate fields at the Earth's surface in NZ between 1991 and 1998. They adopted the method proposed by Haines et al. [26], which enhanced NZ's known deformations and discovered new features in both the North and South Islands.

John Beavan (1950–2012) directed most of the GNSS data acquisition in NZ [14]. His last shear strain rate model of NZ was based on survey-mode GNSS data collected in NZ between January 1996 and February 2011. This model was based on the method of Beavan and Haines [30] and was submitted to Land Information New Zealand (LINZ) to be used in updating the geodetic and cadastral systems following the 4 September 2010 Darfield earthquake. This was the shear strain rate (BSS) used by Rhoades et al. [11].

The Global Strain Rate Model (GSRM v.2.1) of Kreemer et al. [13] is an improved version of the GSRM-1 Kreemer et al. model in [2] with a larger input dataset and improved spatial resolution. The maximum shear strain rate from GSRM v.2.1 is referred to here as the GSS model. Like the BSS model, the GSS model was obtained using the Haines and Holt method [25,27,30] to model the strain rate field. GSS is derived from worldwide GNSS data between 1 January 1996 and 31 December 2013, including velocities from 233 published and unpublished regionally focused studies. Table A1 lists the NZ-focused studies used in the GNSS compilation of the GSS model [13]. Apart from in the southeast South Island region, the GSS model uses less up-to-date data than the BSS model.

The NZ strain rate model of Haines and Wallace [14] was produced from GNSS inter-seismic velocities at 918 stations between 1995 and 2013 [31]. This model was based on the novel method of Haines et al. [32]. They used a physics-based approach to invert the GNSS displacement fields for Vertical Derivatives of Horizontal Stress (VDoHS) rates. This method uses constraints from linear elasticity in place of artificial smoothing. This is to ensure physical plausibility, higher resolution, and an improved signal-to-noise ratio compared to the commonly used methods of Haines and Holt [25] and Beavan and Haines [30]. The one modification made to the published Haines and Wallace model of [14] is one implemented by the NSHM Geodesy Working Group: a sill model of rapid magmatic contraction in the Taupo Volcanic Zone (TVZ) was removed from the GNSS velocities, and then the velocities were re-inverted using the Haines et al. [32] method [31]. The maximum shear strain rate from this modified model is the scalar quantity that we use in our hybrid modelling and is henceforth referred to as HWS [33].

Table A1. Data from NZ GNSS campaigns used in constructing the GSS rate model.

Reference	Region
Beavan et al. (1999) [34]	Central Southern Alps, NZ, Alpine Fault
Beavan and Haines (2001) [30]	NZ
Darby and Beavan (2001) [35]	Southernmost North Island, NZ
Denys et al. (2014) [36]	Southeast South Island, NZ

Appendix B

Appendix B.1 Proximity to the Plate Interface (PPI)

A PPI covariate was initially introduced by Rhoades et al. [10]. Here we use a modified version of PPI that differs in some minor details ([33]; Figure 1a). PPI is based on the notion of proximity to the plate interface. A map of the interface between the Australian and Pacific plates in the Hikurangi subduction zone [37] is used to define the North Island plate boundary. South of the Hikurangi subduction zone, the plate boundary is defined by the Hope Fault, the Alpine Fault, and the Fiordland subduction interface, as given in the 2010 NZNSHM fault model [38]. In estimating the proximity of a cell to the interface, no allowance is made for the possibility of different slip rates on different patches of the plate boundary. The plate boundary is represented by a set of n_I points on the interface at approximately 1 km spacing, and a cell is represented by a set of n_c points at its middle latitude and longitude coordinates with depths from 1 to 39 km at a spacing of 2 km. The PPI covariate $\rho(j)$ in the jth grid cell is defined by

$$\rho(j) = \frac{1}{n_c} \sum_{l=1}^{n_c} h(j,l), \tag{A1a}$$

where $h(i,l)$ is an index of the proximity of the lth point in the jth cell to ith point on the interface:

$$h(j,l) = \frac{1}{[d_I^2 + \min(\Delta(i,l)^2)_{i=1,\dots,n_I}]^{3/2}}, \tag{A1b}$$

where $\Delta(i,l)$ is the distance in km from $[x_I(i), y_I(i), z_I(i)]$ to $[x_j(l), y_j(l), z_j(l)]$ and d_I is a constant smoothing distance, taken as 10 km.

Appendix B.2 Proximity to Mapped Faults (PMF)

The concept of proximity to mapped faults was initially introduced by Rhoades and Stirling [39]. It takes account of the distance from the *j*th cell to all fault elements and their associated slip rates, assumed uniform on each fault segment. As in [39], each fault plane is divided into numerous point sources closely spaced at intervals of 1 km. Thus the set of fault segment planes and associated slip rates is expanded into a much larger set of point sources, at longitude, latitude, and depth coordinates $[x_F(i), y_F(i), z_F(i)]$, $i = 1, \cdots, n_F$, and with associated slip rates $r_i, i = 1, \cdots, n_F$. Similarly, a cell with dimensions of 0.1 degrees in latitude and longitude and 40 km in depth is expanded into a representative set of points within it $[x_j(l), y_j(l), z_j(l)]$, $l = 1, \cdots, n_c$ on a cuboidal grid spaced at 0.033 degrees in latitude and longitude and 5 km in depth. Following [10], the PMF covariate $\rho(j)$ in the *j*th grid cell is defined by

$$\rho(j) = \frac{1}{n_c n_f} \sum_{i=1}^{n_f} \sum_{l=1}^{n_c} h(i,l), \tag{A2}$$

where $h(i,l)$ is an index of the proximity of the *l*th point in the *j*th cell to slip on the *i*th fault point source:

$$h(i,l) = \frac{r_i}{\left[d_F^2 + \Delta(i,l)^2\right]^{\frac{3}{2}}}, \tag{A3}$$

where r_i is the slip rate in mm/yr, $\Delta(i,l)$ is the distance in km from $[x_F(i), y_F(i), z_F(i)]$ to $[x_j(l), y_j(l), z_j(l)]$, and d_F is a constant smoothing distance. The value of $\rho(j)$ is, therefore, high for cells near to, or intersected by, mapped faults and low far away from mapped faults. Further, $\rho(j)$ is greater when the slip rates of nearby mapped faults are greater.

Here we used a modified version [33]. In the modified PMF fault, sources are planes rather than linear traces and distances are measured in three dimensions rather than two. Rhoades and Stirling [39] found an optimal smoothing distance of 1.0 km for a PMF model defined in continuous space. Rhoades et al. used a smoothing distance of 2.5 km, but their PMF has since been found to have small-scale variations between adjacent cells due to the smoothing distance being too small [10]. Here we use a larger smoothing distance of $d_F = 10$ km, so that the denominators in Equation (5) are not overly sensitive to the exact positions of points on the fault surfaces and within the cells. The PMF covariate used here (Figure 1b) also uses updated fault information from the New Zealand Community Fault Model [18].

Appendix B.3 Fault in Cell (FLT)

The fault in cell (FLT) covariate [10] is defined to have the value 1 if a fault intersects the cell anywhere between the surface and a depth of 40 km, given the surface trace of the fault, its dip angle and estimated width. Otherwise, it takes the value 0. The FLT covariate is shown in Figure 1c [33] and has been updated using all faults from the New Zealand Community Fault Model [18].

References

1. Ward, S.N. On the consistency of earthquake moment rates, geological fault data, and space geodetic strain: The United States. *Geophys. J. Int.* **1998**, *134*, 172–186. [CrossRef]
2. Kreemer, C.; Holt, W.E.; Haines, A.J. The global moment rate distribution within plate boundary zones. In *Plate Boundary Zones*; Stein, S., Freymueller, J.T., Eds.; AGU: Washington, DC, USA, 2002; pp. 173–190.
3. Bird, P.; Kreemer, C.; Holt, W.E. A Long-term Forecast of Shallow Seismicity Based on the Global Strain Rate Map. *Seismol. Res. Lett.* **2010**, *81*, 184–194. [CrossRef]
4. Bird, P.; Jackson, D.D.; Kagan, Y.Y.; Kreemer, C.; Stein, R.S. GEAR1: A Global Earthquake Activity Rate Model Constructed from Geodetic Strain Rates and Smoothed Seismicity. *Bull. Seismol. Soc. Am.* **2015**, *105*, 2538–2554. [CrossRef]

5. Mazzotti, S.; Leonard, L.J.; Cassidy, J.F.; Rogers, G.C.; Halchuk, S. Seismic hazard in western Canada from GPS strain rates versus earthquake catalog. *J. Geophys. Res. Earth Surf.* **2011**, *116*, B12310. [CrossRef]
6. Rhoades, D.A.; Gerstenberger, M.C.; Christophersen, A.; Zechar, J.D.; Schorlemmer, D.; Werner, M.; Jordan, T.H. Regional Earthquake Likelihood Models II: Information Gains of Multiplicative Hybrids. *Bull. Seismol. Soc. Am.* **2014**, *104*, 3072–3083. [CrossRef]
7. Field, E.H. Overview of the Working Group for the Development of Regional Earthquake Likelihood Models (RELM). *Seismol. Res. Lett.* **2007**, *78*, 7–16. [CrossRef]
8. Schorlemmer, D.; Gerstenberger, M.C. RELM Testing Center. *Seismol. Res. Lett.* **2007**, *78*, 30–36. [CrossRef]
9. Zechar, J.D.; Schorlemmer, D.; Werner, M.; Gerstenberger, M.C.; Rhoades, D.; Jordan, T.H. Regional Earthquake Likelihood Models I: First-Order Results. *Bull. Seismol. Soc. Am.* **2013**, *103*, 787–798. [CrossRef]
10. Rhoades, D.A.; Christophersen, A.; Gerstenberger, M.C. Multiplicative Earthquake Likelihood Models Based on Fault and Earthquake Data. *Bull. Seismol. Soc. Am.* **2015**, *105*, 2955–2968. [CrossRef]
11. Rhoades, D.; Christophersen, A.; Gerstenberger, M. Multiplicative earthquake likelihood models incorporating strain rates. *Geophys. J. Int.* **2017**, *208*, 1764–1774. [CrossRef]
12. Beavan, J.; Wallace, L.M.; Palmer, N.; Denys, P.; Ellis, S.; Fournier, N.; Hreinsdottir, S.; Pearson, C.; Denham, M. New Zealand GPS velocity field: 1995–2013. *N. Z. J. Geol. Geophys.* **2016**, *59*, 5–14. [CrossRef]
13. Kreemer, C.; Blewitt, G.; Klein, E.C. A geodetic plate motion and Global Strain Rate Model. *Geochem. Geophys. Geosyst.* **2014**, *15*, 3849–3889. [CrossRef]
14. Haines, A.J.; Wallace, L.M. New Zealand-Wide Geodetic Strain Rates Using a Physics-Based Approach. *Geophys. Res. Lett.* **2020**, *47*, e2019GL084606. [CrossRef]
15. Rhoades, D.A.; Schorlemmer, D.; Gerstenberger, M.C.; Christophersen, A.; Zechar, J.D.; Imoto, M. Efficient testing of earthquake forecasting models. *Acta Geophys.* **2011**, *59*, 728–747. [CrossRef]
16. Hurvich, C.M.; Tsai, C.-L. Regression and time series model selection in small samples. *Biometrika* **1989**, *76*, 297–307. [CrossRef]
17. Gerstenberger, M.C.; Rhoades, D.A. New Zealand Earthquake Forecast Testing Centre. *Pure Appl. Geophys.* **2010**, *167*, 877–892. [CrossRef]
18. Van Dissen, R.; Seebeck, H.; Litchfield, N.; Barnes, P.; Nicol, A.; Langridge, R.; Barrell, D.; Villamor, P.; Ellis, S.; Rattenbury, M.; et al. Development of the New Zealand Community Fault Model version 1.0. In Proceedings of the NZSEE Annual Technical Conference, Christchurch, New Zealand, 15 April 2021; p. 92.
19. Molchan, G. Strategies in strong earthquake prediction. *Phys. Earth Planet. Inter.* **1990**, *61*, 84–98. [CrossRef]
20. Molchan, G.M. Structure of optimal strategies in earthquake prediction. *Tectonophysics* **1991**, *193*, 267–276. [CrossRef]
21. Green, D.M.; Swets, J.A. *Signal Detection Theory and Psychophysics*; John Wiley: Oxford, UK, 1966.
22. Fawcett, T. An introduction to ROC analysis. *Pattern Recognit. Lett.* **2006**, *27*, 861–874. [CrossRef]
23. Zechar, J.D.; Jordan, T.H. The Area Skill Score Statistic for Evaluating Earthquake Predictability Experiments. *Pure Appl. Geophys.* **2010**, *167*, 893–906. [CrossRef]
24. Cleveland, W.S. Robust Locally Weighted Regression and Smoothing Scatterplots. *J. Am. Stat. Assoc.* **1979**, *74*, 829–836. [CrossRef]
25. Haines, A.J.; Holt, W.E. A procedure for obtaining the complete horizontal motions within zones of distributed deformation from the inversion of strain rate data. *J. Geophys. Res. Earth Surf.* **1993**, *98*, 12057–12082. [CrossRef]
26. Haines, A.J.; Jackson, J.A.; Holt, W.E.; Agnew, D.C. *Representing Distributed Deformation by Continuous Velocity Fields*; Sci. Rep.98/5; GNS Science: Lower Hutt, New Zealand, 1998; p. 110.
27. Holt, W.E.; Shen-Tu, B.; Haines, J.; Jackson, J. On the determination of self-consistent strain rate fields within zones of distributed continental deformation. *Geophys. Monogr. Ser.* **2000**, *121*, 113–141. [CrossRef]
28. Iezzi, F.; Roberts, G.; Walker, J.F.; Papanikolaou, I.; Ganas, A.; Deligiannakis, G.; Beck, J.; Wolfers, S.; Gheorghiu, D. Temporal and spatial earthquake clustering revealed through comparison of millennial strain-rates from 36Cl cosmogenic exposure dating and decadal GPS strain-rate. *Sci. Rep.* **2021**, *11*, 23320. [CrossRef] [PubMed]
29. Bondár, I.; Engdahl, E.R.; Villaseñor, A.; Harris, J.; Storchak, D. ISC-GEM: Global instrumental earthquake catalogue (1900–2009), II. Location and seismicity patterns. *Phys. Earth Planet. Inter.* **2015**, *239*, 2–13. [CrossRef]
30. Beavan, J.; Haines, J. Contemporary horizontal velocity and strain rate fields of the Pacific-Australian plate boundary zone through New Zealand. *J. Geophys. Res. Earth Surf.* **2001**, *106*, 741–770. [CrossRef]
31. Johnson, K.; Wallace, L.M.; Maurer, J.; Hamling, I.; Williams, C.; Rollins, C.; Gerstenberger, M. The 2022 New Zealand National Seismic Hazard Model: Geodetic Deformation Model, GNS Science Report. In Proceedings of the 2021 New Zealand Society for Earthquake Engineering Annual Technical Conference, Christchurch, New Zealand, 15 April 2021.
32. Haines, A.J.; Dimitrova, L.L.; Wallace, L.M.; Williams, C.A. Enhanced Surface Imaging of Crustal Deformation: Obtaining Tectonic Force Fields Using GPS Data. Springer Briefs in Earth Sciences. 2015. 99p. Available online: https://link.springer.com/book/10.1007/978-3-319-21578-5 (accessed on 1 April 2022).
33. Rastin, S.J.; Rhoades, D.A.; Rollins, C.; Gerstenberger, M.C.; Christophersen, A. *Spatial Distribution of Earthquake Occurrence for the New Zealand National Seismic Hazard Model Revision*; GNS Science report, SR 2021/51; GNS Science: Lower Hutt, New Zealand, 2022. [CrossRef]

34. Beavan, J.; Moore, M.; Pearson, C.; Henderson, M.; Parsons, B.; Bourne, S.; England, P.; Walcott, D.; Blick, G.; Darby, D.; et al. Crustal deformation during 1994–1998 due to oblique continental collision in the central Southern Alps, New Zealand, and implications for seismic potential of the Alpine fault. *J. Geophys. Res. Earth Surf.* **1999**, *104*, 25233–25255. [CrossRef]
35. Darby, D.; Beavan, J. Evidence from GPS measurements for contemporary interplate coupling on the southern Hikurangi subduction thrust and for partitioning of strain in the upper plate. *J. Geophys. Res. Earth Surf.* **2001**, *106*, 30881–30891. [CrossRef]
36. Denys, P.; Norris, R.; Pearson, C.; Denham, M. A Geodetic Study of the Otago Fault System of the South Island of New Zealand. In *Earth on the Edge: Science for a Sustainable Planet*; Rizos, C., Willis, P., Eds.; International Association of Geodesy Symposia; Springer: Berlin/Heidelberg, Germany, 2014; Volume 139. [CrossRef]
37. Williams, C.A.; Eberhart-Phillips, D.; Bannister, S.; Barker, D.H.N.; Henrys, S.; Reyners, M.; Sutherland, R. Revised Interface Geometry for the Hikurangi Subduction Zone, New Zealand. *Seismol. Res. Lett.* **2013**, *84*, 1066–1073. [CrossRef]
38. Stirling, M.; McVerry, G.; Gerstenberger, M.; Litchfield, N.; Van Dissen, R.; Berryman, K.; Barnes, P.; Wallace, L.; Villamor, P.; Langridge, R.; et al. National Seismic Hazard Model for New Zealand: 2010 Update. *Bull. Seismol. Soc. Am.* **2012**, *102*, 1514–1542. [CrossRef]
39. Rhoades, D.A.; Stirling, M.W. An Earthquake Likelihood Model Based on Proximity to Mapped Faults and Cataloged Earthquakes. *Bull. Seismol. Soc. Am.* **2012**, *102*, 1593–1599. [CrossRef]

Article

Probabilistic and Scenario-Based Seismic Hazard Assessment on the Western Gulf of Corinth (Central Greece)

George Kaviris [1,*], Angelos Zymvragakis [1], Pavlos Bonatis [2], Vasilis Kapetanidis [1] and Nicholas Voulgaris [1]

1   Department of Geophysics and Geothermics, National and Kapodistrian University of Athens (NKUA), Panepistimiopolis, Zografou, 15784 Athens, Greece
2   Geophysics Department, Aristotle University of Thessaloniki, 54124 Thessaloniki, Greece
*   Correspondence: gkaviris@geol.uoa.gr

Abstract: The Gulf of Corinth (Central Greece) is one of the most rapidly extending rifts worldwide, with its western part being the most seismically active, hosting numerous strong (M $\geq$ 6.0) earthquakes that have caused significant damage. The main objective of this study was the evaluation of seismic hazard through a probabilistic and stochastic methodology. The implementation of three seismotectonic models in the form of area source zones via a logic tree framework revealed the expected level of peak ground acceleration and velocity for return periods of 475 and 950 years. Moreover, PGA values were obtained through the stochastic simulation of strong ground motion by adopting worst-case seismic scenarios of potential earthquake occurrences for known active faults in the area. Site-specific analysis of the most populated urban areas (Patras, Aigion, Nafpaktos) was performed by constructing uniform hazard spectra in terms of spectral acceleration. The relative contribution of each selected fault segment to the seismic hazard characterizing each site was evaluated through response spectra obtained for the adopted scenarios. Almost all parts of the study area were found to exceed the reference value proposed by the current Greek National Building Code; however, the three urban areas are covered by the Eurocode 8 regulations.

Keywords: PGA; PGV; return period; PSHA; stochastic seismic hazard assessment

Citation: Kaviris, G.; Zymvragakis, A.; Bonatis, P.; Kapetanidis, V.; Voulgaris, N. Probabilistic and Scenario-Based Seismic Hazard Assessment on the Western Gulf of Corinth (Central Greece). *Appl. Sci.* **2022**, *12*, 11152. https://doi.org/10.3390/app122111152

Academic Editor: José A. Peláez

Received: 14 October 2022
Accepted: 1 November 2022
Published: 3 November 2022

Publisher's Note: MDPI stays neutral with regard to jurisdictional claims in published maps and institutional affiliations.

1. Introduction

The Gulf of Corinth, located in Central Greece, is a continental rift with an extension rate of 6–16 mm/yr [1], one of the highest values known worldwide for this type of tectonic structure. Its main axis is aligned in an approximately E–W direction, flanked by major normal faults at its southern and northern shoulders. Major earthquakes (M$_w$ $\geq$ 6.0) have occurred in the past, both at the eastern [2] and western parts of the gulf. However, the Western Gulf of Corinth (WGoC) is the most active in terms of observed microseismicity [3,4], as it exhibits a higher extension rate—i.e., ~10.8 mm/yr near Aigion—than its eastern counterpart (~5.5 mm/yr) [5]. This has led to the WGoC being studied extensively during past decades, and it is closely monitored by the Corinth Rift Laboratory (CRL) local seismological network [6]. The CRL has also obtained the status of an international EPOS Near Fault Observatory (NFO), providing high-resolution multidisciplinary data and products [7–9].

The WGoC is bordered by major north-dipping normal faults in the south, with measured dipping angles on the surface ranging from 50° (e.g., Psathopyrgos fault [10]) to ~60° (e.g., Aigion fault [11]) and up to 65–70° (e.g., Lakka fault [12]), with average slip rates in the Late Quaternary of the order of 2.4–5.0 mm/yr for the Aigion and East and West Helike faults and 2.0–3.5mm/yr for the Psathopyrgos fault [10]. On the northern coast, steep south-dipping antithetic normal faults have been mapped, such as the Marathias fault, dipping at 55° [13], and the Trizonia fault, dipping at 64–72°, the latter with an average slip rate of 0.36–0.44 mm/yr over the last ~130 ka [14]. Offshore E–W-trending normal faults, along with smaller structures, oblique to the main axis of the rift with significant strike-slip components, have also been documented [10,14,15]. Average slip rates during the Quaternary for major offshore faults in the WGoC—i.e., the North and South Eratini and West and East Channel faults—are estimated to be between 0.4 mm/yr and over 1.4 mm/yr [10].

Seismicity in the WGoC is mainly concentrated at focal depths between 5 and 10 km [6], with the vast majority of fault plane solutions indicating dominant E–W normal faulting [4,16]. However, oblique-normal slip has also been reported, usually related to seismic swarms [7,17] or to the activation of older structures [18]; i.e., the External Hellenides, which crosscut the gulf in a roughly N–S direction. Although such faults are not optimally oriented to the regional crustal stress field [19], slip can be facilitated by the intrusion of high pore-pressure fluids in the fault network, which play a significant role in the evolution of swarms in the area [20,21]. Towards the western edge of the rift, the crustal stress field changes [19], favoring a dextral strike-slip regime along the SW–NE-trending Rion–Patras fault. The relocated seismicity presented in Figure 1 shows that the bulk of activity mainly occurs along the rift axis, between Trizonia Island and Aigion, with hypocenters deepening towards the north, developing a north-dipping low-angle detachment zone in which the major faults of the WGoC are rooting [22]. Several earthquakes have been located at intermediate depths of 50–60 km, with resolved focal mechanisms representing reverse faulting (Figure 1). These are indicative of seismic activity related to the oceanic slab at the northwestern end of the Hellenic subduction zone [23], with ascending fluids due to its dehydration possibly playing a role in the triggering of earthquake swarms at shallow depths [22,24].

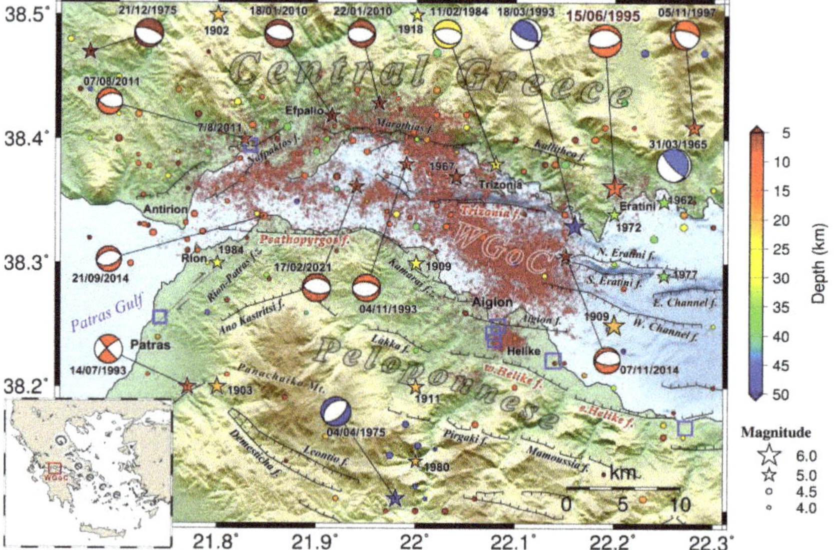

Figure 1. Seismotectonic map of the western Gulf of Corinth (WGoC). Solid circles and stars (for $M_w \geq 5.0$) denote earthquakes with $M_w \geq 4.0$ from the 1900–2009 catalogue of Makropoulos et al. [25], extended in this study up to 2019, with symbol size proportional to the magnitude (M_w) and color according to the focal depth. Selected microseismicity, presented with small, hollow, red circles, is adopted from the 2000–2015 relocated catalogue of Duverger et al. [3]. Focal mechanisms of significant events, presented as beachballs, are adopted from [26–32], as well as the databases of the NKUA, NOA, CMT, and ISC. Blue squares mark the macroseismic epicenters of historical earthquakes (1000–1899) acquired from the SHEEC database [33]. Fault lines are after [10,14,34,35]. Faults marked with red labels were used for the stochastic seismic hazard modeling in this study.

Despite the constant presence of microseismicity, certain parts of major faults in the WGoC appear to be locked, accumulating stress that is released in major events. The last destructive earthquake in the area was the 15 June 1995 $M_s = 6.2$ event, which severely damaged the town of Aigion [29]. The source of this earthquake was a north-dipping, low-angle (33°) normal fault, with more recent data excluding its association with the East Helike fault [22]. The rupture surface of the 1995 event appears to be locked, with the observed microseismicity stopping abruptly at the eastern end of the study area (Figure 1). Other major earthquakes have also occurred in the past at the eastern end of the WGoC, with epicenters towards the northern coasts. Significant earthquakes have struck the WGoC in the historical era (i.e., before 1900). One of the more notable was the 373 BCE earthquake, with an estimated minimum magnitude of M = 6.6 [36], which caused a tsunami that destroyed the ancient city of Helike [37]. An earthquake of similar macroseismic characteristics struck the same area on 26 December 1861 ($M_w = 6.5$ according to Boiselet [38]), triggering a tsunami and causing a strip of plain to submerge under the sea [37]. Both earthquakes are attributed to the north-dipping Helike normal fault, most likely its eastern branch [22]. Recently, the contemporary town of Helike was in the epicentral area of a swarm that started in May 2013, persisting for several months and involving a few earthquakes with magnitudes ranging from 3.3 to 3.7, but which was attributed to activity near the root of Pirgaki fault [17,39]. Other historical earthquakes that struck the town of Aigion occurred in 1748, 1817, and 1888, with estimated magnitudes of 6.3–6.6 [33]. The area of Patras was affected by historical earthquakes in 1785, 1804, and 1806, with magnitudes between 6.1 and 6.4, while the town of Nafpaktos was struck by earthquakes in 1462, 1703, 1714, 1756, and 1831, with magnitudes in the range from 5.9 to 6.5 [33]. In the instrumental era (1900 to today), before the 15 June 1995 earthquake, the previous major event with a magnitude ~6 occurred in the WGoC on 30 May 1909, while intermediate-depth events with magnitudes of 6.4 and 5.9 occurred in 1965 and 1993, respectively; the epicenters of these events were located at the eastern end of the study area.

Although no $M_w \geq 6.0$ earthquake has struck the WGoC since 1995, moderate magnitude events ($M_w \approx 5$) have occurred in recent years. On 14 July 1993, an $M_w = 5.5$ event with a strike-slip focal mechanism [30,40] struck near Patras, causing considerable damage. Although the locally mapped structures trend SW–NE, it has been suggested, based on the aftershock distribution, that the earthquake ruptured a NW–SE sinistral-slip fault [41]. A notable case is the earthquake doublet of January 2010 near Efpalio, with an $M_w = 5.3$ event being followed by an $M_w = 5.2$ four days later, both related to blind E–W-trending normal faults [32,42,43]. This activity likely played a role in the triggering of another $M_w = 5.1$ earthquake on 7 August 2011 to the west, near Nafpaktos, characterized by a similar normal faulting style. Later, in 2013–2014, the WGoC presented strong signs of microseismic activity, with an earthquake swarm initiating in September 2013 at the offshore region between Nafpaktos and Psathopyrgos that slowly migrated eastwards, triggering earthquake clusters and culminating in an $M_w = 5.0$ event, offshore of Aigion, on 7 November 2014 [21,44]. A second seismic swarm excitation, characterized by bilateral spatiotemporal migration in the offshore area between Nafpaktos and Psathopyrgos, involved an $M_w = 4.9$ earthquake on 21 September 2014, activating a patch related to the westernmost edge of the Psathopyrgos fault, near its junction with the Rion–Patras fault zone [3]. The most recent significant activity in the WGoC was the 2020–2021 seismic sequence [7,45,46], which started on 23 December 2020 with a moderate event near Marathias and evolved in three stages, first triggering earthquake clusters towards Trizonia Island involving some strike-slip events and later producing an $M_w = 5.3$ earthquake offshore of Psathopyrgos that exhibited peculiarities in terms of its complicated rupture characteristics [47].

In this study, we reassess the seismic hazard in the WGoC utilizing a well-established methodology and incorporating recent data. The term "seismic hazard" describes the expected ground motion level for potential earthquake occurrences within a study area. The methodologies used to evaluate seismic hazard are grouped into two major categories: probabilistic and scenario-based. The first exploits data from historical and instrumental earthquake catalogues and integrates a seismotectonic model for its application. Subsequently, given an earthquake occurrence model, results for the maximum expected ground motion for selected return periods are generated. The second methodology is based on individual earthquake scenarios, with which synthetic waveforms are produced and analyzed. Strong ground motion may be a catalyst for other secondary hazards, the most important of which are rock falls, landslides, and liquefaction phenomena. Therefore, the visualization of the spatial distribution of peak ground motion parameters can be of great importance for scientists and decision makers. The advantage that the study area offers is the wealth of earthquake data. The latter is critical because seismic hazard can be assessed without adopting algorithms that cope with this difficulty, such as the one proposed by Meirova et al. [48]. A plethora of seismic hazard studies have been conducted for the Greek territory, either as a whole or for specific areas, such as [49–58]. The applied methods for the assessment of seismic hazard are described in detail in Section 2.

Patras, the third largest city in Greece and one of its largest harbors, is situated at the western edge of the study area, on the southeastern shores of the Patras Gulf. The nearby town of Rion hosts important infrastructure, such as the University of Patras and the University General Hospital of Patras, while the Rion–Antirrion bridge connects Peloponnese with Central Greece. Other important towns on the shores of the WGoC include Aigion and Nafpaktos. As the seismic risk strongly depends on the exposure to seismically prone areas, the aforementioned city, towns, and infrastructure may suffer increased economic losses from potentially destructive earthquakes, both in terms of human lives and structural damage. As a consequence, the proper assessment of seismic hazard in the WGoC area, both probabilistic and with specific earthquake scenarios, is required for the subsequent estimation of seismic risk and consideration of means for its mitigation.

2. Materials and Methods

Seismic hazard in the study area was evaluated using two approaches: (a) a Probabilistic Seismic Hazard Assessment (PSHA) and (b) a stochastic approach to worst-case seismic scenarios for known active faults. PSHA remains the primary framework for assessing seismic hazard and is based on prior knowledge of seismicity at a given place. On the other hand, during past decades, finite-fault stochastic ground motion simulations have been proven to be a powerful tool to reliably estimate strong ground motion parameters, such as the Peak Ground Acceleration (PGA) and Peak Ground Velocity (PGV). The combination of these methodologies can provide a holistic evaluation of seismic hazard, which in turn can contribute significantly to the assessment of seismic risk and its mitigation at a given area.

2.1. PSHA

Concerning PSHA, the approach proposed by Cornell and McGuire [59,60] was followed, utilizing the instrumental earthquake catalogue from Makropoulos et al. [25]. Given that this catalogue ends in 2009, it was deemed necessary to extend the period up to 2019 in a homogenous manner, meaning that only reviewed events from the International Seismological Centre (ISC) earthquake catalogue were taken into consideration. Therefore, we refrained from expanding the database for post-2019 events in order to retain homogeneity. It should be noted that only few $M_w \geq 4$ magnitude earthquakes have been recorded in the study area since 2019, the largest being the $M_w = 5.3$ earthquake of 17 February 2021 [7,47]. Thus, we considered that the annual rate of exceedance of the magnitude of completeness would not be altered significantly and would not impact the results of PSHA. According to the Cornell–McGuire approach to PSHA, the results follow a normal distribution, being essentially time-independent. Given that there are both aftershocks and foreshocks included in the utilized earthquake catalogue, a declustering procedure was conducted to retain only the mainshocks. However, it must be noted that ground motion exceedances can be caused randomly by any earthquake occurrence and a significant amount of data could be omitted by applying a declustering procedure. This would have an impact on the obtained results because annual rates of exceedance for the magnitude of completeness would be lower, resulting to the underestimation of the maximum expected ground motion parameters [58,61]. The latter is strengthening the implementation of a non–declustered earth-quake catalogue in a Poissonian process, as ground motion at a site caused either by foreshocks or aftershocks may exceed a certain level [61].

Three seismotectonic models were considered herein; namely, the Euro-Mediterranean Seismic Hazard Model 2013 (ESHM13) [62,63], the updated Euro-Mediterranean Seismic Hazard Model 2020 (ESHM20) [64], and the model from Vamvakaris et al. [65] proposed for the Greek territory. The boundaries of the Area Source Zones are selected to include areas with similar seismological and tectonic characteristics, avoiding dividing fault systems, whenever possible. Each model divides the study area into polygons with common seismotectonic characteristics (Area Source Zones, ASZ). Individual faults were not implemented in the PSHA, as a significant part of the total seismicity of the WGoC is related to offshore faults for which certain characteristics are not known; i.e., mean slip rates. The decision to employ more than one seismotectonic model was made in order to acquire differently parameterized earthquake occurrence models so that the epistemic uncertainties would be reduced in a qualitative manner.

The Modified Gutenberg–Richter (MG-R) earthquake occurrence model was applied to characterize the seismic potential of every area source zone for each seismotectonic model [66]. The MG-R was parameterized by importing the following seismicity data for each zone: (a) the b-value, (b) the threshold magnitude (M_0; also considered as the magnitude of completeness (M_c)), (c) the average annual exceedance rate of M_c ($\lambda(M_c)$), and (d) the maximum expected magnitude (M_u). The M_c and b-values were estimated for each zone using the maximum curvature (MAXC) method from Wiemer and Wyss [67] and the maximum likelihood function from Aki [68], respectively, with ZMAP software [69]. This method has been proved to be stable, even in cases when a small number of earthquakes within each zone are used [70,71]. The $\lambda(M_c)$ was calculated through the analysis of the earthquake catalogue and M_u was estimated via the Robson–Whitlock–Cooke (R-W-C) procedure [72,73]. Regarding the geometry data, the depth of each seismic zone was estimated through depth histograms constructed using the focal depths reported in the earthquake catalogue. It must be noted that ESHM13, ESHM20 and VAM16 provide values concerning the aforementioned parameters, each using a different earthquake catalogue. However, in the present study the b-value, M_c, $\lambda(M_c)$ and M_u were recomputed for each ASZ, im-plementing the updated earthquake catalogue of Makropoulos et al. [25] to retain ho-mogeneity regarding the seismicity and geometry data. Consequently, only the ASZ were adopted from the three zonation models. The computed seismicity and geometry data are depicted in Table S1.

Ground Motion Prediction Equations (GMPEs) proposed for the Greek territory were adopted for the calculation of the maximum expected ground motion in terms of peak ground acceleration and velocity (PGA, PGV). In particular, for PGA, the GMPEs from [74–78] were employed. Sakkas [77] has not proposed a GMPE for PGV estimation. Furthermore, the model from Skarlatoudis et al. [75] was replaced with an upgraded GMPE for PGV, which was taken into account. As area source zones were herein adopted, it was only feasible to use epicentral distances. As a result, more recent GMPEs [79] that utilize different distance metrics were not considered. In addition to PGA and PGV, Spectral acceleration (S_a) values for different periods (T) were calculated for the three most important urban areas of the WGoC—namely, Patras, Nafpaktos, and Aigion—using the GMPE from Danciu and Tselentis [76]. All GMPEs (except for the one from Margaris et al. [74]) include a term related to the type of focal mechanism, taking a value of 0 or 1 for normal and strike-slip/thrust ruptures, respectively. As a seismic zone could potentially include any type of focal mechanism, it was deemed necessary to compute their respective participation rates. Thus, each GMPE was calibrated with the exact percentage of focal mechanism types in each zone, generating weight-specific GMPEs (hereafter, w-sGMPEs). The final results (PGA, PGV for return periods of 475 and 950 years, and S_a for a return period of 475 years) were computed via an equal-weighted logic tree, where each secondary branch was a w-sGMPE (except [74]) and each primary branch was a different seismotectonic model (Figure 2). The reason for assigning equal weights to the primary branches was that none of the three zonation models outweighs the others. There are uncertainties that would be increased if a higher weighting factor would be assigned to the ASZ of ESHM13, ESHM20 or VAM16. For example, there are zones including both the western and eastern Gulf of Corinth in ESHM13 and ESHM20 (zones 13 and 1 in Figures S1a and S1b, respectively). In addition, certain small zones of the VAM16 model (zones 5, 13 in Figure S1c), result in an insufficient number of earthquakes, not adequate to reliably compute the seismicity parameters. The software used for PSHA was R-CRISIS and, specifically, its newest version (V20.0) [66].

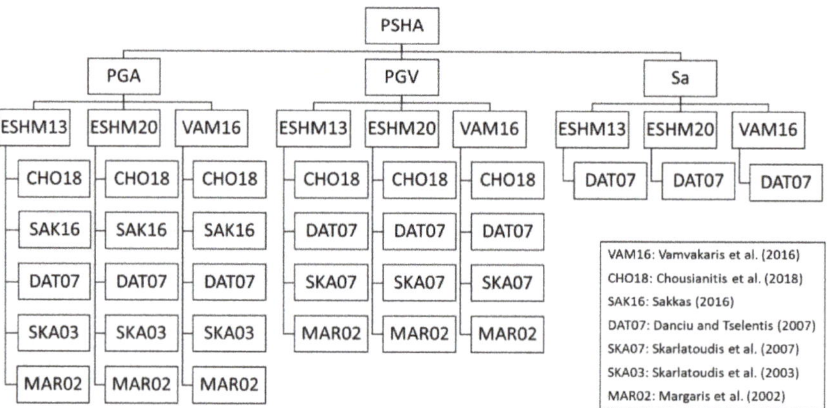

Figure 2. The logic tree constructed to calculate PGA, PGV, and S_a. All branches have the same weighting coefficient. Results were generated for return periods of 475 and 950 years for both PGA and PGV and 475 years for S_a. Abbreviations: VAM16: Vamvakaris et al. [65], CHO18: Chousianitis et al. [78], SAK16: Sakkas [77], DAT07: Danciu and Tselentis [76], SKA07: Skarlatoudis et al. [75], SKA03: Skarlatoudis et al. [75], MAR02: Margaris et al. [74].

2.2. Seismic Scenarios

We compiled a set of seismic scenarios to assess the seismic hazard in the broader WGoC area in terms of maximum expected strong ground motion from specific seismic sources. These scenarios were generated through stochastic simulations for a predicted maximum magnitude (M_{max}) for well-studied known faults in the area. Simulations were performed using a stochastic finite-fault model based on a dynamic corner frequency approach with the EXSIM code [80,81]. The modeling strategy was based upon the discretization of the earthquake fault plane into smaller subfaults, each of which was considered as a potential earthquake source. The point-source stochastic method [82] was employed to generate synthetic time series for each subfault. The summation of the individual contributions of each subfault, along with a suitable time delay, led to the final ground motion parameters at the sites of interest. This approach has been widely implemented worldwide [83], as well as in Greece [84–88].

The anticipated strong ground motion at a given site is a result of a complex physical process that includes the relative contribution of source, path, and local site effects. Source effects describe the characteristics of the accumulated strain release from the fault, which, when released, results in the generation of an earthquake. These include, among others, the source dimensions (length, width), the moment magnitude (M_w), the slip distribution for the causative fault, and the stress parameter ($\Delta\sigma$). After the earthquake nucleation and rupture, seismic waves travel through the Earth's crust; therefore, path properties must also be taken into account. Their propagation is most strongly affected by two path parameters; i.e., the geometrical spreading and the anelastic attenuation, which is controlled by the quality factor (Q). Ultimately, the local site properties play an important role in the resulting surface ground motions, given that the impact of the shallower layers may lead to significant amplification of the seismic waves. These effects are treated by the EXSIM code using the kappa (κ_0) parameter [89] and user-defined soil amplification factors.

As a first step to obtain high-resolution model parameters for the desired earthquake scenarios, we performed a stochastic simulation of the most recent strong ($M_w \geq 6.0$) earthquake in the study area; namely, the 1995 Aigion M_S = 6.2 (M_w = 6.5 according to the Global Centroid Moment Tensor (CMT) project) mainshock [29]. Modeling of past earthquakes in the study area can significantly help in constraining the path and site components of strong ground motion through comparison with GMPEs, as well as with real strong motion data. In the case of the Aigion 1995 mainshock, 17 recordings from seismic stations up to an epicentral distance of 140 km are available, allowing the calibration of the synthetic results through an iterative procedure to achieve the best fit. The final model parameters are summarized in Table 1. The causative fault's dimensions and geometry (strike and dip) were adopted following [29] and the upper edge of the fault plane was set according to the model from Console et al. [90], which is in compliance with the geometry of the seismogenic layer of the study area. The slip distribution onto the fault plane was modeled through a random slip pattern, given that detailed finite-fault slip inversions are not available for this case. This approach has been also followed by other authors for this mainshock [86,88]. Finite-fault discretization into smaller subfaults was performed using the empirical relationship proposed by Beresnev and Atkinson [91] (Equation (1)):

$$\log\Delta l = 0.4M - 2, \tag{1}$$

where Δl is the length of each subfault and M is the moment magnitude of the mainshock.

The $\Delta\sigma$ is a parameter closely related to the actual stress drop and slip velocity due to an earthquake occurrence, but it does not include the natural context of stress drop [92]. It is generally considered to weakly scale with moment magnitude, but in a more regional extent their interconnection appears to be more profound [93]. It was determined by employing an iterative procedure, comparing the synthetic ground motions with recorded ones, as well as with estimated peak values from well-established GMPEs [94,95] appropriate for the study area. The $\Delta\sigma$ value of 56 bars, which is routinely used as a mean value for earthquakes in Greece [96,97], was used initially and, afterwards, various other values were tested to find the best fit. Ultimately, $\Delta\sigma$ = 30 bars was deemed appropriate and was adopted in the simulations. This deviation from the mean value was foreseen, given that low stress drop has been documented for this earthquake [98], as well as for the study area, especially in comparison with the eastern part of the gulf where stress drop values appear to be larger [99]. Regarding the path effects, the geometrical spreading model, as described by Atkinson [100], was adopted, which divides the slope of the attenuation relation in three distance intervals (<70 km, 70–130 km, and >130 km), reflecting the dominant type of wave in the seismic signal. The anelastic attenuation, on the other hand, is described by the quality factor proposed by Hatzidimitriou [101] for the broader Aegean region (Equation (2)):

$$Q(f) = 100f^{0.8}, \tag{2}$$

Lastly, to account for the local site conditions for synthetic ground motions, the average κ_0 value of 0.044 for class C (NEHRP) sites in Greece [102] was adopted, along with the corresponding amplification factors. The selection of class C soil characterization was based on local soil data (e.g., Vs30 profiles) for the major cities in the area and geology data from the locations of the available seismic stations. In order to calibrate and validate our results, the GMPEs of A14 and B14 were employed. Both were derived using strong motion datasets from Italy, Greece, and Turkey.

After validating the basic stochastic simulation parameters for the study area, three well-studied seismogenic faults (Psathopyrgos Fault (PT); Helike Fault (HF); Trizonia Fault (TF)) were selected to assess the worst-case earthquake scenarios (Table 2, Figure 1). Given the limited extent of the area under study and the consistency in the seismotectonic regime, we assumed that path and local site effects were properly constrained from the Aigion 1995 simulations. It was, therefore, only necessary to define the seismic source parameters and, particularly, the fault dimensions and geometry, as well as the M_{max}. Fault characteristics were adopted from various studies (see Table 2 caption) and the M_{max} was calculated following the approach described by Kourouklas et al. [103], which is a slightly different version of the method proposed by Pace et al. [104]. The adopted approach combined various scaling relationships between magnitude and rupture length with the maximum observed magnitude obtained from historical data. The finally adopted M_{max} (Table 2) was weighted and acquired through the combination of all relative M_{max} values, along with their standard deviations. In our assessment, the scaling laws from [105–108] were used.

Table 1. Modeling parameters used for the stochastic ground motion simulation of the $M_w = 6.5$ Aigion 1995 mainshock performed with the EXSIM code.

Parameter	Value	References
Strike	277°	
Dip	33°	
Burial depth	5 km	[29,90]
Fault dimensions (Length × Width)	15 km × 9 km	
M_w	6.5	
Stress parameter ($\Delta\sigma$)	30 bars	This study
Kappa parameter (κ_0)	0.044	[102]
Quality factor	$100f^{0.8}$	[101]
Geometrical spreading as a factor of distance (R^n)	N = −1.0, R < 70 km n = 0.0, 70 km ≤ R < 130 km n = −0.5, R ≥ 130 km	[100]
Site amplification	Empirical amplification factors from Klimis et al. [102]	

Table 2. Stochastic modeling source parameters for the three selected seismic scenarios. Fault orientation and dimensions were adopted from various sources [6,38,90]. Associated past strong events were adopted from Boiselet [38].

Source Parameters	Psathopyrgos Fault (PF)	Helike Fault (HF)	Trizonia Fault (TF)
Strike	270°	281°	95°
Dip	40°	34°	65°
Dimensions (length × width)	16 km × 8 km	22 km × 12.5 km	10.5 km × 8.5 km
M_{max}	6.3 ± 0.3	6.4 ± 0.3	6.0 ± 0.4
Associated strong events	1462 ($M_w = 6.4$) 1703 ($M_w = 6.1$) 1714 ($M_w = 6.1$) 1756 ($M_w = 5.9$)	373 B.C.E. ($M_w \approx 6.6$) 61 C.E. ($M_w = 6.3$) 1758 ($M_w = 5.9$) 1817 ($M_w = 6.4$) 1861 ($M_w = 6.5$) 1888 ($M_w = 6.3$)	23 C.E. ($M_w = 6.3$)

Furthermore, the definition of the hypocenter position and the slip distribution onto the fault plane is a crucial step in stochastic simulations. For the purposes of the present study, we divided each fault segment into subfaults using Equation (1), and the hypocenter position was placed randomly among them (minimum of 10 iterations). Different slip rupture patterns (minimum of 5 iterations) were examined, applying random slip weights to each subfault.

3. Results

3.1. PSHA

The spatial distribution of the PGA, computed via the logic tree approach (Figure 2) for a return period of 475 years, is illustrated in Figure 3a. The highest values were identified in close proximity to the coastline, near the towns of Nafpaktos and Aigion. The PGA decreased towards the north and reached its minimum of about 250 cm/s^2 at the NE edge of the study area. A maximum of ~325 cm/s^2 was obtained approximately 5 km SSE of Aigion. The difference between the highest and the lowest values was 75 cm/s^2, which indicates that their variation was low to intermediate. The spatial distribution pattern of the PGA for a return period of 950 years was nearly identical (Figure 3b). A maximum of ~400 cm/s^2 was found SSE of Aigion, meaning that there was an increase of 75 cm/s^2 compared to RP = 475 yr. The lowest value was about 300 cm/s^2, increased by 50 cm/s^2 from the respective value obtained for the return period of 475 years.

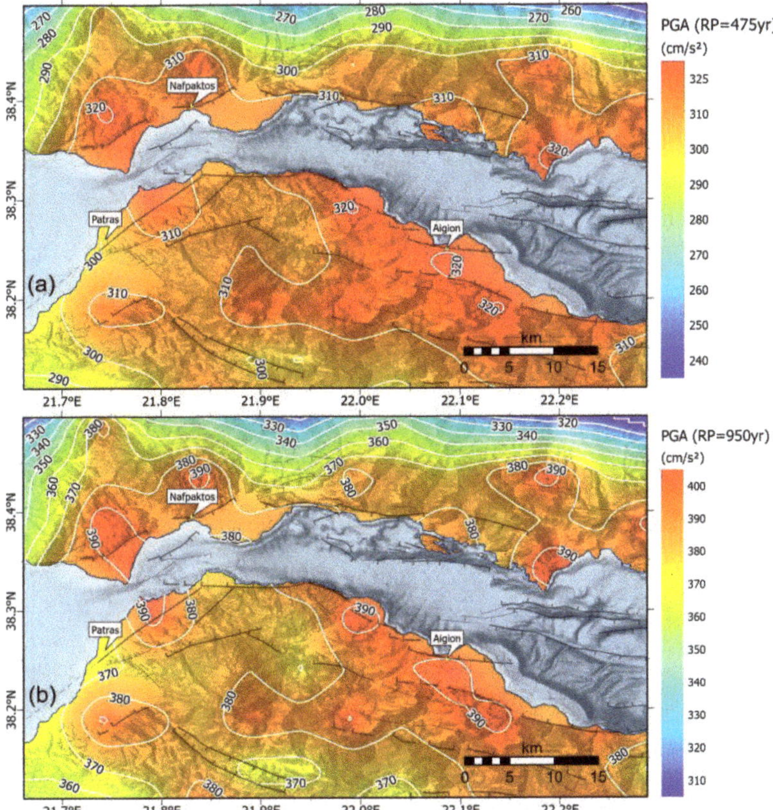

Figure 3. Spatial distribution of PGA values for a return period (RP) of (**a**) 475 yr and (**b**) 950 yr.

Regarding the seismic hazard intensity measure of the PGV, the spatial distribution was similar to the PGA for a return period of 475 years (Figure 4a). A maximum of ~18.5 cm/s was found toward the southern edge of Aigion, where the highest PGA was also determined. The lowest value persistently remained at the NE end of the study area, with an approximate PGV of 14 cm/s. In Nafpaktos, the PGVs were lower than in Aigion but higher than in Patras. The variation in the values was small, as the difference between the highest and lowest PGV was only 4 cm/s. In Figure 4b, the spatial distribution of PGVs for a return period of 950 years is depicted. The highest PGV was about 24 cm/s, south of Aigion, and the lowest was 18 cm/s at the NE tip. The increase between PGVs for the two return periods was only 5 cm/s for the highest and 4 cm/s for the lowest values.

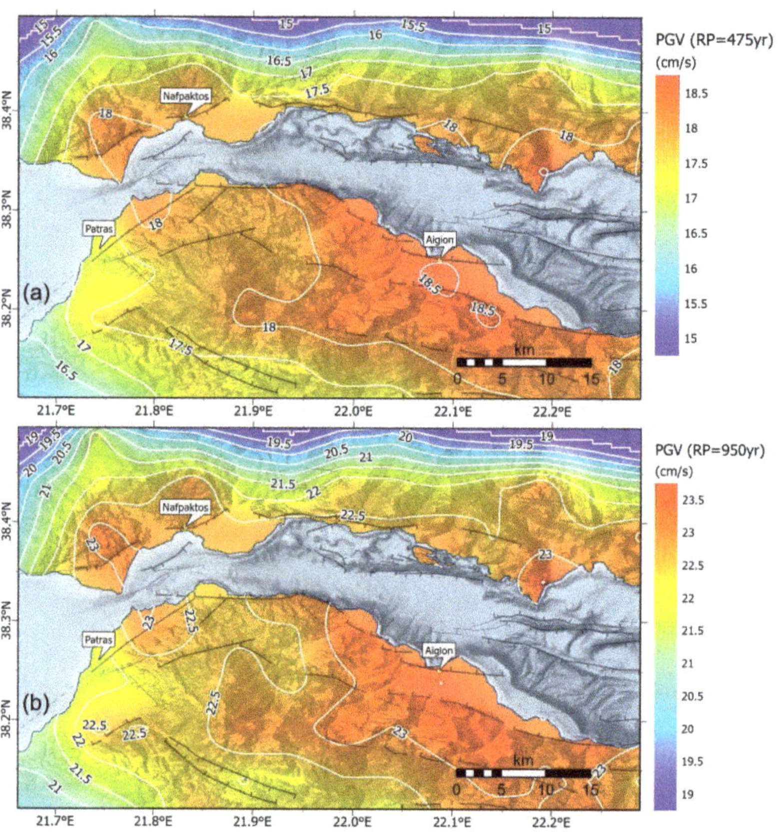

Figure 4. Spatial distribution of PGV values for (**a**) RP = 475 yr and (**b**) RP = 950 yr.

Hazard curves were also generated for the three main WGoC city and towns; namely, Patras, Nafpaktos, and Aigion (Figure 5a). PGA values were calculated for a range of exceedance probabilities in 50 years to observe the increase rate for the PGA as a function of the return period. The hazard curves were proximal to each other, indicating that the level of seismic hazard was similar among the three sites. Patras presented a slightly lower PGA than Nafpaktos and Aigion. It could not be determined whether Aigion or Nafpaktos exhibited higher seismic hazard, as the PGAs were almost identical between the two cities. It is worth noting that the value of 1000 cm/s^2 was exceeded only for very small probabilities of exceedance, which implies that such extreme ground motions are unlikely to be reached. The last output of the PSHA calculations was the computation of Spectral acceleration (S$_a$) values for periods ranging from 0.1 s to 2.0 s in the three aforementioned sites in order to construct uniform hazard spectra (Figure 5b). The S$_a$ curves for Patras, Nafpaktos, and Aigion were very similar to each other throughout all periods. This was reasonable in light of the close geographical distance among them. As illustrated in both PGA and PGV maps, the city and the two towns were adjacent to regions of comparable maximum expected ground motions. The maximum of the three spectra was about 500 cm/s^2 at a period of 0.25 s. The curves seemed to be nearly identical for periods above 0.9 s. Thus, at high periods, there was almost no difference between the three sites. Minor deviations were detected in the period range [0.3, 0.5] s, for which Nafpaktos seemed to have slightly lower S_a values than Patras and Aigion. The elastic design spectra proposed by the Current National Building Code (EAK2003) [109] and Eurocode 8 (Ec8) [110] were also plotted to investigate their relation with our results. The seismicity was defined as high (type I) and the soil as bedrock (type A) to match the input data of the herein proposed model. The Ec8 spectrum overlay the spectra of the three urban areas for all periods, while that of EAK2003 covered the spectra for the period range [1.4, 2.0] s.

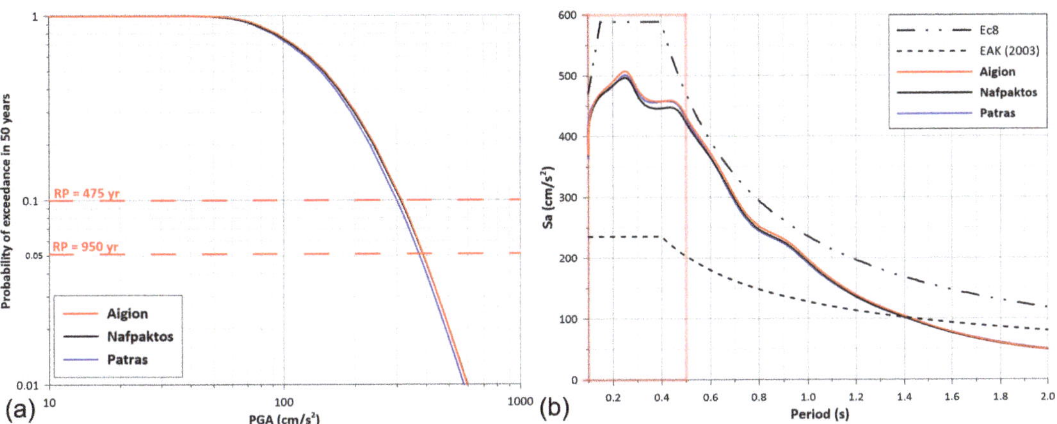

Figure 5. Hazard curves in terms of (**a**) PGA for Aigion, Nafpaktos, and Patras and (**b**) UHS for the aforementioned sites, alongside the Ec8 and EAK2003 response spectra. The red rectangle indicates the range of eigenperiods of most buildings in the study area.

3.2. Seismic Scenarios

3.2.1. Validation of the Aigion 1995 Mainshock Simulation

In order to assess the ground motion variability caused by the Aigion 1995 mainshock, PGA values were computed on a grid enclosing the area under study, with nodes at a spacing of 0.03° in latitude and longitude. The spatial distribution of synthetic PGA values is depicted in Figure 6a, along with the surface projection of the activated fault [29] and the available recorded PGA. The highest values estimated onshore exceeded 500 cm/s^2, whereas in the city of Aigion, where the highest observed PGA was reported, they were close to 300 cm/s^2. The spatial distribution of simulated values was similar to those presented in other studies [88,111], despite using diverse input parameters. The validation of the final parameters used in the simulation is shown in Figure 6b, where simulated PGA values are plotted against those derived from the selected GMPEs and the recorded ones as a function of Joyner and Boore distances (R_{jb}). The functional form of the GMPEs used corresponded to a normal faulting style and type C (NEHRP) soil condition. As shown in Figure 6b, the PGAs obtained from EXSIM were in good agreement with the GMPE curves throughout the entire R_{jb} range, except for the points that lay very close to the surface projection of the fault plane ($R_{jb} \leq 2$–3 km). This was, however, anticipated, given that GMPEs are generally not fully capable of reliably reproducing the ground motion in very short distances due to the limited availability of near-fault strong motion datasets, which affects their formulation. Moreover, directivity effects may have a strong influence on near-fault ground motion that is not fully captured by the GMPEs used. In this case, for example, the high PGA value recorded in Aigion (AIGA; Figure 6b) has been attributed to forward rupture directivity, in addition to local soil and topographic characteristics [29,111]. Nevertheless, the overall PGA variability lay inside the $\pm 1\sigma$ range for both GMPEs in the entire R_{jb} range. Regarding the attenuation pattern, PGAs prescribed by the A14 and B14 models appeared to decay slightly faster than the simulated ones; however, taking into account the fact that the same finding also applied to the recorded ones (red triangles;

Figure 6b), this highlights the importance of retrieving region-specific parameters that can be incorporated into GMPEs. The finally adopted synthetic PGA values were obtained based on the desire to achieve the right balance between keeping the lowest misfit among synthetic, GMPE-derived, and real values. Consequently, as shown in Figure 6b, our simulations were capable of reproducing the PGAs recorded from the two closest stations (AIGA, AMIA) more closely than the GMPEs, whereas at larger distances the general trend of the recorded PGAs (which presented a relatively high variability in certain cases) was captured by both GMPEs and synthetic values.

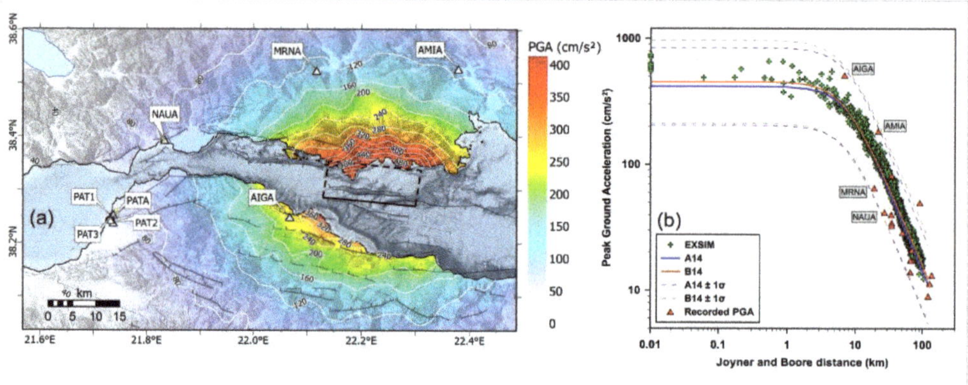

Figure 6. (**a**) Spatial distribution of simulated PGA values (cm/s^2) calculated with the EXSIM code for the case of the Aigion 1995 mainshock. (**b**) Comparison between the recorded (red triangles), GMPE-derived (lines), and simulated (crosses) PGAs, plotted as a function of the Joyner and Boore distance (R_{jb}). The simulated values on the surface projection of the fault plane ($R_{jb} = 0$) were assigned a very small positive R_{jb} value (~0.01 km) to make them visible on the logarithmic axis.

3.2.2. Spatial Distribution of Simulated PGA for Selected Faults and Site-Specific Analysis

In the finite-fault stochastic scenarios of the present study, a worst-case seismic scenario was adopted in contrast to the PSHA, where peak ground motion parameters were obtained from a large number of possible earthquakes with a certain probability of exceedance. Figure 7 presents the spatial distribution of the simulated PGA values generated by the three selected faults (Table 2). The color scales retain a common, fixed range of PGA values to enable a comparison between each scenario. Simulated acceleration time series for Patras, Nafpaktos, and Aigion are also illustrated along with the resulting PGA values.

In all cases, the maximum PGAs, located at the surface projection of the faults or nearby, appeared to be relatively close, ranging from approximately 350 to 450 cm/s^2. The HF may have had larger dimensions among the three cases, but PF produced the highest acceleration values onshore (Figure 7a), even though it was assigned a slightly lower M_{max} (6.3) than HF (6.4; Table 2).

Similar to the PSHA framework, hazard response spectra for Aigion, Nafpaktos, and Patras for hypothetical future earthquakes originating from the selected faults (PF, HF, TF) were constructed for 5% damping (Figure 8). In this case, however, it became plausible to gain insight into which seismic sources are more capable of causing damage to each urban area. In all cases, the response spectra of simulated ground motions exhibited a spike behavior at short periods of [0.1–0.3] s, followed by a relatively sharp decline. As shown in Figure 8a, the town of Aigion is mostly threatened by HF and TF, whereas PF poses the greatest threat to Nafpaktos and Patras. The city of Patras (Figure 8b) is exposed to a lower level of seismic hazard compared to the other cases when taking into account the major active faults of the WGoC. Maximum S_a values exceeded 300 cm/s^2 for a hypothetical rupture of PF, whereas HF produced the lowest expected values (approximately half of the corresponding ones for PF). Moreover, the town of Nafpaktos, located in between the aforementioned sites, is highly susceptible to high levels of seismic hazard due to the PF, with maximum expected S_a values approaching almost 900 cm/s^2. TF and HF, which are located quite far from the town, do not pose significant hazards, given that the predicted maximum S_a values were within the ranges appointed by the National Building Code [109].

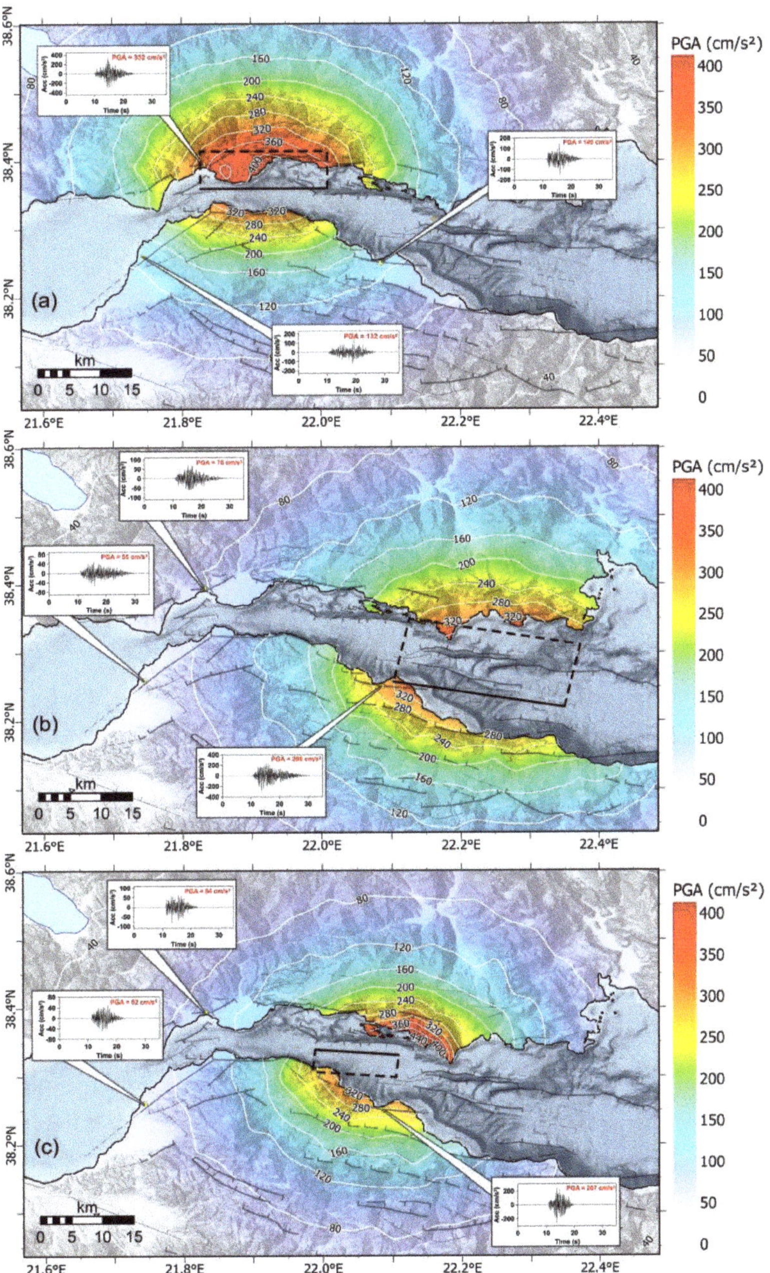

Figure 7. Spatial distribution of simulated PGA values (cm/s^2) calculated with the EXSIM code for the cases of the (**a**) Psathopyrgos, (**b**) Helike, and (**c**) Trizonia Faults. Synthetic acceleration time series for Patras, Nafpaktos, and Aigion are also displayed along with the maximum TF obtained values.

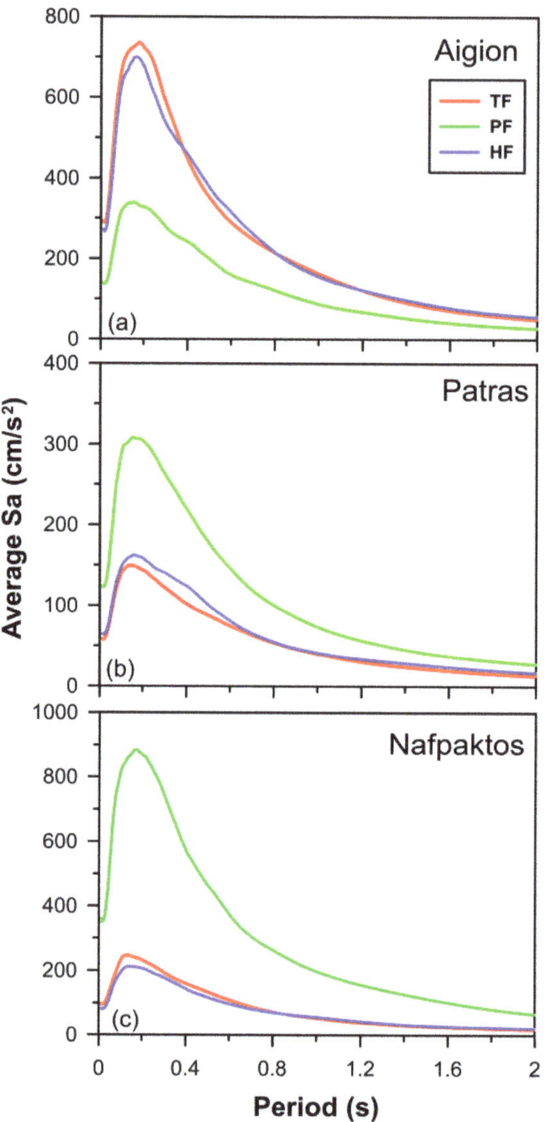

Figure 8. Response spectra of the selected earthquake scenarios for (**a**) Aigion, (**b**) Patras, and (**c**) Nafpaktos corresponding to 5% damping. TF: Trizonia Fault, HF: Helike Fault, PF: Psathopyrgos Fault.

4. Discussion

The high seismicity of the study area, both onshore and offshore, is due to the high average extension rate of the tectonic rift [5]. It is expressed through both large earthquakes and seismic swarms [7,17,20,39,112–114], which confirm the great importance of seismic hazard assessment as a means to quantify anticipated levels of ground shaking.

When performing PSHA, there are two types of uncertainties, i.e., aleatory uncertainty and epistemic uncertainty [115]. The first one accounts for random variations in PGA, PGV and Sa values due to the implementation of a GMPE, whereas the second one accounts for the accuracy of the values [115]. Epistemic uncertainty is usually handled by the incorporation of a logic tree approach to account for the implementa-tion of more than one

GMPE [115–117]. This is the procedure followed in the present study, given that by taking into account alternative models the uncertainties can be reduced [115].

The PGA results for a return period of 475 years, obtained through the Cornell–McGuire approach, can be directly compared with the reference value proposed by the current Greek Building Code [109]. EAK2003 divides Greece into three zones and defines the maximum expected ground acceleration for a return period of 475 years. The Western Gulf of Corinth is within zone II, with a reference value of 0.24 g (~240 cm/s^2). However, in the herein proposed model, the entire study area exceeded 240 cm/s^2. The computed PGA results in the northeastern part of the area were close to the EAK provisions, even though they still slightly exceeded them. The highest value was about 325 cm/s^2, surpassing EAK2003 by 74%. The calculated PGA for a return period of 950 years can be utilized for the construction of buildings of greater importance, such as schools and medical centers. The spatial patterns of PGA and PGV for both return periods were similar; i.e., the highest values were found close to the Gulf, where seismicity is higher, whereas the lowest values were onshore, far from the coastline.

A comparison of the PGA results obtained to those of Banitsiotou et al. [49] and Tselentis et al. [54] was attempted, both of which are for the whole Greek territory. Banitsiotou et al.'s [49] computed PGA values were for certain cities, one of them being Patras, which was assigned a PGA of 0.26 g for a return period of 475 years. In this study, the corresponding PGA was ≈300 cm/s^2, which is slightly higher. Tselentis et al. [54] computed PGA and PGV for the same return period for Greece. PGA values varied between 0.40 g and 0.50 g close to Patras and Nafpaktos and between 0.50 g and 0.60 g close to Aigion. Concerning PGV, Tselentis et al. [54] proposed values in the broad range of [20.0–105.0] cm/s. In the present study, PGA and PGV values were in the range of [0.25–0.33] g and [15.0–18.5] cm/s, respectively. The existing deviations can be attributed to the different seismic and geometry data used in each case.

Hazard curves were constructed for the most populated sites of the study area, the city of Patras and the towns of Nafpaktos and Aigion. All three are located close to the coastline and are characterized by high expected ground motion values. The distances between them are relatively small and they belong in the same or neighboring (according to the model) area source zones, parameterized with similar geometry and seismicity data. For this reason, their corresponding hazard curves were quite close to each other, whereas strong ground motions (>1 g) occurred only at low exceedance probabilities.

Uniform hazard spectra were obtained by adopting the GMPE from Danciu and Tselentis [76]. The final results were derived using the same logic tree but without the minor branches, as they were replaced with this GMPE. However, the adopted model was, again, parameterized to take into account all types of focal mechanisms for each zone. The values for the spectral acceleration were almost the same for all three cases, as was also observed in the hazard curves. Differences from the elastic design spectra of EAK2003 [109] and Ec8 [110] were evident. The fundamental periods of interest had ranges of [0.1, 0.5] s (Figure 5) for the majority of buildings in Greece. In this range of values, the spectra for Patras, Nafpaktos, and Aigion significantly surpassed EAK2003. However, they were fully overlaid by the design spectra from Ec8. Moreover, given that the probability of exceeding the respective S_a value was equal for all periods, and since all future earthquake events were taken into account, it is reasonable that higher values were computed. Hence, it is suggested that the Ec8 regulations should be respected for the majority of buildings in the study area.

In addition to the traditional PSHA approach, finite-fault stochastic simulations were performed by generating earthquake scenarios for a predicted maximum magnitude (M_{max}), calculated by taking into account past seismicity and fault data. The three modeled active faults were the Psathopyrgos, Helike, and Trizonia Faults. Final results were obtained by testing both random earthquake nucleation points and co-seismic slip distribution on the surface. Prior to the computations, stochastic modeling parameters regarding the path and local site effects were calibrated by performing strong ground motion simulation of the largest recent event; namely, the Aigion earthquake that struck the study area in 1995. Comparison with real strong motion data and GMPEs proved that this methodology is reliable for the reproduction of the ground motion, regardless of the uncertainties that are involved, which can be ascribed mainly to source complexity and local site condition variability. As a result, this methodology can be treated as a powerful tool for the simulation of the expected strong ground motion of future strong earthquakes, especially for cases where recordings are sparse or not available. It, therefore, provides a unique opportunity to comprehensively evaluate seismic hazard with the synergy between PSHA outcomes and specific scenarios.

Site-specific analysis at the three selected sites was performed through a comparison between response spectra for each seismic scenario. S_a values constitute a good indicator of seismic loading for a variety of structures, as they describe the absolute maximum response of a single degree-of-freedom oscillator to an enforced ground motion. The mean response spectra exceeded the current EAK2003 in some cases, as in Nafpaktos for the hypothetical rupture of the Psathopyrgos Fault. The city of Patras was revealed to be the site less exposed to seismic hazard among those examined for the selected scenarios. However, it should be noted that other faults outside the WGoC may be more dangerous for Patras, such as the Rion–Patras fault zone (Figure 1); the local NW–SE sinistral-slip fault, which was related to the 1993 earthquake [41]; or even the Andravida fault, further southwest, which has caused strong earthquakes in the past [118]. The latter two faults were not taken into account, as the present study assessed seismic hazard in the WGoC area.

Overall, the PGA variability obtained from the specific fault scenarios lay inside the PGA values calculated through the PSHA. Figure 3a,b indicate that the western part of the study area exhibited a similar level of seismic hazard under the PSHA evaluation. The expected peak ground motions in this part, however, were considerably affected by the nearby seismic sources to the west (the Ionian Islands and the westernmost end of the Hellenic Arc), which are among the most active and productive in terms of seismic potential [119]. PSHA provided valuable information reflecting the combined effects of all potential seismic sources, making it possible to assess seismic hazard in terms of statistical likelihood of occurrence. On the other hand, the employment of seismic scenarios made it possible to acquire more realistic results concerning specific strong events through the definition of local site effects and path properties but without taking into account the time frame of occurrence.

5. Conclusions

A holistic seismic hazard assessment, including both probabilistic and scenario-based methods, was conducted for the highly active Western Gulf of Corinth area in Central Greece. Initially, seismic hazard was evaluated using a probabilistic approach, taking into account all the $M_w \geq 4.0$ earthquakes recorded during 1900–2019, in order to determine values for PGA, PGV, and S_a in a dense grid. For this purpose, the Cornell–McGuire method was utilized. Aiming to incorporate a multitude of seismotectonic models in order to qualitatively cope with epistemic uncertainties, three models—namely, ESHM13, ESHM20, and VAM16—were considered, and GMPEs were introduced for each zone. The PGA and PGV results were obtained through an equal-weight logic tree approach in which each major branch was a seismotectonic model and each minor branch was a modified GMPE. The logic tree procedure incorporated with the five GMPEs that were developed using Greek data qualitatively reduced epistemic uncertainties, which simple PSHA approaches may carry. In addition, PGA values were computed for the most populated urban areas (Patras, Nafpaktos, and Aigion) for various exceedance probabilities in 50 years.

The results obtained herein highlight the great significance of seismic hazard assessment using a combined approach that takes into account not only the evaluation of ground parameters in terms of probabilities of occurrence in a given time frame but also the anticipated effects of deterministic worst-case scenarios. Future work could include disaggregation of PSHA results so that the parameter pair of magnitude and epicentral distance that contributes most to seismic hazard can be identified. In this way, scenarios could be generated based on this result. Furthermore, a study of the Peak Ground Rotational Acceleration (PGRA) and Velocity (PGRV) values could be undertaken, as it has been proven that their results provide aid for engineering purposes [58,77,120]. In addition, the incorporation of a comprehensive microzonation scheme could provide valuable insight into the impact of future strong earthquakes on the major cities of the study area by identifying the possible amplification trends with respect to the structural response. Lastly, the results of the present seismic hazard study can be exploited in the future to assess seismic risk at urban centers in the WGoC area after incorporating structural vulnerability data.

Supplementary Materials: The following supporting information can be downloaded at: https://www.mdpi.com/article/10.3390/app122111152/s1, Figure S1. The three zonation models that were implemented for the WGoC, (a) the ESHM13 model [62,63], (b) the ESHM20 model [64] and (c) the model VAM16 from Vamvakaris et al. [65] proposed for the Greek territory; Table S1: The seismicity and geometry data that was computed for each Area Source Zone (ASZ).

Author Contributions: Conceptualization, G.K.; Data curation, G.K. and A.Z.; Formal analysis, G.K. and A.Z.; Funding acquisition, G.K.; Investigation, G.K., A.Z., P.B. and V.K.; Methodology, G.K.; Resources, N.V.; Software, A.Z. and P.B.; Supervision, G.K. and N.V.; Validation, G.K. and A.Z.; Visualization, P.B. and V.K.; Writing—original draft, G.K., A.Z., P.B. and V.K.; Writing—review & editing, G.K., A.Z., V.K. and N.V. All authors have read and agreed to the published version of the manuscript.

Funding: This research received no external funding.

Institutional Review Board Statement: Not applicable.

Informed Consent Statement: Not applicable.

Data Availability Statement: Publicly available datasets were analyzed in this study. ESHM13,20 ASZ can be found here: http://hazard.efehr.org/en/home/, accessed on 30 October 2022. In accordance with the earthquake catalogue, the earthquakes for the period 1900-2009 were adopted from Makropoulos et al. [25]. For the temporal expansion (2010-1019) of the catalogue, the reviewed events from International Seismological Centre (http://www.isc.ac.uk/iscbulletin/search/bulletin/, accessed on 30 October 2022) were adopted.

Acknowledgments: We express our gratitude to the three anonymous reviewers for their constructive comments that helped improve the manuscript, as well as to the editor and the guest editors for providing us with the opportunity to publish this study. We would like to thank the personnel of the following institutions: (a) the CRL team (CL network, data hosted at RESIF, https://doi.org/10.15778/RESIF.CL, accessed on 30 October 2022); (b) the National and Kapodistrian University of Athens (NKUA) (HA network, data hosted at NOA, https://doi.org/10.7914/SN/HA, accessed on 30 October 2022); (c) the University of Patras (HP network, data hosted at NOA, https://doi.org/10.7914/SN/HP, accessed on 30 October 2022), which operates 11 stations jointly with Charles University, Prague; (d) the National Observatory of Athens (NOA) (HL network, data hosted at NOA, https://doi.org/10.7914/SN/HL, accessed on 02 November 2022); and (e) the Institute of Engineering Seismology and Earthquake Engineering (ITSAK) (HI network, data hosted at NOA, https://doi.org/10.7914/SN/HI, accessed on 30 October 2022). Stations operated by the last four institutes are also part of the Hellenic Unified Seismological Network (HUSN, https://www.gein.noa.gr/en/networks-equipment/hellenic-unified-seismic-network-h-u-s-n/, accessed on 30 October 2022). Figure 1 was drawn using the Generic Mapping Tools (GMT) software [121]. Grapher version 15 (http://www.GoldenSoftware.com, accessed on 30 October 2022) was used for some of the figures.

Conflicts of Interest: The authors declare no conflict of interest.

References

1. Avallone, A.; Briole, P.; Agatza-Balodimou, A.M.; Billiris, H.; Charade, O.; Mitsakaki, C.; Nercessian, A.; Papazissi, K.; Paradissis, D.; Veis, G. Analysis of Eleven Years of Deformation Measured by GPS in the Corinth Rift Laboratory Area. *Comptes Rendus Geosci.* **2004**, *336*, 301–311. [CrossRef]
2. Kaviris, G. Study of Seismic Source Properties of the Eastern Gulf of Corinth. Ph.D. Thesis, Geophysics-Geothermics Department, Faculty of Geology, University of Athens, Zografou, Greece, 2003. [CrossRef]

3. Duverger, C.; Lambotte, S.; Bernard, P.; Lyon-Caen, H.; Deschamps, A.; Nercessian, A. Dynamics of Microseismicity and Its Relationship with the Active Structures in the Western Corinth Rift (Greece). *Geophys. J. Int.* **2018**, *215*, 196–221. [CrossRef]

4. Mesimeri, M.; Karakostas, V.; Papadimitriou, E.; Tsaklidis, G.; Jacobs, K. Relocation of Recent Seismicity and Seismotectonic Properties in the Gulf of Corinth (Greece). *Geophys. J. Int.* **2018**, *212*, 1123–1142. [CrossRef]

5. Briole, P.; Ganas, A.; Elias, P.; Dimitrov, D. The GPS Velocity Field of the Aegean. New Observations, Contribution of the Earthquakes, Crustal Blocks Model. *Geophys. J. Int.* **2021**, *226*, 468–492. [CrossRef]

6. Bernard, P.; Lyon-Caen, H.; Briole, P.; Deschamps, A.; Boudin, F.; Makropoulos, K.; Papadimitriou, P.; Lemeille, F.; Patau, G.; Billiris, H.; et al. Seismicity, Deformation and Seismic Hazard in the Western Rift of Corinth: New Insights from the Corinth Rift Laboratory (CRL). *Tectonophysics* **2006**, *426*, 7–30. [CrossRef]

7. Kaviris, G.; Elias, P.; Kapetanidis, V.; Serpetsidaki, A.; Karakonstantis, A.; Plicka, V.; De Barros, L.; Sokos, E.; Kassaras, I.; Sakkas, V.; et al. The Western Gulf of Corinth (Greece) 2020–2021 Seismic Crisis and Cascading Events: First Results from the Corinth Rift Laboratory Network. *Seism. Rec.* **2021**, *1*, 85–95. [CrossRef]

8. Chiaraluce, L.; Festa, G.; Bernard, P.; Caracausi, A.; Carluccio, I.; Clinton, J.; Di Stefano, R.; Elia, L.; Evangelidis, C.; Ergintav, S.; et al. The Near Fault Observatory Community in Europe: A New Resource for Faulting and Hazard Studies. *Ann. Geophys.* **2022**, *65*, DM316. [CrossRef]

9. Wessels, R.; ter Maat, G.; Del Bello, E.; Cacciola, L.; Corbi, F.; Festa, G.; Funiciello, F.; Kaviris, G.; Lange, O.; Lauterjung, J.; et al. Transnational Access to Research Facilities: An EPOS Service to Promote Multi-domain Solid Earth Sciences in Europe. *Ann. Geophys.* **2022**, *65*, DM214. [CrossRef]

10. Bell, R.E.; McNeill, L.C.; Bull, J.M.; Henstock, T.J.; Collier, R.E.L.; Leeder, M.R. Fault Architecture, Basin Structure and Evolution of the Gulf of Corinth Rift, Central Greece. *Basin Res.* **2009**, *21*, 824–855. [CrossRef]

11. Cornet, F.H.; Doan, M.L.; Moretti, I.; Borm, G. Drilling through the Active Aigion Fault: The AIG10 Well Observatory. *Comptes Rendus Geosci.* **2004**, *336*, 395–406. [CrossRef]

12. Ghisetti, F.; Vezzani, L. Inherited Structural Controls on Normal Fault Architecture in the Gulf of Corinth (Greece). *Tectonics* **2005**, TC4016. [CrossRef]

13. Gallousi, C.; Koukouvelas, I.K. Quantifying Geomorphic Evolution of Earthquake-Triggered Landslides and Their Relation to Active Normal Faults. An Example from the Gulf of Corinth, Greece. *Tectonophysics* **2007**, *440*, 85–104. [CrossRef]

14. Beckers, A.; Hubert-Ferrari, A.; Beck, C.; Bodeux, S.; Tripsanas, E.; Sakellariou, D.; De Batist, M. Active Faulting at the Western Tip of the Gulf of Corinth, Greece, from High-Resolution Seismic Data. *Mar. Geol.* **2015**, *360*, 55–69. [CrossRef]

15. Bell, R.E.; McNeill, L.C.; Bull, J.M.; Henstock, T.J. Evolution of the Offshore Western Gulf of Corinth. *GSA Bull.* **2008**, *120*, 156–178. [CrossRef]

16. Rigo, A.; Lyon-Caen, H.; Armijo, R.; Deschamps, A.; Hatzfeld, D.; Makropoulos, K.; Papadimitriou, P.; Kassaras, I. A Microseismic Study in the Western Part of the Gulf of Corinth (Greece): Implications for Large-Scale Normal Faulting Mechanisms. *Geophys. J. Int.* **1996**, *126*, 663–688. [CrossRef]

17. Kapetanidis, V.; Deschamps, A.; Papadimitriou, P.; Matrullo, E.; Karakonstantis, A.; Bozionelos, G.; Kaviris, G.; Serpetsidaki, A.; Lyon-Caen, H.; Voulgaris, N.; et al. The 2013 Earthquake Swarm in Helike, Greece: Seismic Activity at the Root of Old Normal Faults. *Geophys. J. Int.* **2015**, *202*, 2044–2073. [CrossRef]

18. Pacchiani, F.; Lyon-Caen, H. Geometry and Spatio-Temporal Evolution of the 2001 Agios Ioanis Earthquake Swarm (Corinth Rift, Greece). *Geophys. J. Int.* **2010**, *180*, 59–72. [CrossRef]

19. Kapetanidis, V.; Kassaras, I. Contemporary Crustal Stress of the Greek Region Deduced from Earthquake Focal Mechanisms. *J. Geodyn.* **2019**, *123*, 55–82. [CrossRef]

20. Duverger, C.; Godano, M.; Bernard, P.; Lyon-Caen, H.; Lambotte, S. The 2003–2004 Seismic Swarm in the Western Corinth Rift: Evidence for a Multiscale Pore Pressure Diffusion Process along a Permeable Fault System. *Geophys. Res. Lett.* **2015**, *42*, 7374–7382. [CrossRef]

21. Kapetanidis, V.; Michas, G.; Kaviris, G.; Vallianatos, F. Spatiotemporal Properties of Seismicity and Variations of Shear-Wave Splitting Parameters in the Western Gulf of Corinth (Greece). *Appl. Sci.* **2021**, *11*, 6573. [CrossRef]

22. Lambotte, S.; Lyon-Caen, H.; Bernard, P.; Deschamps, A.; Patau, G.; Nercessian, A.; Pacchiani, F.; Bourouis, S.; Drilleau, M.; Adamova, P. Reassessment of the Rifting Process in the Western Corinth Rift from Relocated Seismicity. *Geophys. J. Int.* **2014**, *197*, 1822–1844. [CrossRef]

23. Halpaap, F.; Rondenay, S.; Perrin, A.; Goes, S.; Ottemöller, L.; Austrheim, H.; Shaw, R.; Eeken, T. Earthquakes Track Subduction Fluids from Slab Source to Mantle Wedge Sink. *Sci. Adv.* **2019**, *5*, eaav7369. [CrossRef] [PubMed]

24. Pearce, F.D.; Rondenay, S.; Sachpazi, M.; Charalampakis, M.; Royden, L.H. Seismic Investigation of the Transition from Continental to Oceanic Subduction along the Western Hellenic Subduction Zone. *J. Geophys. Res. Solid Earth* **2012**, B07306. [CrossRef]

25. Makropoulos, K.; Kaviris, G.; Kouskouna, V. An Updated and Extended Earthquake Catalogue for Greece and Adjacent Areas since 1900. *Nat. Hazards Earth Syst. Sci.* **2012**, *12*, 1425–1430. [CrossRef]

26. McKenzie, D. Some Remarks on the Development of Sedimentary Basins. *Earth Planet. Sci. Lett.* **1978**, *40*, 25–32. [CrossRef]

27. Papadopoulos, G.A.; Kondopoulou, D.P.; Leventakis, G.-A.; Pavlides, S.B. Seismotectonics of the Aegean Region. *Tectonophysics* **1986**, *124*, 67–84. [CrossRef]

28. Baker, C.; Hatzfeld, D.; Lyon-Caen, H.; Papadimitriou, E.; Rigo, A. Earthquake Mechanisms of the Adriatic Sea and Western Greece: Implications for the Oceanic Subduction-Continental Collision Transition. *Geophys. J. Int.* **1997**, *131*, 559–594. [CrossRef]

29. Bernard, P.; Briole, P.; Meyer, B.; Lyon-Caen, H.; Gomez, J.-M.; Tiberi, C.; Berge, C.; Cattin, R.; Hatzfeld, D.; Lachet, C.; et al. The Ms = 6.2, June 15, 1995 Aigion Earthquake (Greece): Evidence for Low Angle Normal Faulting in the Corinth Rift. *J. Seismol.* **1997**, *1*, 131–150. [CrossRef]

30. Louvari, E. A Detailed Seismotectonic Study in the Aegean Sea and the Surrounding Area with Emphasis on the Information Obtained from Microearthquakes. Ph.D. Thesis, Aristotle University of Thessaloniki, Thessaloniki, Greece, 2000.

31. Kiratzi, A.; Louvari, E. Focal Mechanisms of Shallow Earthquakes in the Aegean Sea and the Surrounding Lands Determined by Waveform Modelling: A New Database. *J. Geodyn.* **2003**, *36*, 251–274. [CrossRef]

32. Sokos, E.; Zahradník, J.; Kiratzi, A.; Janský, J.; Gallovič, F.; Novotny, O.; Kostelecký, J.; Serpetsidaki, A.; Tselentis, G.-A. The January 2010 Efpalio Earthquake Sequence in the Western Corinth Gulf (Greece). *Tectonophysics* **2012**, *530–531*, 299–309. [CrossRef]

33. Stucchi, M.; Rovida, A.; Gomez Capera, A.A.; Alexandre, P.; Camelbeeck, T.; Demircioglu, M.B.; Gasperini, P.; Kouskouna, V.; Musson, R.M.W.; Radulian, M.; et al. The SHARE European Earthquake Catalogue (SHEEC) 1000–1899. *J. Seismol.* **2013**, *17*, 523–544. [CrossRef]

34. Armijo, R.; Meyer, B.; King, G.C.P.; Rigo, A.; Papanastassiou, D. Quaternary Evolution of the Corinth Rift and Its Implications for the Late Cenozoic Evolution of the Aegean. *Geophys. J. Int.* **1996**, *126*, 11–53. [CrossRef]

35. Flotté, N.; Sorel, D.; Müller, C.; Tensi, J. Along Strike Changes in the Structural Evolution over a Brittle Detachment Fault: Example of the Pleistocene Corinth–Patras Rift (Greece). *Tectonophysics* **2005**, *403*, 77–94. [CrossRef]

36. Papadopoulos, G. A Reconstruction of the Great Earthquake of 373 B.C. in the Western Gulf of Corinth. In Proceedings of the 2nd International Conference "Ancient Helike and Aigialeia", Aighion, Greece, 1–3 December 1995; 1998; pp. 479–494.

37. Mouyaris, N.; Papastamatiou, D.; Vita-Finzi, C. The Helice Fault? *Terra Nova* **1992**, *4*, 124–128. [CrossRef]

38. Boiselet, A. Cycle Sismique et Aléa Sismique d'un Réseau de Failles Actives: Le Cas Du Rift de Corinthe (Grèce). Ph.D. Thesis, Sciences de La Terre, Ecole Normale Superieure de Paris-ENS Paris, Paris, France, 2014.

39. Kaviris, G.; Spingos, I.; Kapetanidis, V.; Papadimitriou, P.; Voulgaris, N.; Makropoulos, K. Upper Crust Seismic Anisotropy Study and Temporal Variations of Shear-Wave Splitting Parameters in the Western Gulf of Corinth (Greece) during 2013. *Phys. Earth Planet. Inter.* **2017**, *269*, 148–164. [CrossRef]

40. Plicka, V.; Sokos, E.; Tselentis, G.A.; Zahradník, J. The Patras Earthquake (14 July 1993): Relative Roles of Source, Path and Site Effects. *J. Seismol.* **1998**, *2*, 337–349. [CrossRef]

41. Karakostas, B.; Papadimitriou, E.; Hatzfeld, D.; Makaris, D.; Makropoulos, K.; Diagourtas, D.; Papaioannou, C.; Stavrakakis, G.; Drakopoulos, J.; Papazachos, B. The Aftershock Sequence and Focal Properties of the July 14, 1993 (M = 5.4) Patras Earthquake. *Bull. Geol. Soc. Greece* **1994**, *30*, 167–194.

42. Ganas, A.; Chousianitis, K.; Batsi, E.; Kolligri, M.; Agalos, A.; Chouliaras, G.; Makropoulos, K. The January 2010 Efpalion Earthquakes (Gulf of Corinth, Central Greece): Earthquake Interactions and Blind Normal Faulting. *J. Seismol.* **2013**, *17*, 465–484. [CrossRef]

43. Leptokaropoulos, K.M.; Papadimitriou, E.E.; Orlecka–Sikora, B.; Karakostas, V.G. An Evaluation of Coulomb Stress Changes from Earthquake Productivity Variations in the Western Gulf of Corinth, Greece. *Pure Appl. Geophys.* **2016**, *173*, 49–72. [CrossRef]

44. Kaviris, G.; Millas, C.; Spingos, I.; Kapetanidis, V.; Fountoulakis, I.; Papadimitriou, P.; Voulgaris, N.; Makropoulos, K. Observations of Shear-Wave Splitting Parameters in the Western Gulf of Corinth Focusing on the 2014 Mw = 5.0 Earthquake. *Phys. Earth Planet. Inter.* **2018**, *282*, 60–76. [CrossRef]

45. Mesimeri, M.; Ganas, A.; Pankow, K.L. Multisegment Ruptures and Vp/Vs Variations during the 2020–2021 Seismic Crisis in Western Corinth Gulf, Greece. *Geophys. J. Int.* **2022**, *230*, 334–348. [CrossRef]

46. Papadimitriou, E.; Bonatis, P.; Bountzis, P.; Kostoglou, A.; Kourouklas, C.; Karakostas, V. The Intense 2020–2021 Earthquake Swarm in Corinth Gulf: Cluster Analysis and Seismotectonic Implications from High Resolution Microseismicity. *Pure Appl. Geophys.* **2022**, *179*, 3121–3155. [CrossRef]

47. Zahradník, J.; Aissaoui, E.M.; Bernard, P.; Briole, P.; Bufféral, S.; De Barros, L.; Deschamps, A.; Elias, P.; Evangelidis, C.P.; Fountoulakis, I.; et al. An Atypical Shallow Mw 5.3, 2021 Earthquake in the Western Corinth Rift (Greece). *J. Geophys. Res. Solid Earth* **2022**, *127*, e2022JB024221. [CrossRef]

48. Meirova, T.; Shapira, A.; Eppelbaum, L. PSHA in Israel by Using the Synthetic Ground Motions from Simulated Seismicity: The Modified SvE Procedure. *J. Seismol.* **2018**, *22*, 1095–1111. [CrossRef]

49. Banitsiotou, I.D.; Tsapanos, T.M.; Margaris, V.N.; Hatzidimitriou, P.M. Estimation of the Seismic Hazard Parameters for Various Sites in Greece Using a Probabilistic Approach. *Nat. Hazards Earth Syst. Sci.* **2004**, *4*, 399–405. [CrossRef]

50. Burton, P.W.; Xu, Y.; Tselentis, G.-A.; Sokos, E.; Aspinall, W. Strong Ground Acceleration Seismic Hazard in Greece and Neighboring Regions. *Soil Dyn. Earthq. Eng.* **2003**, *23*, 159–181. [CrossRef]

51. Makropoulos, K.C.; Burton, P.W. Seismic Hazard in Greece. II. Ground Acceleration. *Tectonophysics* **1985**, *117*, 259–294. [CrossRef]

52. Tsapanos, T.M. Seismicity and Seismic Hazard Assessment in Greece. In *Earthquake Monitoring and Seismic Hazard Mitigation in Balkan Countries*; Husebye, E.S., Ed.; NATO Science Series: IV: Earth and Environmental Sciences; Springer: Dordrecht, The Netherlands, 2008; Volume 81, pp. 253–270, ISBN 9781402068133.

53. Tselentis, G.-A.; Danciu, L. Probabilistic Seismic Hazard Assessment in Greece—Part 1: Engineering Ground Motion Parameters. *Nat. Hazards Earth Syst. Sci.* **2010**, *10*, 25–39. [CrossRef]

54. Tselentis, G.-A.; Danciu, L.; Sokos, E. Probabilistic Seismic Hazard Assessment in Greece—Part 2: Acceleration Response Spectra and Elastic Input Energy Spectra. *Nat. Hazards Earth Syst. Sci.* **2010**, *10*, 41–49. [CrossRef]

55. Papadopoulou-Vrynioti, K.; Bathrellos, G.D.; Skilodimou, H.D.; Kaviris, G.; Makropoulos, K. Karst Collapse Susceptibility Mapping Considering Peak Ground Acceleration in a Rapidly Growing Urban Area. *Eng. Geol.* **2013**, *158*, 77–88. [CrossRef]

56. Pavlou, K.; Kaviris, G.; Chousianitis, K.; Drakatos, G.; Kouskouna, V.; Makropoulos, K. Seismic Hazard Assessment in Polyphyto Dam Area (NW Greece) and Its Relation with the "Unexpected" Earthquake of 13 May 1995 (M_s = 6.5, NW Greece). *Nat. Hazards Earth Syst. Sci.* **2013**, *13*, 141–149. [CrossRef]

57. Pavlou, K.; Kaviris, G.; Kouskouna, V.; Sakkas, G.; Zymvragakis, A.; Sakkas, V.; Drakatos, G. Minor Seismic Hazard Changes in the Broader Area of Pournari Artificial Lake after the First Filling (W. Greece). *Results Geophys. Sci.* **2021**, *7*, 100025. [CrossRef]

58. Kaviris, G.; Zymvragakis, A.; Bonatis, P.; Sakkas, G.; Kouskouna, V.; Voulgaris, N. Probabilistic Seismic Hazard Assessment for the Broader Messinia (SW Greece) Region. *Pure Appl. Geophys.* **2022**, *179*, 551–567. [CrossRef]

59. Cornell, C.A. Engineering Seismic Risk Analysis. *Bull. Seismol. Soc. Am.* **1968**, *58*, 1583–1606. [CrossRef]

60. McGuire, R.K. *FORTRAN Computer Program for Seismic Risk Analysis*; Open-File Report; Geological Survey: Reston, VA, USA, 1976.

61. Taroni, M.; Akinci, A. Good practices in PSHA: Declustering, b-value estimation, foreshocks and aftershocks inclusion; a case study in Italy. *Geophys. J. Int.* **2021**, *224*, 1174–1187. [CrossRef]

62. Giardini, D.; Wössner, J.; Danciu, L. Mapping Europe's Seismic Hazard. *Eos Trans. Am. Geophys. Union* **2014**, *95*, 261–262. [CrossRef]

63. Woessner, J.; Laurentiu, D.; Giardini, D.; Crowley, H.; Cotton, F.; Grünthal, G.; Valensise, G.; Arvidsson, R.; Basili, R.; Demircioglu, M.B.; et al. The 2013 European Seismic Hazard Model: Key Components and Results. *Bull. Earthq. Eng.* **2015**, *13*, 3553–3596. [CrossRef]

64. Danciu, L.; Nandan, S.; Reyes, C.; Basili, R.; Weatherill, G.; Beauval, C.; Rovida, A.; Vilanova, S.; Sesetyan, K.; Bard, P.-Y.; et al. *ESHM20—EFEHR Technical ReportThe 2020 Update of the European Seismic Hazard Model—ESHM20: Model Overview; EFEHR European Facilities of Earthquake Hazard and Risk*; EFEHR European Facilities of Earthquake Hazard and Risk: Zürich, Switzerland, 2021.

65. Vamvakaris, D.A.; Papazachos, C.B.; Papaioannou, C.A.; Scordilis, E.M.; Karakaisis, G.F. A Detailed Seismic Zonation Model for Shallow Earthquakes in the Broader Aegean Area. *Nat. Hazards Earth Syst. Sci.* **2016**, *16*, 55–84. [CrossRef]

66. Ordaz, M.; Salgado-Gálvez, M.A.; Giraldo, S. R-CRISIS: 35 Years of Continuous Developments and Improvements for Probabilistic Seismic Hazard Analysis. *Bull Earthq. Eng* **2021**, *19*, 2797–2816. [CrossRef]

67. Wiemer, S.; Wyss, M. Minimum Magnitude of Completeness in Earthquake Catalogs: Examples from Alaska, the Western United States, and Japan. *Bull. Seismol. Soc. Am.* **2000**, *90*, 859–869. [CrossRef]

68. Aki, K. Maximum Likelihood Estimate of b in the Formula Log N = A-BM and Its Confidence Limits. *Bull. Earthq. Res. Inst.* **1965**, *43*, 237–239.

69. Wiemer, S. A Software Package to Analyze Seismicity: ZMAP. *Seismol. Res. Lett.* **2001**, *72*, 373–382. [CrossRef]

70. Wiemer, S.; Wyss, M. Mapping the Frequency-Magnitude Distribution in Asperities: An Improved Technique to Calculate Recurrence Times? *J. Geophys. Res. Solid Earth* **1997**, *102*, 15115–15128. [CrossRef]

71. Zhou, Y.; Zhou, S.; Zhuang, J. A Test on Methods for MC Estimation Based on Earthquake Catalog. *Earth Planet. Phys.* **2018**, *2*, 150–162. [CrossRef]

72. Robson, D.S.; Whitlock, J.H. Estimation of a Truncation Point. *Biometrika* **1964**, *51*, 33–39. [CrossRef]

73. Cooke, P. Statistical Inference for Bounds of Random Variables. *Biometrika* **1979**, *66*, 367–374. [CrossRef]

74. Margaris, B.; Papazachos, C.; Papaioannou, C.; Theodulidis, N.; Kalogeras, I.; Skarlatoudis, A. Ground Motion Attenuation Relations for Shallow Earthquakes in Greece. In Proceedings of the 12th European Conference on Earthquake Engineering 12ECEE, London, UK, 9–13 September 2002; pp. 318–327.

75. Skarlatoudis, A.A.; Papazachos, C.B.; Margaris, B.N.; Theodulidis, N.; Papaioannou, C.; Kalogeras, I.; Scordilis, E.M.; Karakostas, V. Empirical Peak Ground-Motion Predictive Relations for Shallow Earthquakes in Greece. *Bull. Seismol. Soc. Am.* **2003**, *93*, 2591–2603, Erratum in *Bull. Seismol. Soc. Am.* **2007**, *97*, 2219–2221. [CrossRef]

76. Danciu, L.; Tselentis, G.-A. Engineering Ground-Motion Parameters Attenuation Relationships for Greece. *Bull. Seismol. Soc. Am.* **2007**, *97*, 162–183. [CrossRef]

77. Sakkas, G. Calculation and Analysis of the Seismic Motion Rotational Components in Greece. Ph.D. Thesis, National and Kapodistrian University of Athens, Athens, Greece, 2016; 278p. [CrossRef]

78. Chousianitis, K.; Del Gaudio, V.; Pierri, P.; Tselentis, G.-A. Regional Ground-Motion Prediction Equations for Amplitude-, Frequency Response-, and Duration-Based Parameters for Greece. *Earthq. Engng. Struct Dyn.* **2018**, *47*, 2252–2274. [CrossRef]

79. Boore, D.M.; Stewart, J.P.; Skarlatoudis, A.A.; Seyhan, E.; Margaris, B.; Theodoulidis, N.; Scordilis, E.; Kalogeras, I.; Klimis, N.; Melis, N.S. A Ground-Motion Prediction Model for Shallow Crustal Earthquakes in Greece. *Bull. Seismol. Soc. Am.* **2021**, *111*, 857–874. [CrossRef]

80. Motazedian, D.; Atkinson, G. Stochastic Finite-Fault Modeling Based on a Dynamic Corner Frequency. *Bull. Seismol. Soc. Am.* **2005**, *95*, 995–1010. [CrossRef]

81. Boore, D.M. Comparing Stochastic Point-Source and Finite-Source Ground-Motion Simulations: SMSIM and EXSIM. *Bull. Seismol. Soc. Am.* **2009**, *99*, 3202–3216. [CrossRef]

82. Boore, D.M. Simulation of Ground Motion Using the Stochastic Method. *Pure Appl. Geophys.* **2003**, *160*, 635–676. [CrossRef]

83. Akinci, A.; Aochi, H.; Herrero, A.; Pischiutta, M.; Karanikas, D. Physics-Based Broadband Ground-Motion Simulations for ProbableMw$\geq$7.0 Earthquakes in the Marmara Sea Region (Turkey). *Bull. Seismol. Soc. Am.* **2016**, *107*, 1307–1323. [CrossRef]

84. Roumelioti, Z.; Kiratzi, A.; Margaris, B.; Chatzipetros, A. Simulation of Strong Ground Motion on Near-Fault Rock Outcrop for Engineering Purposes: The Case of the City of Xanthi (Northern Greece). *Bull. Earthq. Eng.* **2017**, *15*, 25–49. [CrossRef]

85. Kassaras, I.; Kazantzidou-Firtinidou, D.; Ganas, A.; Kapetanidis, V.; Tsimi, C.; Valkaniotis, S.; Sakellariou, N.; Mourloukos, S. Seismic Risk and Loss Assessment for Kalamata (SW Peloponnese, Greece) from Neighbouring Shallow Sources. *Boll. Di Geofis. Teor. Ed Appl.* **2018**, *59*, 1–26. [CrossRef]

86. Giannaraki, G.; Kassaras, I.; Roumelioti, I.; Roumelioti, Z.; Kazantzidou-Firtinidou, A.; Ganas, A. Scenario-Based Seismic Risk Assessment in the City of Aigion (Greece). *Bull. Earthq. Eng.* **2019**, *17*, 603–634. [CrossRef]

87. Bonatis, P.; Akinci, A.; Karakostas, V.; Papadimitriou, E.; Kaviris, G. Near-Fault Broadband Ground Motion Simulation Applications at the Central Ionian Islands, Greece. *Pure Appl. Geophys.* **2021**, *178*, 3505–3527. [CrossRef]

88. Ravnalis, M.; Kkallas, C.; Papazachos, C.; Margaris, B.; Papaioannou, C. Semi-Automated Stochastic Simultaneous Simulation of Macroseismic Information and Strong Ground Motion Records for Recent Strong Earthquakes of the Aegean Area. *Ann. Geophys.* **2022**, *65*, SE425. [CrossRef]

89. Anderson, J.G.; Hough, S.E. A Model for the Shape of the Fourier Amplitude Spectrum of Acceleration at High Frequencies. *Bull. Seismol. Soc. Am.* **1984**, *74*, 1969–1993. [CrossRef]

90. Console, R.; Carluccio, R.; Murru, M.; Papadimitriou, E.; Karakostas, V. Physics-Based Simulation of Spatiotemporal Patterns of Earthquakes in the Corinth Gulf, Greece, Fault System. *Bull. Seismol. Soc. Am.* **2021**, *112*, 98–117. [CrossRef]

91. Beresnev, I.A.; Atkinson, G.M. Generic Finite-Fault Model for Ground-Motion Prediction in Eastern North America. *Bull. Seismol. Soc. Am.* **1999**, *89*, 608–625.

92. Atkinson, G.M.; Beresnev, I. Don't Call It Stress Drop. *Seismol. Res. Lett.* **1997**, *68*, 3–4. [CrossRef]

93. Oth, A.; Bindi, D.; Parolai, S.; Di Giacomo, D. Earthquake Scaling Characteristics and the Scale-(in)Dependence of Seismic Energy-to-Moment Ratio: Insights from KiK-Net Data in Japan. *Geophys. Res. Lett.* **2010**, *37*, L19304. [CrossRef]

94. Akkar, S.; Sandıkkaya, M.A.; Bommer, J.J. Empirical Ground-Motion Models for Point- and Extended-Source Crustal Earthquake Scenarios in Europe and the Middle East. *Bull Earthq. Eng* **2014**, *12*, 359–387. [CrossRef]

95. Bindi, D.; Massa, M.; Luzi, L.; Ameri, G.; Pacor, F.; Puglia, R.; Augliera, P. Pan-European Ground-Motion Prediction Equations for the Average Horizontal Component of PGA, PGV, and 5 %-Damped PSA at Spectral Periods up to 3.0 s Using the RESORCE Dataset. *Bull Earthq. Eng.* **2014**, *12*, 391–430. [CrossRef]

96. Margaris, B.N.; Boore, D.M. Determination of $\Delta\sigma$ and K0 from Response Spectra of Large Earthquakes in Greece. *Bull. Seismol. Soc. Am.* **1998**, *88*, 170–182. [CrossRef]

97. Margaris, B.; Hatzidimitriou, P. Source Spectral Scaling and Stress Release Estimates Using Strong-Motion Records in Greece. *Bull. Seismol. Soc. Am.* **2002**, *92*, 1040–1059. [CrossRef]

98. Chouliaras, G.; Stavrakakis, G.N. Seismic Source Parameters from a New Dial-up Seismological Network in Greece. *Pure Appl. Geophys.* **1997**, *150*, 91–111. [CrossRef]

99. Mangira, O.; Vasiliadis, G.; Papadimitriou, E. Application of a Linked Stress Release Model in Corinth Gulf and Central Ionian Islands (Greece). *Acta Geophys.* **2017**, *65*, 517–531. [CrossRef]

100. Atkinson, G.M. Empirical Attenuation of Ground-Motion Spectral Amplitudes in Southeastern Canada and the Northeastern United States. *Bull. Seismol. Soc. Am.* **2004**, *94*, 1079–1095.

101. Hatzidimitriou, P.M. *S*-Wave Attenuation in the Crust in Northern Greece. *Bull. Seismol. Soc. Am.* **1995**, *85*, 1381–1387. [CrossRef]

102. Klimis, N.S.; Margaris, B.N.; Koliopoulos, P.K. Site-dependent amplification functions and response spectra in greece. *J. Earthq. Eng.* **1999**, *3*, 237–270. [CrossRef]

103. Kourouklas, C.; Console, R.; Papadimitriou, E.; Murru, M.; Karakostas, V. Strong Earthquakes Recurrence Times of the Southern Thessaly, Greece, Fault System: Insights from a Physics-Based Simulator Application. *Front. Earth Sci.* **2021**, *9*, 596854. [CrossRef]

104. Pace, B.; Visini, F.; Peruzza, L. *FiSH*: MATLAB Tools to Turn Fault Data into Seismic-Hazard Models. *Seismol. Res. Lett.* **2016**, *87*, 374–386. [CrossRef]

105. Wells, D.L.; Coppersmith, K.J. New Empirical Relationships among Magnitude, Rupture Length, Rupture Width, Rupture Area, and Surface Displacement. *Bull. Seismol. Soc. Am.* **1994**, *84*, 974–1002.

106. Papazachos, B.C.; Scordilis, E.; Panagiotopoulos, D.; Papazachos, C.; Karakaisis, G. Global Relations between Seismic Fault Parameters and Moment Magnitude of Earthquakes. *Bull. Geol. Soc. Greece* **2004**, *36*, 1482–1489. [CrossRef]

107. Leonard, M. Earthquake Fault Scaling: Self-Consistent Relating of Rupture Length, Width, Average Displacement, and Moment Release. *Bull. Seismol. Soc. Am.* **2010**, *100*, 1971–1988. [CrossRef]

108. Thingbaijam, K.; Mai, P.; Goda, K. New Empirical Earthquake Source-Scaling Laws. *Bull. Seismol. Soc. Am.* **2017**, *107*, 2225–2246. [CrossRef]

109. EAK. *Greek Seismic Code Edited by: Earthquake Planning and Protection Organization*; Earthquake Planning and Protection Organization: Athens, Greece, 2003.

110. EC8. *European Committee for Standardization, EN-1998-1, Eurocode 8, "Design of Structures for Earthquake Resistance", Part 1: General Rules, Seismic Actions and Rules for Buildings*; European Committee: Brussels, Belgium, 2004.

111. Mavroeidis, G.; Ding, Y.; Moharrami, N. Revisiting the 1995 MW 6.4 Aigion, Greece, Earthquake: Simulation of Broadband Strong Ground Motion and Site Response Analysis. *Soil Dyn. Earthq. Eng.* **2018**, *104*, 156–173. [CrossRef]

112. Mesimeri, M.; Karakostas, V.; Papadimitriou, E.; Schaff, D.; Tsaklidis, G. Spatio-Temporal Properties and Evolution of the 2013 Aigion Earthquake Swarm (Corinth Gulf, Greece). *J. Seismol.* **2016**, *20*, 595–614. [CrossRef]

113. Michas, G.; Vallianatos, F. Stochastic Modeling of Nonstationary Earthquake Time Series with Long-Term Clustering Effects. *Phys. Rev. E* **2018**, *98*, 042107. [CrossRef]

114. De Barros, L.; Cappa, F.; Deschamps, A.; Dublanchet, P. Imbricated Aseismic Slip and Fluid Diffusion Drive a Seismic Swarm in the Corinth Gulf, Greece. *Geophys. Res. Lett.* **2020**, *47*, e2020GL087142. [CrossRef]

115. Atkinson, G.M.; Bommer, J.J.; Abrahamson, N.A. Alternative Approaches to Modeling Epistemic Uncertainty in Ground Motions in Probabilistic Seismic-Hazard Analysis. *Seismol. Res. Lett.* **2014**, *85*, 1141–1144. [CrossRef]

116. Bommer, J.J.; Scherbaum, F. The Use and Misuse of Logic Trees in Probabilistic Seismic Hazard Analysis. *Earthq. Spectra* **2008**, *24*, 997–1009. [CrossRef]

117. Marzocchi, W.; Taroni, M.; Selva, J. Accounting for Epistemic Uncertainty in PSHA: Logic Tree and Ensemble Modeling. *Bull. Seismol. Soc. Am.* **2015**, *105*, 2151–2159. [CrossRef]

118. Karakostas, V.; Mirek, K.; Mesimeri, M.; Papadimitriou, E.; Mirek, J. The Aftershock Sequence of the 2008 Achaia, Greece, Earthquake: Joint Analysis of Seismicity Relocation and Persistent Scatterers Interferometry. *Pure Appl. Geophys.* **2017**, *174*, 151–176. [CrossRef]

119. Bonatis, P.; Karakostas, V.G.; Papadimitriou, E.E.; Kaviris, G. Investigation of the Factors Controlling the Duration and Productivity of Aftershocks Following Strong Earthquakes in Greece. *Geosciences* **2022**, *12*, 328. [CrossRef]

120. Lee, W.H.K.; Çelebi, M.; Todorovska, M.I.; Igel, H. Introduction to the Special Issue on Rotational Seismology and Engineering Applications. *Bull. Seismol. Soc. Am.* **2009**, *99*, 945–957. [CrossRef]

121. Wessel, P.; Smith, W.H.F. New, Improved Version of Generic Mapping Tools Released. *Eos Trans. Am. Geophys. Union* **1998**, *79*, 579. [CrossRef]

applied sciences

Article

A Non-Extensive Statistical Physics View of the Temporal Properties of the Recent Aftershock Sequences of Strong Earthquakes in Greece

Sophia-Ekaterini Avgerinou [1,2], Eleni-Apostolia Anyfadi [1,2], Georgios Michas [1,2] and Filippos Vallianatos [1,2,*]

1   Section of Geophysics-Geothermics, Department of Geology and Geoenvironment, National and Kapodistrian University of Athens, Panepistimiopolis, 15784 Athens, Greece
2   Institute of Physics of Earth's Interior and Geohazards, UNESCO Chair on Solid Earth Physics and Geohazards Risk Reduction, Hellenic Mediterranean University Research Center, Crete, 73133 Chania, Greece
*   Correspondence: fvallian@geol.uoa.gr

Abstract: Greece is one of Europe's most seismically active areas. Seismic activity in Greece has been characterized by a series of strong earthquakes with magnitudes up to $M_w = 7.0$ over the last five years. In this article we focus on these strong events, namely the $M_w6.0$ Arkalochori (27 September 2021), the $M_w6.3$ Elassona (3 March 2021), the $M_w7.0$ Samos (30 October 2020), the $M_w5.1$ Parnitha (19 July 2019), the $M_w6.6$ Zakynthos (25 October 2018), the $M_w6.5$ Kos (20 July 2017) and the $M_w6.1$ Mytilene (12 June 2017) earthquakes. Based on the probability distributions of interevent times between the successive aftershock events, we investigate the temporal evolution of their aftershock sequences. We use a statistical mechanics model developed in the framework of Non-Extensive Statistical Physics (NESP) to approach the observed distributions. NESP provides a strictly necessary generalization of Boltzmann–Gibbs statistical mechanics for complex systems with memory effects, (multi)fractal geometries, and long-range interactions. We show how the NESP applicable to the temporal evolution of recent aftershock sequences in Greece, as well as the existence of a crossover behavior from power-law ($q \neq 1$) to exponential ($q = 1$) scaling for longer interevent times. The observed behavior is further discussed in terms of superstatistics. In this way a stochastic mechanism with memory effects that can produce the observed scaling behavior is demonstrated. To conclude, seismic activity in Greece presents a series of significant earthquakes over the last five years. We focus on strong earthquakes, and we study the temporal evolution of aftershock sequences of them using a statistical mechanics model. The non-extensive parameter q related with the interevent times distribution varies between 1.62 and 1.71, which suggests a system with about one degree of freedom.

Keywords: aftershocks sequences; Tsallis entropy; interevent times; power-law scaling; complexity; Greek seismicity

Citation: Avgerinou, S.-E.; Anyfadi, E.-A.; Michas, G.; Vallianatos, F. A Non-Extensive Statistical Physics View of the Temporal Properties of the Recent Aftershock Sequences of Strong Earthquakes in Greece. *Appl. Sci.* **2023**, *13*, 1995. https://doi.org/10.3390/app13031995

Academic Editor: Alexey Zavyalov

Received: 12 December 2022
Revised: 23 January 2023
Accepted: 30 January 2023
Published: 3 February 2023

1. Introduction

Due to the fact that a strong mainshock immediately after its occurrence can induce a high number of aftershocks in the broader epicentral area, aftershock sequences are typically regarded as an important component of the earthquake occurrence. Following the mainshock, many aftershocks typically occur in and around the fault rupture regions. In the larger framework of seismic activity analysis research, understanding the temporal characteristics of these earthquake sequences is a crucial first step. Time-correlated structures that determine the time series of observed earthquakes can provide usable data about the dynamic features of earthquake activities and the associated geodynamic mechanisms [1]. In this paper, we investigate the temporal properties of seven recent aftershock sequences that occurred in Greece between 2017 and 2021. Greece is located at the limits of contact and convergence of the Eurasian and African plates, which gives rise to intense geodynamic processes and seismicity, with several large magnitude events reported in both historic and modern times [2]. In terms of seismic energy release, Greece is ranked first in the Mediterranean and Europe, and sixth in the world [3]. This high seismic activity is commonly linked to the following geotectonic features: (a) the continental convergence, which consists of the oceanic component of the North African plate subduction beneath the European plate. Due to the accretion of African plate sediments beneath the underlying Aegean plate, this movement was coupled with severe crustal shortening and an uplift rate of a few millimeters per year throughout the Hellenic Arc, (b) the rollback of the subducting African slab causes high-rate extension in the back-arc area and last (c) the most prominent tectonic feature of the North Aegean Sea, the North Aegean Trough (NAT) and the Cephalonia Transform Zone (CTFZ) [4].

Based on the recent earthquake activity over the last four years, this area of Greece is characterized by strong earthquakes. More specifically, we focus on the recent strong earthquakes such as that of $M_w6.0$ Arkalochori (27 September 2021), the $M_w6.3$ Elassona (3 March 2021), the $M_w7.0$ Samos (30 October 2020), the $M_w5.1$ Parnitha (19 July 2019), the $M_w6.6$ Zakynthos (25 October 2018), the $M_w6.5$ Kos (20 July 2017) and the $M_w6.1$ Mytilene (12 June 2017) earthquakes. These events generated intense and prolonged aftershock sequences.

Herein, we study the temporal properties of these aftershock sequences that occurred in the area of Greece, with particular emphasis on the probability distribution of the interevent times T between successive aftershocks, in view of the ideas of non-extensive statistical physics [5,6]. The Non-Extended Statistical Physics (NESP) is a generalization of Boltzmann–Gibbs (BG) statistical physics and is used to estimate the probability distribution of T and to determine its non-additive entropic parameter q [7], which is estimated to vary in the range 1.62–1.71. In all analyzed aftershock sequences, we recognize a crossover behavior from power-law ($q \neq 1$) to exponential ($q = 1$) scaling for larger interevent times.

2. Principles of Non-Extensive Statistical Physics

In this work, we use a generalized formulation of Boltzmann–Gibbs (BG) statistical physics, termed non-extensive statistical physics (NESP) [8–12], to investigate the distribution of the interevent times between the successive aftershocks. The fundamental benefit of NESP, is that it takes into account correlations on all length scales between system elements, resulting in asymptotic power-law behavior. NESP has been used in a wide variety of fields such as non-linear dynamical systems, including aftershock sequences [13], seismicity [5–7,14–16], natural hazards [17], and complexity in volcanic areas [12], among others [8]. Such characteristics can be described in fracture-related phenomena. Non-extensive statistical physics is concerned with precisely such phenomena.

Initially, NESP begins by defining entropy by Tsallis [18]. This entropic functional is appropriate for characterizing complex systems with finite degrees of freedom, self-organized critically, and non-Markovian characteristics with long-range memory, properties as that commonly occur in geosciences [5,19–21]. The present application of Tsallis entropy introduces the variable of T (i.e., the interevent times) between two successive aftershocks, where $p(T)\,dT$ indicates the number of the parameter between T and $T+dT$. An earthquake complex system, in a non-equilibrium state, can be described by an entropic functional S_q introduced by Tsallis [18]

$$S_q = k_B \frac{1 - \sum_i p^q(T_i)}{q - 1} \tag{1}$$

where k_B is Boltzmann's constant, $p(T)$ is the probability distribution of interevent times T and the index q expresses the degree of non-additivity of the system. The index q may violate the additivity principle of classical BG entropy [8,18]. In [18] it was demonstrated that in the limit of $q \to 1$, the non-extensive entropy S_q recovers the Boltzmann–Gibbs (BG) one.

In earth sciences, the cumulative distribution function is traditionally used in the framework of NESP [20,21]. This expression is derived by maximizing S_q while imposing appropriate constraints and employing the Lagrange multipliers method, yielding to [8]:

$$p(T) = \frac{\left[1 - (1-q)\left(\frac{T}{T_q}\right)\right]^{\frac{1}{1-q}}}{Z_q} = \frac{1}{Z_q} exp_q\left(-\frac{T}{T_q}\right) \tag{2}$$

whith Z_q the so-called q-partition function

$$Z_q = \int_0^\infty exp_q(-T/T_q)dT, \tag{3}$$

and T_q denotes the generalized scaled interevent time. With respect to Equation (2) q-exponential function appears, defined as [8]

$$exp_q(X) = [1 + (1-q)X]^{\frac{1}{1-q}}, \tag{4}$$

for $1 + (1-q)X \geq 0$, while in other cases $exp_q(X) = 0$.

Equation (2) is further used to estimate the cumulative distribution function (CDF) of the interevent times:

$$P(> T) = \frac{N(> T)}{N_0}, \tag{5}$$

with $N(>T)$, is the number of the interevent times with value greater than T and N_0 their total number [22,23]. By using Equation (2), $P(>T)$ equals to Equation (6) which has the form of a q-exponential function, hereafter calles Q-exponential one:

$$P(> T) = exp_Q(-T/T^*), \tag{6}$$

with:

$$T^* = T_q Q \tag{7}$$

$$q = 2 - \left(\frac{1}{Q}\right) \tag{8}$$

The Q-logarithmic function is the inverse function of the Q-exponential and it is defined as:

$$ln_Q P(> T) = \frac{P(> T)^{1-Q} - 1}{1 - Q}, \tag{9}$$

Equation (9) demonstrates that the Q-logarithm [8,24] of CDF of interevent times, is linearly scaled with T with an expression:

$$ln_Q P(> T) = -\frac{T}{T^*}, \tag{10}$$

with slope $1/T^*$ [14].

According to the different values that the parameter q can take, three particular cases arise. More specifically, in the limit $q \rightarrow 1$, the q-exponential leads to the ordinary exponential function. For $q > 1$, the q-exponential function exhibits an asymptotic power-law behavior with slope $-1/(q-1)$, whereas for $0 < q < 1$, the q-exponential function presents a cut-off [22].

The Tsallis entropy S_q (with $q \neq 1$) is non-additive, whereas the BG entropy is additive, which means that in the merged system's $(A + B)$, BG entropy is equal to the sum of the constituent BG entropies of the systems A and B respectively [19,24–27]. In NESP approach, in the case where A and B are probabilistically independent, we have [19]:

$$S_q(A + B) = S_q(A) + S_q(B) + \frac{(q-1)}{k_B} S_q(A) S_q(B), \tag{11}$$

When $q = 1$, the Tsallis entropy S_q coincides with the BG one. Despite having several characteristics in common, such as non-negativity, expansibility, and concavity, S_q and S_{BG} differ significantly from one another. Particularly, there are three types of additivity: $q < 1$ represents super-additivity, $q > 1$ represents sub-additivity and the right-hand side of Equation (11) vanishes at $q = 1$, leading to additivity features [7,8].

3. Data Analysis and Results

In this paragraph, we present the findings based on the previously described methodology. The study is focused on the scaling properties of the aftershock sequences' temporal evolution, for the seven strong shallow earthquakes that took place over the previous five years in Greece (Table 1). The epicenters and focal mechanisms of these strong events are illustrated in Figure 1, with event numbers corresponding to the ones dictated in Table 1.

Table 1. Results of all analyzed aftershock sequences, where M_c is the completeness magnitude of the catalogue used, q is the Tsallis entropic parameter of the interevent time distribution, T_q denotes the generalized scaled interevent time, and T_c is the cross-over point at which the transition from Tsallis to BG statistical mechanics occurs.

Event No.	Date	Time (G.M.T.)	Lat. (°N)	Long. (°E)	Depth (km)	M_w	M_c	Database	No. of Aftershocks	Duration (Days)	q	T_q (s)	T_c (s)
1. Arkalochori	27/09/21	06:17:21	35.15	25.27	10	6.0	2.5	H.U.S.N.	700	95	1.62	774	7.8×10^3
2. Elassona	03/03/21	10:16:08	39.73	22.22	10	6.3	2.5	A.U.TH.	676	33	1.62	231	4.6×10^3
3. Samos	30/10/20	11:51:25	37.91	26.84	10	7.0	2.5	E.M.S.C.	1158	64	1.63	220	9.8×10^3
4. Parnitha	19/07/19	11:13:15	38.13	23.53	14.97	5.1	1.0	S.L.-N.K.U.A.	436	400	1.71	942	175×10^3
5. Zakynthos	25/10/18	22:54:50	37.35	20.49	13	6.6	2.1	H.U.S.N	1668	366	1.68	1629	5.6×10^3
6. Kos	20/07/17	22:31:10	36.97	27.41	7.1	6.5	1.6	K.O.E.R.I.	6492	530	1.63	201	15×10^3
7. Mytilene	12/06/17	12:28:37	38.85	26.31	9	6.3	2.0	E.M.S.C.	1610	365	1.68	418	11×10^3

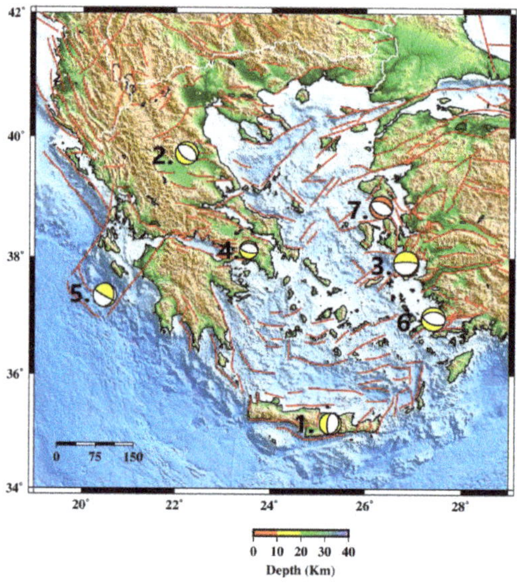

Figure 1. Geographical distribution of the seven studied mainshocks. The index numbers depicted in this figure correspond to the event indices given in Table 1. The colors and sizes of the focal mechanisms (beachballs) are related to the depth and the magnitude of each event. The faults are visualized according to the GEM database [28].

For the purpose of the definition of the aftershock zone, an elliptical region was initially picked for every main shock based on the distribution of its aftershocks and consequently the catalogue for each aftershock sequence was obtained. In order to test the stability of our results, we examined creating catalogues of earthquakes with 20% greater major axes of the ellipse. Considering the spatial distribution of aftershocks, small changes do not affect the parameter estimations that we will consider below, since the majority of the aftershock events included in the elliptic area first selected. Subsequently, the catalogue was updated to include all earthquake event occurrences inside this zone and for a period of two to four months after the main shock (see Table 1). Following the creation of the catalogues, we estimated the magnitude of completeness (M_c) for each aftershock sequence using the frequency–magnitude distribution [29,30]. It is worth noting that aftershock sequences can be depicted in terms of the modified parameters of the Gutenberg–Richter law [31,32].

The locations of the seven shallow mainshocks are illustrated in Figure 1. The earthquake numbers are presented chronologically from most recent to oldest in the event indexes. Along with the entropic parameters q and T_c, which represent the transition point from the non-additive to additive range in every aftershock series, the parameters of each mainshock and its aftershock sequence are summarized in Table 1.

Next, the interevent time distribution is calculated for each aftershock sequence, and a Q-exponential function fitting up to a value T_c, yielding to the Q and q parameters, respectivelly. In all cases that we study, we observe a deviation from the Q-exponential function for high values of time T, with $T > T_c$. Additionally, using the estimated Q value from the prior analysis, the Q-logarithmic function of $P(>T)$ as a function of T is constructed. The range of interevent times, provided by Equation (12), where $ln_Q P(>T)$ vs. T is a straight line, is then specified along with its correlation coefficient. The transition from NESP to BG statistical physics is indicated by the deviation from linearity at T_c. This demonstrates that in the immediate aftermath of the mainshock, the system is controlled by NESP, whereas as the aftershock sequence develops at $T > T_c$, the system is controlled by BG statistical mechanics.

Figure 2 presents a flowchart of the process behind the non-extensive statistical physics flowchart implemented in the present work.

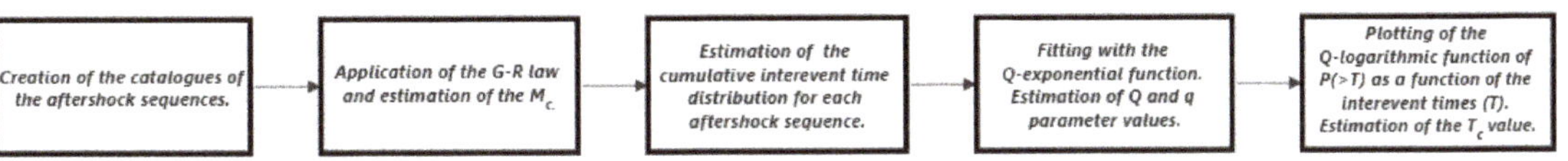

Figure 2. This flowchart summarizes the process behind the non-extensive statistical physics model implemented in the present work.

3.1. The Arkalochori Aftershock Sequence

In this section, we investigate the space–time distribution of the main event's aftershock sequence, which struck the Greek island of Crete at a depth of about 10 km on 27 September 2021 [33,34] The earthquake's epicenter was located southeast of Heraklion. The mainshock had a magnitude $M_w 6.0$. Based on a detailed examination of the aftershock sequence, as located by the Hellenic Unified Seismological Network (HUSN) station network (http://www.gein.noa.gr/en/networks/husn, accessed on 27 September 2022), the aftershock area encompasses the region between longitudes 25.17° E–25.40° E and latitudes 35.03° N–35.24° N. The aftershocks' catalogue includes events characterized by magnitudes $2.5 \leq M_w \leq 5.8$, with a completeness magnitude of $M_c = 2.5$. According to different networks and catalogues, M_c-value varies systematically in space and time. However, we should be cautious because commonly this value may lead to inaccurate estimations in statistical analyses due to it being higher in the early part of an earthquake sequence.

We study the probability distribution of interevent times in the aftershock series of the Arkalochori, 2021 event using the NESP framework, as described previously. This approach results from the generalized expression of entropy (Equation (1)), which is characteristic for complex systems with finite degrees of freedom and long-range memory [5,8,22].

Figure 3a shows a typical Q-exponential pattern in the log–log plot of the cumulative distribution function (CDF), $P(>T) = N(>T)/N_0$ of aftershocks' interevent times. Figure 3a shows that for values of T greater than a critical interevent time T_c (i.e., when $T > T_c$), there is a divergence from the Q-exponential function. Furthermore, fitting the Q-exponential to the instances up to a value near to T_c yields $q = 1.62$, as shown by Equations (7) and (8).

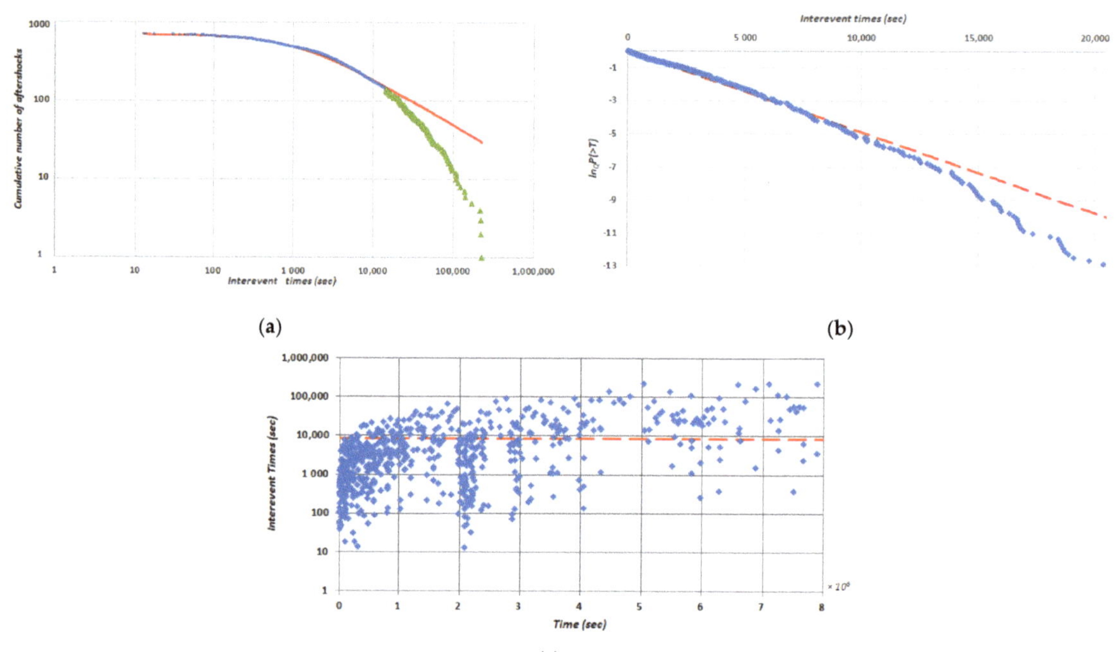

(a)

(b)

(c)

Figure 3. (**a**) The interevent times CDF of the 2021 $M_w6.0$ Arkalochori earthquake, for the aftershocks with $M > M_c$. The scarlet solid line is the Q-exponential operation with $q = 1.62$. The change of colors indicates the crossover between the NESP (blue circles) and BG statistics (green triangles). (**b**) The $ln_Q P(>T)$ as a function of interevent times T, where the scarlet line presents the fitting with $q = 1.62$ and correlation coefficient 0.9953 up to T_c. T_c value close to 7750 s is suggested by the deviation from linearity. (**c**) Interevent time T evolution with time t since the main shock. The red line illustrates the T_c value.

Next, we show $ln_Q P(>T)$ (see Equation (10)) as a function of interevent times T for $q = 1.62$ in Figure 3b. We estimate T_c to be $\approx$7750 s based on the divergence from predicted linearity during the transition from one system to another.

Figure 3c illustrates that the T evolves as a function of time t since the main shock. The T_c value indicates that the majority of interevent times have a value of T below T_c (Figure 3c) in the early aftershock time, supporting the idea that the NESP mechanism is predominant in the beginning of the aftershock evolution, indicating finite degrees of freedom and long-range memory effects. As time passes, these traits of the aftershock sequence are not prevalent anymore and BG statistics are restored (i.e., $q = 1$) [14].

3.2. The Elassona Aftershock Sequence

Here we focus on the aftershock sequence of the main earthquake that took place near the capital city of Larissa in Thessaly on 3 March 2021. The mainshock had a magnitude of $M_w6.0$ (from the Geophysical Laboratory of the Aristotle University of Thessaloniki

(GL-AUTH), http://geophysics.geo.auth.gr/, accessed on 27 September 2022) (Table 1), and generated a prolonged aftershock sequence in a general SE–NW direction [34]. The aftershock area, for a 1-month time interval from 3 March 2021 to 4 April 2021, covers the region between longitudes 21.47° E–23.13° E and latitudes 39.01° N–40.56° N. The aftershocks' catalogue includes 676 aftershocks characterized by magnitudes $2.5 \leq M_w \leq 5.8$, with a completeness magnitude of M_c = 2.5 [33,35].

The CDF of the interevent times for the Elassona 2021 aftershock sequence, based on the fitting of the Q-exponential function (Equations (7) and (8)) to the values of T, up to a value approaching T_c, reaches to q = 1.62 (see Figure 4). Next, in Figure 4b we present the $ln_Q P(>T)$ as an operation of interevent times T for q = 1.62. From its deviation from the expected linearity, the approximated value of $T_c \approx 4587$ s is extracted. A graph of the evolution of interevent time T over time t since the main shock is shown in Figure 4c.

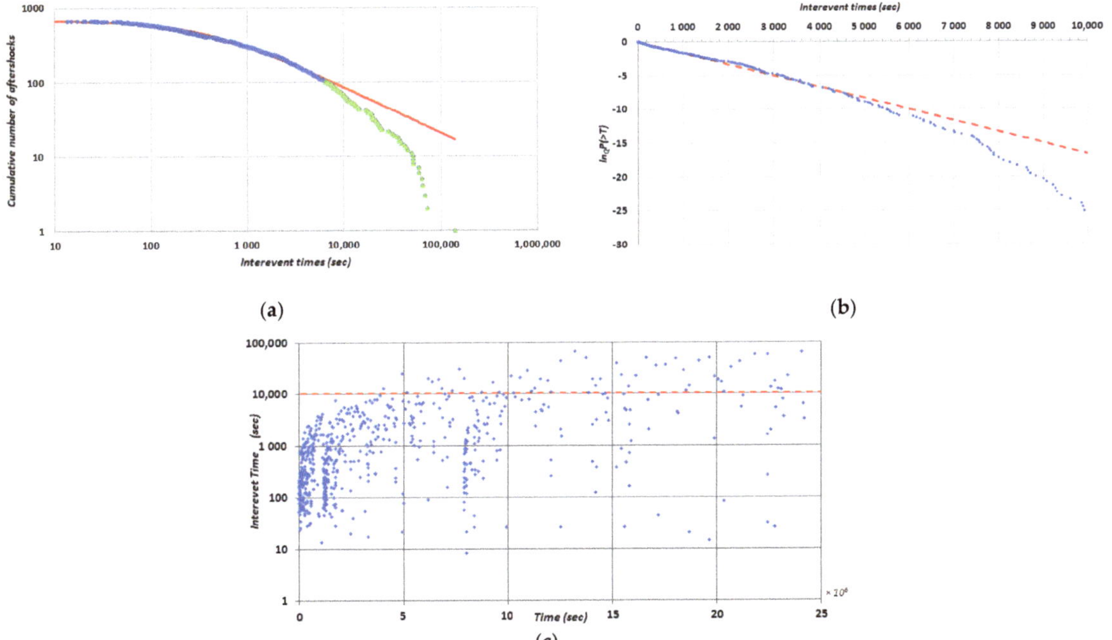

Figure 4. (**a**) The interevent times CDF of the 2021 M_w6.3 Elassona Earthquake, for the aftershocks with $M > M_c$. The scarlet stroke is the Q-exponential fitting with q = 1.62. The change of colors indicates the crossover between the NESP (blue circles) and BG statistics (green triangles). (**b**) The $ln_Q P(>T)$ as a function of T, where the scarlet line presents the fitting with q = 1.62 and correlation coefficient 0.9936 up to T_c. T_c value close to 4587 s is suggested by the deviation from linearity. (**c**) Interevent time T evolution with time t since the main shock. The red line illustrates the T_c value.

3.3. The Samos Aftershock Sequence

A strong and shallow earthquake of M_w = 7.0 struck Samos Island on the Aegean Sea (Figure 1), on 30 October 2020. Its aftershocks area covers the region between longitudes 26.10° E–26.99° E and latitudes 37.64° N–37.98° N. The catalogue, for a completeness magnitude of M_c = 2.5, includes 1158 aftershocks (Table 1).

Using the same methodology as previously, fitting the Q-exponential function to the noticed data up to a value near to T_c yields q = 1.63 (Figure 5a). We estimate T_c to be 9761 s, based on the deviation from predicted linearity (Figure 5b). In the early time aftershock part, most of the interevent times, with T values less than T_c exist, which forces us to conclude that the Tsallis entropy mechanism is dominant in this early part of the aftershock evolution. With the progress of time, the pattern of the aftershock sequence, such as finite degrees of freedom and long-range memory, are notpredominant anymore and the BG statistical physics controls the aftershocks evolution (i.e., $q = 1$) (Figure 5c) [14].

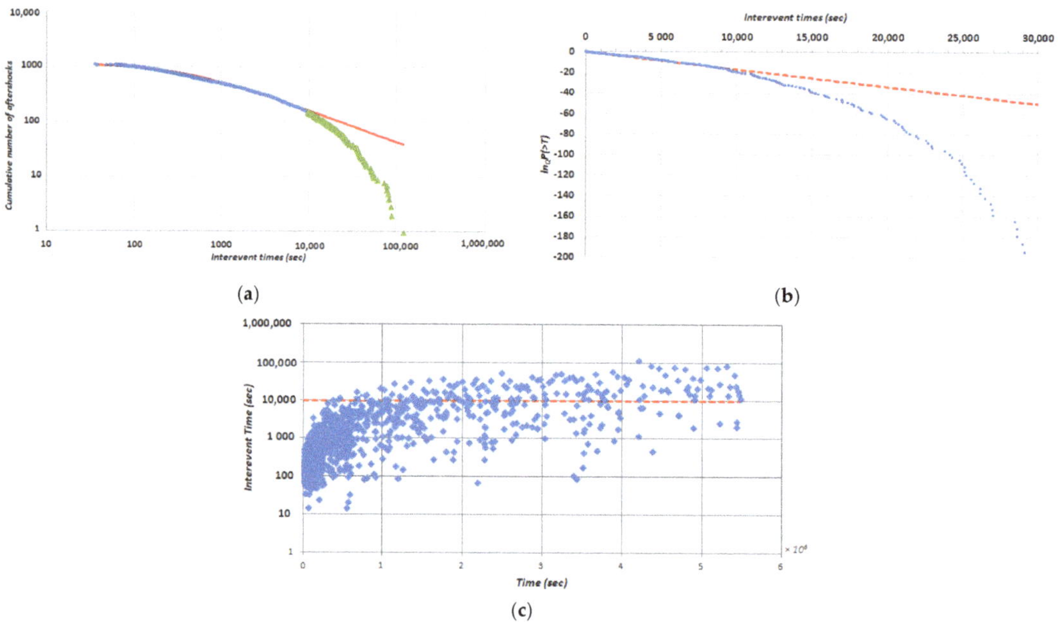

Figure 5. (**a**) The interevent times CDF of the 2020 M_w7.0 Samos Earthquake, for the aftershocks with $M > M_c$. The scarlet solid stroke is the Q-exponential fitting with $q = 1.63$. The change of colors indicates the crossover between the NESP (blue circles) and BG statistics (green triangles). (**b**) The $ln_Q P(>T)$ as a function of T, where the red line presents the fitting with $q = 1.63$ and correlation coefficient 0.9942 up to T_c. A T_c value close to 9761 s is suggested by the deviation from linearity. (**c**) Interevent time T evolution with time t since the main shock. The scarlet line illustrates the T_c value.

3.4. The Parnitha Aftershock Sequence

On 19 July 2019, at 11:13:15 GMT (Greenwich Mean Time), an earthquake of $M_w = 5.1$ struck Athens, the Capital of Greece. The mainshock's location parameters were obtained from the catalogue of Kapetanidis et al. (2020) [36], summarized in Table 1. The event took place NW of the Thriassio basin. The aftershock distribution of the 436 events covers the region between longitudes 23.47° E–23.67° E and latitudes 38.05° N–38.18° N and characterizes aftershocks with magnitudes $1.0 \leq M_w \leq 4.2$. The catalogue of this earthquake with completeness magnitude $M_c = 1.0$, covers the period from the day of the main event up to 21 August 2020.

In Figure 6a, $q = 1.71$ is obtained by fitting the Q-exponential function to the observed data up to a value close to T_c. In the present aftershock sequence, there is a slight increase in the parameter q compared to the previous ones (see also Table 1). The deviation of the Q-logarithmic operation from the expected linearity is observed at a T_c value of $\approx$175,460 s (Figure 6b). In the early aftershock period, there are more interevent times with values lower than T_c indicating that the Tsallis entropy description dominates the aftershock evolution in the immediate to the main shock time. As time passes, some of the traits of the early aftershock sequence related to NESP are insignificant and BG statistical physics are recovered.

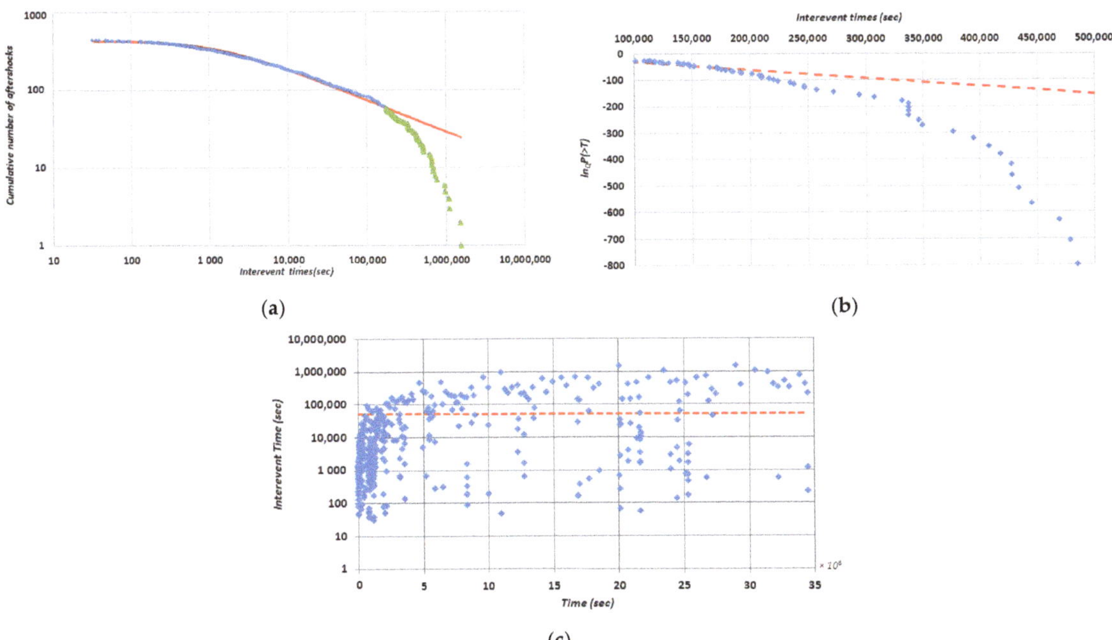

Figure 6. (**a**) The interevent times CDF of the 2019 M_w5.1 Parnitha Earthquake, for the aftershocks with $M > M_c$. The scarlet stroke is the Q-exponential fitting with $q = 1.71$. The change of colors indicates the crossover between the NESP (blue circles) and BG statistics (green triangles). (**b**) The $ln_Q P(>T)$ as a function of T, where the red line is the fitting with $q = 1.71$ and correlation coefficient 0.9912 up to T_c. T_c value close to 175,460 s is suggested by the deviation from linearity. (**c**) Interevent time T evolution with time t since the main shock. The scarlet line illustrates the T_c value.

3.5. The Zakynthos Aftershock Sequence

Herein, we study the seismic sequence that began on 25 October 2018 with a shallow $M_w = 6.6$ earthquake off the coast Zakynthos (Ionian Sea, Greece) (Figure 1). Based on detailed examination of the 1668 aftershock sequence, which were located by the station network of the Hellenic Unified Seismological Network (HUSN), we conclude that the duration corresponds to the 1-year time period i.e., from 25 October 2018 up to 19 October 2019. According to the catalogue used, aftershocks occurred with a magnitude greater than M_w2.1.

Plotting the CDF, $P(>T) = N(>T)/N_0$ of aftershocks interevent times on a double-log scale, a typical Q-exponential pattern presents for $T < T_c$, with $q = 1.68$ (Figure 7a). The transition from NESP to BF statistics is estimated to be at about $T_c \approx 5607$ s (Figure 7b). The T_c value (red dashed line in Figure 7c) suggests that the Tsallis entropy controls the early stages of the aftershocks' evolution. Certain traits of the early aftershock sequence related to the NESP become less significant as time passes by, and the statistical physics of BG is recovered.

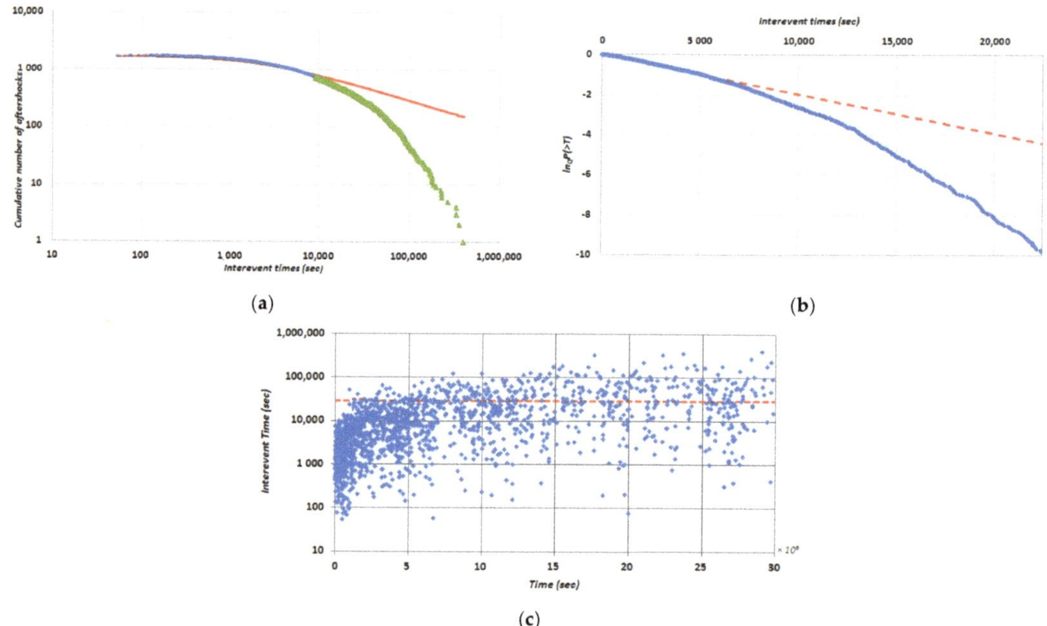

Figure 7. (**a**) The interevent times CDF of the 2018 M_w6.6 Zakynthos Earthquake, for the events with $M > M_c$. The scarlet stroke is the Q-exponential fitting with $q = 1.68$. The change of colors indicates the crossover between the NESP (blue circles) and BG statistics (green triangles). (**b**) The $ln_Q P(>T)$ as a function of T, where the red line is the fitting with $q = 1.68$ and correlation coefficient 0.9966 up to T_c. T_c value close to 5607 s is suggested by the deviation from linearity. (**c**) Interevent time T evolution with time t since the main shock. The scarlet line illustrates the T_c value.

3.6. The Kos Aftershock Sequence

An earthquake with magnitude $M_w = 6.5$ at a depth of 7.1 km, which had a normal faulting mechanism striking about east–west (Figure 1), happened on 20 July 2017 in Gökova Bay, in the Aegean Sea, at 22:31:10 GMT between Bodrum town, Turkey, and Kos Island, Greece. As stated in the data, the mainshock epicenter was given as 27.41° E and 36.97° N located 12 km ENE to Kos in Greece and 8 km SE to Bodrum in Muğla in Turkey. The earthquake generated a tsunami that affected the coast of the Bodrum peninsula and the northeast coast of Kos. A tide gauge in Bodrum, close to the earthquake's epicenter, recorded the tsunami [37].

The data were obtained from the Boun Koeri Regional Earthquake-Tsunami Monitoring Center, Kandilli Observatory and Earthquake Research Institute (RETMC) (the Turkish Disaster and Emergency Management Presidency, AFAD; Boğaziçi University (KOERI), http://www.koeri.boun.edu.tr/, accessed on 27 September 2022). This study's goal is to give a thorough region-time analysis with a variety of aftershock attributes such as the parameter q by Tsallis for 6492 aftershocks identified in six months after the mainshock.

In terms of Tsallis Entropy the value of q is equal to $q = 1.63$ (Figure 8a). Following, the transition estimated to be at $T_c \approx 15{,}005$ s (Figure 8b). The parameter T_c (red dashed line in Figure 7c) shows that the NESP describes the early part of the aftershocks while as time goes on, BG statistics are revealed.

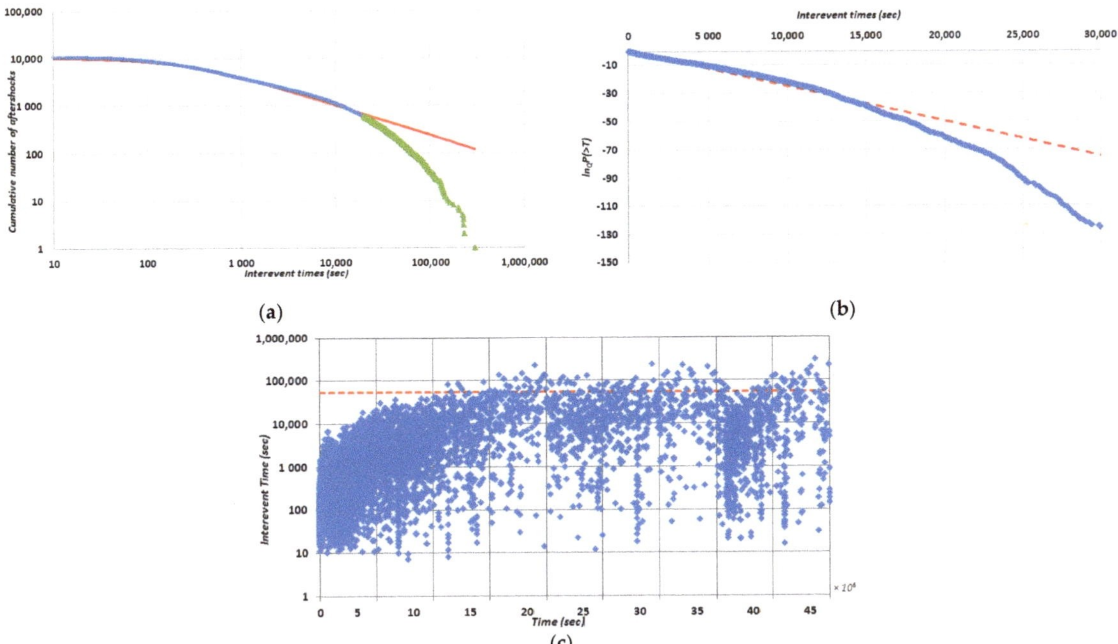

(a)

(b)

(c)

Figure 8. (**a**) The interevent times CDF of the 2017 $M_w6.5$ Kos Earthquake, for the aftershocks with $M > M_c$. The scarlet line is the Q-exponential fitting with $q = 1.63$. The change of colors indicates the crossover between the NESP (blue circles) and BG statistics (green triangles). (**b**) The $ln_Q P(>T)$ as a function of T, where the red line is the fitting with $q = 1.63$ and correlation coefficient 0.9935 up to T_c. T_c value close to 15,005 s is suggested by the deviation from linearity. (**c**) Interevent time T evolution with time t since the main shock. The scarlet line illustrates the T_c value.

3.7. The Mytilene Aftershock Sequence

The 2017 Mytilene earthquake of $M_w6.3$, took place at the coordinates (26.31, 38.85) (see more for location parameters in the Table 1) on 12 June at 12:28:37 GMT.

This destructive offshore event occurred northeast of Chios and almost 15 km south of the southeast coast of Lesbos. In Vrissa village, a collapsed building killed one person and injured 15 others due to a collapsed building and falling debris. Damage was reported in at least 12 villages across the southeast region of Lesvos, and there was additional impact along the Turkish coast [38]. Regarding the environmental impact of the earthquake, slope displacement and ground cracks occurred in many places in the disaster area. Also, tsunamis were reported in Plomari Port [38].

A total of 1610 aftershocks were detected over the period between 12 June 2017 and 11 June 2018 (European Mediterranean Seismological Centre, EMSC). The aftershock area covers the region between the coordinates by longitudes 25.22° E–27.30° E and latitudes 38.23° N–39.22° N.

In line with the analysis of the previous aftershock sequences, for the earthquake of Mytilene, we study the distribution of the interevent times. The value of q is equal to $q = 1.68$ (Figure 9a). The transition estimated to be at $T_c \approx 10,761$ s (Figure 9b). Since the most interevent times have T values $T < T_c$, the parameter T_c (red dashed line in Figure 9c) confirms that the Tsallis entropy description dominates the early stages of the aftershocks' evolution.

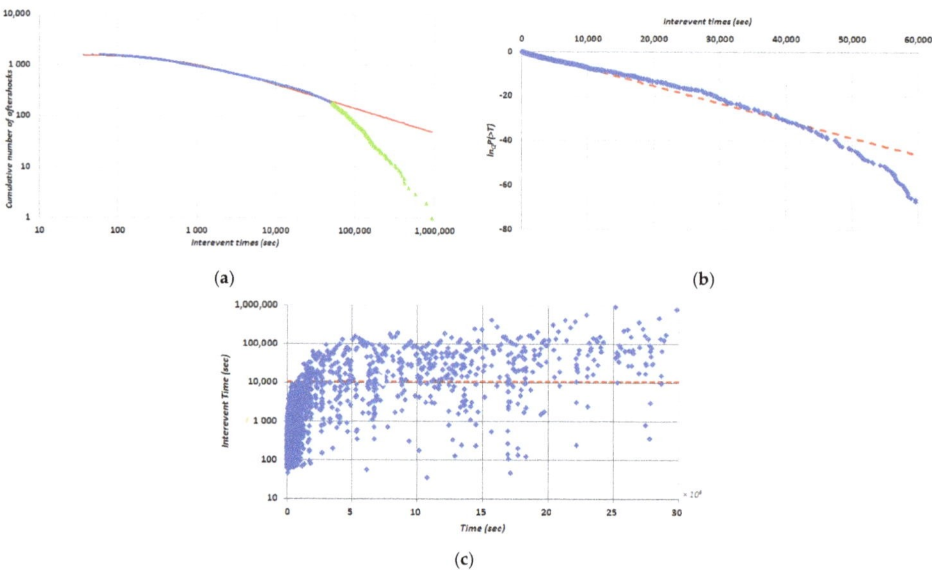

(a)

(b)

(c)

Figure 9. (**a**) The interevent times CDF of the 2017 $M_w6.3$ Mytilene Earthquake, for the aftershocks with $M > M_c$. The scarlet line is the Q-exponential fitting with $q = 1.68$. The change of colors indicates the crossover between the NESP (blue circles) and physical BG statistics (green triangles). (**b**) The $ln_Q P(>T)$ as a function of T, where the red line is the fitting with $q = 1.68$ and correlation coefficient 0.9927 up to T_c. T_c value close to 10,761 s is suggested by the deviation from linearity. (**c**) Interevent time T evolution with time t since the main shock. The scarlet line illustrates the T_c value.

In addition, and for all the aftershock sequences that were studied, we introduce a normalized parameter, $x = T/T_c$, where $x < 1$ indicates the range where the Tsallis entropy describes the evolution of aftershocks sequence interevent times, while $x > 1$ is related to the Boltzmann–Gibbs (BG) process. This is because $P(>T) = exp_Q(-T/T^*)$ for $T < T_c$. A deviation from the Q-exponential is present for all of the examined aftershock sequences when $x \gg 1$ (i.e., $T \gg T_c$). An inspection of Figure 10, where all aftershock sequences are plotted together, suggests that for $0.01 < x < 1$, power-law scaling emerges for all the aftershock sequences, with a slope in the range 0.40–0.60, conforming to the q values calculated from the analysis (Table 1). The latter is expected since the asymptotic expression of Equation (6) is $P(> T) \sim (T/T^*)^{-\frac{1}{(Q-1)}}$ is a typical expression of a power law.

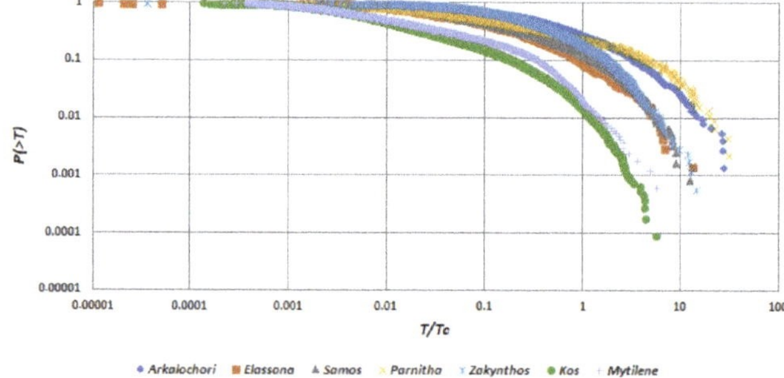

Figure 10. The interevent times distribution $P(>x)$ for all the studied aftershock sequences as a function of $x = T/T_c$. Deviations from the Q-exponential operation are pronounced at $T/T_c \gg 1$, for all sequences.

It is important to mention, that in all estimations the accuracy of the estimated values is in the order of ± 0.01.

4. Discussion

All shallow earthquakes are followed by an aftershock sequence. The statistical properties of aftershock sequences are associated with scaling relations such as that extracted in view of non-extensive statistical physics (NESP). In this study, we used a detailed temporal assessment of the aftershock sequences over the last five years of significant earthquakes in Greece with magnitudes that reach up to $M_w = 7.0$. We studied the strong events, such as the $M_w6.0$ Arkalochori (27 September 2021), the $M_w6.3$ Elassona (3 March 2021), the $M_w7.0$ Samos (30 October 2020), the $M_w5.1$ Parnitha (19 July 2019), the $M_w6.6$ Zakynthos (25 October 2018), the $M_w6.5$ Kos (20 July 2017) and the $M_w6.1$ Mytilene (12 June 2017) earthquakes. Based on non-expansive statistical physics, we analyzed the distribution of interevent times for each aftershock sequence for each main shock.

In all cases, the cumulative distribution function $P(>T)$ is defined by a Q-exponential in the early stage of the aftershock sequence where interevent times less than T_c are observed, where T_c is the crossover point between the non-additive and additive behavior. By fitting a Q-exponential function to the data up to a value close to T_c, the parameter q is estimated for each aftershock sequence. In all the cases analyzed, the applicability of non-extensive statistical physics to the interevent times CDF is demonstrated, as well as, the existence of transition behavior from the power-law to exponential scaling for larger interevent times. Since the q entropic parameter is greater than one ($q > 1$) a sub-additive process is implied, supporting the conclusion that long-range memory exists in the early state of temporal evolution of aftershocks where mainly $T < T_c$. Additionally, for aftershock sequences analyzed, the estimated Tsallis entropic q-values that describe the observed CDF are within the range of 1.62–1.71.

In addition, the superposition of two aftershock mechanisms can be used to explain the observed scaling behavior and the deviation from the Q-exponential function for greater interevent times. For $T > T_c$, a second mechanism—characterized by an exponential function—becomes apparent. The first mechanism, as presented by NESP, is dominant for $T < T_c$. We, thus, introduce the generalization described in [20,39,40], to account for a transition from NESP ($q \neq 1$) to BG ($q = 1$) statistical mechanics, where:

$$\frac{dp(T)}{dT} = -\beta_1 p - (\beta_q - \beta_1) p^q, \tag{12}$$

whose solution is

$$p(T) = C \left[1 - \frac{\beta_q}{\beta_1} + \frac{\beta_q}{\beta_1} e^{(q-1)\beta_1 T} \right]^{1/1-q}, \tag{13}$$

In Equation (13) the probability function $p(T)$ decreases monotonically with increasing T for positive β_q and β_1, where C is a normalization factor. As a result, when $(q-1)\beta_1 \ll 1$, a q-exponential, $p(T) \approx C exp_q(-T/T_q)$, where $T_q = 1/\beta_q$, is an approximation of Equation (13), while for $(q-1)\beta_1 \gg 1$, the asymptotic behavior of the probability distribution function $p(T) \propto \left(\frac{\beta_1}{\beta_q} \right)^{1/(q-1)} e^{-\beta_1 T}$, is an exponential one, where $T_c = 1/(q-1)\beta_1$ defines the crossover point from the non-additive to additive behavior [24,41]. The T_c value suggests that the Tsallis entropy is prevalent in the early stages of aftershock evolution, while the traits of aftershock sequences which are associated with a NESP description, become less apparent as time passes and BG statistics are recovered [6,9,17,22,42,43].

The super-statistical theory, which is complementary to NESP, is based on a superposition of ordinary local equilibrium statistical mechanics, with a Gamma distributed intensive parameter that varies over a fairly wide time scale. This approach can be used to explain the q-exponential behavior of the interevent times in aftershock sequences [14].

The super-statistical approach states that the interevent times of an aftershock sequence may be described by a local Poisson process $p(T|\beta) = \beta e^{-\beta T}$, with β an intensive fluctuating parameter. On a long-time scale, β is distributed with a possibility density $f(\beta)$ [25,44–47]. Then the probability distribution $p(T)$ is given as:

$$p(T) = \int_0^\infty f(\beta)\beta e^{-\beta T} d\beta, \tag{14}$$

In the scenario where a Gamma distribution provides the probability density of β [43–46]:

$$f(\beta) = \frac{1}{\Gamma(n/2)} \left(\frac{n}{2\beta_0}\right)^{n/2} \beta^{\frac{n}{2}-1} exp\left(-\frac{n\beta}{2\beta_0}\right), \tag{15}$$

Integration of Equation (14) is analytically calculated [48], obtained $p(T) \approx C(1 + B(q-1)T)^{1/(1-q)}$, which is exactly the result in term of NESP, where $q = 1 + ([2/(n + 2)])$ and $B = 2\beta_0/(2-q)$ [14]. Since in all the analyzed cases q is in the range 1.61–1.71, we conclude that the corresponding number of degrees of freedom is close to one ($n = 1$).

The latter implies that the evolution of an aftershock sequence could be influenced by a stochastic mechanism with memory effects. In accordance with [49,50], the stochastic differential equation for the evolution with time t, of interevent times T of an aftershock sequence:

$$dT = -\gamma(T - <T>)dt + \varphi\sqrt{T}W_t, \tag{16}$$

This stochastic equation is made up of two parts that control how the seismicity evolves. The primary goal of the first deterministic term is to restore the seismic rate R to its usual value of $R = 1/<T>$ based on a constant γ which expresses the rate of relaxation to the mean waiting time $<T>$. Memory effects in seismicity's development are depicted in the second stochastic part. The stochastic term W_t describes a Wiener process following a Gaussian distribution with zero mean and unitary variance that could follow the macroscopic effects in the evolution of interevent times in the aftershock sequence. W_t' random sign causes an increase ($W_t > 0$) or decrease ($W_t < 0$) of T. The construction of this term operates in a way that large values of T cause large amplitude of the stochastic term, which leads to an increase or decrease in T depending on the sign of W_t. The parameter φ introduces a noise component to the process and can be expressed as $\varphi = \sqrt{2\gamma<T>}$ [50].

Equation (16) is a stochastic differential equation that represents a multiplicative noise example, known as the Feller process [48–50].

We write the corresponding Fokker–Planck equation for Equation (16), to ascertain the evolution of the interevent time series T after some time t, given the probability distribution $f(T, t)$, as [51]:

$$\frac{\partial f(T,t)}{\partial t} = \frac{\partial}{\partial T}[\gamma(T - <T>)f(T,t)] + \frac{\partial^2}{\partial T^2}[T<T>\gamma f(T,t)], \tag{17}$$

The latter Fokker–Planck equation's stationary solution, Equation (17), is the distribution [48]:

$$p(T/<T>) = f(T) = \frac{1}{<T>}e^{-\frac{1}{<T>}T}, \tag{18}$$

where Equation (18) presents the conditional probability of T given $<T>$.

It is necessary to account for local variations in the seismic rate $R = 1/<T>$ associated with non-stationarities in the evolution of the earthquake activity over time scales significantly larger than $1/\gamma$ in order to achieve stationarity in Equation (16). In this case, the mean interevent time $<T>$ exhibits local variations, and we assume that these fluctuations adhere to the stationary gamma distribution:

$$f(<T>) = \frac{\left(\frac{1}{\lambda}\right)^\delta}{\Gamma[\delta]}<T>^{-(1+\delta)}e^{-\frac{1}{\lambda<T>}}, \tag{19}$$

where Equation (19) gives the marginal probability of T, [52] as:

$$p(T) = \int_0^\infty p(T/<T>)f(<T>)d<T>,\tag{20}$$

The latter integration leads to:

$$p(T) = \frac{\lambda\Gamma[1+\delta]}{\Gamma[\delta]}(1+\lambda T)^{-(1+\delta)},\tag{21}$$

By continuing to implement the variable changes:

$$\lambda = \frac{q-1}{T_o} \text{ and } \delta = \frac{1}{q-1} - 1 = \frac{2-q}{q-1}\tag{22}$$

and taking into account the form of q-exponential function in Equation (2), Equation (16) can be transformed to [50]:

$$p(T) = \frac{(q-1)\Gamma\left(\frac{1}{q-1}\right)}{T_o\Gamma\left[\frac{1}{q-1}-1\right]}exp_q\left(-\frac{T}{T_o}\right)\tag{23}$$

which is the exact form of the q-exponential function.

5. Concluding Remarks

In summarizing the study's findings, we concentrated on analyzing the distributions of interevent times for each sequence in order to statistically examine its patterns in the most recent aftershock sequences in Greece.

Namely:

- We can state that the aftershock sequences located in Greece follows the statistical mechanics model derived in the framework of Non-Extensive Statistical Physics (NESP);
- Moreover, we should note that the NESP approach is useful for other regions, not only for the Greek territory, such as subduction zones all over the word [53];
- According to the NESP approach used here, it suggests that the system is in an abnormal equilibrium with a transition for large interevent times from abnormal ($q > 1$) to normal ($q = 1$) statistical mechanics;
- The analysis of the interevent times distribution indicates such a system;
- The range of the non-extensive parameter q and for all sequences studied, results in a non-extensive entropic parameter q with a range of between 1.62 and 1.71, which suggests a system with one degree of freedom.

To summarize, the used models fit the noticed distributions reasonably well, and imply the importance of using NESP in evaluating such phenomena.

The main limitations of the work presented in this paper are related to earthquakes in different geotectonic environments along with fault types of the main shock. Studying aftershock sequences as a function of geotectonic environments is a matter of discussion in future studies, and this could be useful for the prediction of damaged aftershocks [54–57].

Author Contributions: Conceptualization, F.V.; methodology, F.V., E.-A.A., S.-E.A., G.M.; software, G.M. and E.-A.A.; validation, E.-A.A. and S.-E.A.; formal analysis, F.V., E.-A.A., S.-E.A., G.M.; investigation, F.V., E.-A.A., S.-E.A., G.M.; resources, F.V., E.-A.A., S.-E.A., G.M.; data curation, F.V., E.-A.A., S.-E.A., G.M.; writing—original draft preparation, F.V., E.-A.A., S.-E.A., G.M.; writing—review and editing, F.V., E.-A.A., G.M.; visualization, F.V., E.-A.A.; supervision, F.V. All authors have read and agreed to the published version of the manuscript.

Funding: This research received no external funding.

Data Availability Statement: Data are available at the Hellenic Unified Seismic Network (H.U.S.N., http://www.gein.noa.gr/en/networks/husn), (A.U.TH.: Aristotle University of Thessaloniki [http://geophysics.geo.auth.gr/ss/], E.M.S.C.: European Mediterranean Seismological Centre [https://www.emsc-csem.org/#2], N.K.U.A.: National and Kapodistrian University of Athens [http://www.geophysics.geol.uoa.gr/], K.O.E.R.I.: Kandilli Observatory and Earthquake Research Institute [http://www.koeri.boun.edu.tr/new/en]. Last visit to the above webpages: 12 September 2022.

Acknowledgments: We would like to thank Andreas Karakonstantis for his valuable assistance in the construction of the seismotectonic map.

Conflicts of Interest: The authors declare no conflict of interest.

References

1. Telesca, L.; Cuomo, V.; Lapenna, V.; Vallianatos, F.; Drakatos, G. Analysis of the temporal properties of Greek aftershock sequences. *Tectonophysics* **2001**, *341*, 163–178. [CrossRef]
2. Papazachos, B.C.; Papazachou, C. *The Earthquakes of Greece*; Ziti Publications: Thessaloniki, Greece, 2003.
3. Tsapanos, T. *Seismicity and Seismic Hazard Assessment in Greece*; Springer Science: Thessaloniki, Greece, 2008.
4. Kassaras, I.; Kapetanidis, V.; Ganas, A.; Tzanis, A.; Kosma, C.; Karakonstantis, A.; Valkaniotis, S.; Chailas, S.; Kouskouna, V.; Papadimitriou, P. The New Seismotectonic Atlas of Greece (v1.0) and Its Implementation. *Geosciences* **2020**, *10*, 447. [CrossRef]
5. Vallianatos, F.; Michas, G.; Papadakis, G. Non-Extensive Statistical Seismology: An Overview. *Complex. Seism. Time Ser.* **2018**, 25–60. [CrossRef]
6. Michas, G.; Vallianatos, F. Scaling properties, multifractality and range of correlations in earthquake timeseries: Are earth-quakes random? In *Statistical Methods and Modeling of Seismogenesis*; Limnios, N., Papadimitriou, E., Tsaklidis, G., Eds.; ISTE John Wiley: London, UK, 2021. [CrossRef]
7. Vallianatos, F.; Sammonds, P. Evidence of non-extensive statistical physics of the lithospheric instability approaching the 2004 Sumatran-Andaman and 2011 Honshu mega-earthquakes. *Tectonophysics* **2013**, *590*, 52–58. [CrossRef]
8. Tsallis, C. *Introduction to Non-Extensive Statistical Mechanics: Approaching a Complex World*; Springer: Berlin/Heidelberg, Germany, 2009; pp. 41–218. [CrossRef]
9. Sobyanin, D.N. Generalization of the Beck-Cohen superstatistics. *Phys. Rev. E* **2011**, *84*, 051128. [CrossRef]
10. Vallianatos, F.; Karakostas, V.; Papadimitriou, E. A Non-Extensive Statistical Physics View in the Spatiotemporal Properties of the 2003 (Mw6.2) Lefkada, Ionian Island Greece, Aftershock Sequence. *Pure Appl. Geophys.* **2013**, *171*, 1443–1449. [CrossRef]
11. Efstathiou, A.; Tzanis, A.; Vallianatos, F. On the nature and dynamics of the seismogenetic systems of North California, USA: An analysis based on Non-Extensive Statistical Physics. *Phys. Earth Planet. Inter.* **2017**, *270*, 46–72. [CrossRef]
12. Vallianatos, F.; Michas, G.; Papadakis, G.; Tzanis, A. Evidence of non-extensivity in the seismicity observed during the 2011–2012 unrest at the Santorini volcanic complex, Greece. *Nat. Hazards Earth Syst. Sci.* **2013**, *13*, 177–185. [CrossRef]
13. Vallianatos, F.; Michas, G.; Papadakis, G.; Sammonds, P. A Non-Extensive Statistical Physics View to the Spatiotemporal Properties of the June 1995, Aigion Earthquake (M6.2) Aftershock Sequence (West Corinth Rift, Greece). *Acta Geophys.* **2012**, *60*, 759–765. [CrossRef]
14. Vallianatos, F.; Pavlou, K. Scaling properties of the Mw7.0 Samos (Greece), 2020 aftershock sequence. *Acta Geophysica* **2021**, *69*, 1067–1084. [CrossRef]
15. Sychev, V.N.; Sycheva, N.A. Nonextensive analysis of aftershocks following moderate earthquakes in Tien Shan and North Pamir. *J. Volcanol. Seismol.* **2021**, *15*, 58–71. [CrossRef]
16. Michas, G.; Papadakis, G.; Vallianatos, F. A Non-Extensive approach in investigating Greek seismicity. *Bull. Geol. Soc. Greece* **2013**, *47*, 1177–1187. [CrossRef]
17. Vallianatos, F.; Telesca, L. Application of Statistical Physics in Earth Sciences and Natural Hazards. *Acta Geophys. (Spec. Issue)* **2012**, *60*, 499. [CrossRef]
18. Tsallis, C. Possible generalization of Boltzmann-Gibbs statistics. *J. Stat. Phys.* **1988**, *52*, 479–487. [CrossRef]
19. Telesca, L. Maximum likelihood estimation of the non-extensive parameters of the earthquake cumulative magnitude distribution. *Bull. Seismol. Soc. Am.* **2012**, *102*, 886–891. [CrossRef]
20. Vallianatos, F.; Michas, G.; Papadakis, G. Non-Extensive Statistical Seismology: An overview. In *Complexity of Seismic Time Series: Measurement and Application*, 1st ed.; Chelidze, T., Vallianatos, F., Telesca, L., Eds.; Elsevier: Amsterdam, The Netherlands, 2018; pp. 25–53.
21. Vallianatos, F.; Papadakis, G.; Michas, G. Generalized statistical mechanics approaches to earthquake and tectonics. *Proc. R. Soc. A* **2016**, *472*, 20160497. [CrossRef]
22. Abe, S.; Suzuki, N. Scale-free statistics of time interval between successive earthquakes. *Physica A* **2005**, *350*, 588–596. [CrossRef]
23. Abe, S.; Suzuki, N. Scale-invariant statistics of period in directed earthquake network. *Eur. Phys. J. B* **2005**, *44*, 115–117. [CrossRef]
24. Tsallis, C. Nonadditive entropy and non-extensive statistical mechanics—An overview after 20 years. *Brazil. J. Phys.* **2009**, *39*, 337–339. [CrossRef]
25. Beck, C. Superstatistical Brownian Motion. *Prog. Theor. Phys. Suppl.* **2006**, *162*, 29–31. [CrossRef]

26. Vallianatos, F.; Benson, P.; Meredith, P.; Sammonds, P. Experimental evidence of a non-extensive statistical physics behavior of fracture in triaxially deformed Etna basalt using acoustic emissions. *EPL* **2012**, *97*, 58002. [CrossRef]
27. Telesca, L. Tsallis-Based Nonextensive Analysis of the Southern California Seismicity. *Entropy* **2011**, *13*, 1267–1280. [CrossRef]
28. Caputo, R.; Pavlides, S. The Greek Database of Seismogenic Sources (GreDaSS), Version 2.0.0: A Compilation of Potential Seismogenic Sources (Mw > 5.5) in the Aegean Region. 2013. Available online: https://gredass.unife.it/ (accessed on 26 October 2022).
29. Gutenberg, B.; Richter, C.F. Magnitude and energy of earthquakes. *Ann. Geophys.* **1956**, *9*, 1–15. [CrossRef]
30. Wyss, M.; Wiemer, S.; Zuniga, R. Zmap: A Tool for Analyses of Seismicity Patterns, Typical Applications and Uses: A Cookbook. Available online: http://:www.researchgate.net/publication/261508570_cookbook (accessed on 26 October 2022).
31. Avgerinou, S.-E.; Anyfadi, E.-A.; Michas, G.; Vallianatos, F. Study of the Frequency Magnitude Distribution of recent Aftershock Sequences of Earthquakes in Greece, in Terms of Tsallis Entropy. In Proceedings of the 16th International Congress of the Geological Society of Greece, Patras, Greece, 17–19 October 2022.
32. Avgerinou, S.-E. A Non-Extensive Statistical Physics View to the Temporal Properties of the Recent Aftershock Sequences of Strong Earthquakes in Greece. Master's Thesis, Earth Science Department of Geology and Geoenviroment, Athens, Greece, 2021.
33. Vallianatos, F.; Michas, G.; Hloupis, G. Seismicity Patterns Prior to the Thessaly (Mw6. 3) Strong Earthquake on 3 March 2021 in Terms of Multiresolution Wavelets and Natural Time Analysis. *Geosciences* **2021**, *11*, 379. [CrossRef]
34. Vallianatos, F.; Karakonstantis, A.; Michas, G.; Pavlou, K.; Kouli, M.; Sakkas, V. On the Patterns and Scaling Properties of the 2021–2022 Arkalochori Earthquake Sequence (Central Crete, Greece) Based on Seismological, Geophysical and Satellite Observations. *Appl. Sci.* **2022**, *12*, 7716. [CrossRef]
35. Michas, G.; Pavlou, K.; Avgerinou, S.-E.; Anyfadi, E.-A.; Vallianatos, F. Aftershock patterns of the 2021 Mw 6.3 Northern Thessaly (Greece) earthquake. *J. Seismol.* **2022**, *26*, 201–225. [CrossRef]
36. Kapetanidis, V.; Karakonstantis, A.; Papadimitriou, P.; Pavlou, K.; Spingos, I.; Kaviris, G.; Voulgaris, N. The 19 July 2019 earthquake in Athens, Greece: A delayed major aftershock of the 1999 Mw = 6.0 event, or the activation of a different structure? *J. Geodyn.* **2020**, *139*, 101766. [CrossRef]
37. Öztürk, S.; Şahin, S. A statistical space-time-magnitude analysis on the aftershocks occurrence of the 21 July 2017 MW = 6.5 Bodrum-Kos, Turkey, earthquake. *J. Asian Earth Sci.* **2018**, *172*, 443–457. [CrossRef]
38. Lekkas, E.; Voulgaris, N.; Karydis, P.; Tselentis, G.-A.; Skourtsos, E.; Antoniou, V.; Andreadakis, E.; Mavroudis, S.; Spirou, N.; Speis, F.; et al. Preliminary Report, Lesvos Earthquake Mw 6.3, 12 June 2017, Environmental, Disaster and Crisis Management Strategies. Available online: https://paleoseismicity.org/ (accessed on 30 September 2022).
39. Vallianatos, F. A non-extensive statistical physics approach to the polarity reversals of the geomagnetic field. *Phys. A Stat. Mech. Appl.* **2011**, *390*, 1773–1778. [CrossRef]
40. Sugiyama, M. Introduction to the topical issue: Nonadditive entropy and non-extensive statistical mechanics. *Continuum Mech. Thermodyn.* **2004**, *16*, 221–222. [CrossRef]
41. Vallianatos, F.; Sammonds, P. Is plate tectonics a case of non-extensive thermodynamics? *Phys. A Stat. Mech. Appl.* **2010**, *389*, 4989–4993. [CrossRef]
42. Chochlaki, K.; Vallianatos, F.; Michas, G. Global regionalized seismicity in view of Non-Extensive Statistical Physics. *Phys. A Stat. Mech. Appl.* **2018**, *493*, 276–285. [CrossRef]
43. Beck, C. Superstatistics: Theory and applications. *Continum Mech. Thermodyn* **2004**, *16*, 294–298. [CrossRef]
44. Beck, C. Recent developments in superstatistics. *Braz. J. Phys.* **2009**, *39*, 357–362. [CrossRef]
45. Beck, C. Dynamical Foundations of Nonextensive Statistical Mechanics. *Phys. Rev. Lett.* **2001**, *87*, 180601. [CrossRef]
46. Beck, C.; Cohen, E.G.D. Superstatistics. *Physica A* **2003**, *322*, 267–269. [CrossRef]
47. Antonopoulos, C.G.; Michas, G.; Vallianatos, F.; Bountis, T.; Physica, A. Evidence of q-exponential statistics in Greek seismici-ty. *Phys. A Stat. Mech. Appl.* **2014**, *409*, 71–77. [CrossRef]
48. Mathai, A.M. A pathway to matrix-variate gamma and normal densities. *Lin. Al. Appl.* **2005**, *396*, 317–328. [CrossRef]
49. Feller, W. Two Singular Diffusion Problems. *An. Math* **1951**, *54*, 173–182. [CrossRef]
50. Michas, G.; Vallianatos, F. Stochastic modeling of nonstationary earthquake time series with long-term clustering effects. *Phys. Rev. E* **2018**, *98*, 042107. [CrossRef]
51. Gardiner, C.W. *Handbook of Stochastic Methods for Physics, Chemistry, and the Natural Sciences*, 1st ed.; Springer: Berlin/Heidelberg, Germany, 1993.
52. Risken, H. *The Fokker-Planck Equation: Methods of Solution and Applications*, 2nd ed.; Springer: Berlin, Heidelberg, 1989. [CrossRef]
53. Anyfadi, E.-A.; Avgerinou, S.-E.; Michas, G.; Vallianatos, F. Universal Non-Extensive Statistical Physics Temporal Pattern of Major Subduction Zone Aftershock Sequences. *Entropy* **2022**, *24*, 1850. [CrossRef] [PubMed]
54. Yeo, G.L.; Allin Cornell, C. A probabilistic framework for quantification of aftershock ground-motion hazard in California: Methodology and parametric study. *Wiley Online Libr. Earthq. Eng. Struct. Dyn.* **2008**, *38*, 45–60. [CrossRef]
55. Jayaram, N.; Baker, J.-W. Correlation model for spatially distributed ground-motion intensities. *Wiley Online Libr. Earthq. Eng. Struct. Dyn.* **2009**, *38*, 1687–1708. [CrossRef]

56. Asim, K.-M.; Martínez-Álvarez, F.; Basit, A.; Iqbal, T. Earthquake magnitude prediction in Hindukush region using machine learning techniques. *SpringerLink Nat. Hazards* **2017**, *85*, 471–486. [CrossRef]
57. Melgarejo-Morales, A.; Vazquez-Becerra, G.E.; Millan-Almaraz, J.-R.; Pérez-Enríquez, R.; Martínez-Félix, C.-A.; Ramon Gaxiola-Camacho, J. Examination of seismo-ionospheric anomalies before earthquakes of Mw $\geq$ 5.1 for the period 2008–2015 in Oaxaca, Mexico using GPS-TEC. *SpringerLink Solid Earth Sci. Acta Geophys.* **2020**, *68*, 1229–1244. [CrossRef]

Article

Incorporating Foreshocks in an Epidemic-like Description of Seismic Occurrence in Italy

Giuseppe Petrillo [1] and Eugenio Lippiello [2,*]

[1] The Institute of Statistical Mathematics, Research Organization of Information and Systems, Tokyo 190-0014, Japan
[2] Department of Mathematics and Physics, University of Campania "L. Vanvitelli", 81100 Caserta, Italy
* Correspondence: eugenio.lippiello@unicampania.it

Abstract: The Epidemic Type Aftershock Sequence (ETAS) model is a widely used tool for cluster analysis and forecasting, owing to its ability to accurately predict aftershock occurrences. However, its capacity to explain the increase in seismic activity prior to large earthquakes—known as foreshocks—has been called into question due to inconsistencies between simulated and experimental catalogs. To address this issue, we introduce a generalization of the ETAS model, called the Epidemic Type Aftershock Foreshock Sequence (ETAFS) model. This model has been shown to accurately describe seismicity in Southern California. In this study, we demonstrate that the ETAFS model is also effective in the Italian catalog, providing good agreement with the instrumental Italian catalogue (ISIDE) in terms of not only the number of aftershocks, but also the number of foreshocks—where the ETAS model fails. These findings suggest that foreshocks cannot be solely explained by cascades of triggered events, but can be reasonably considered as precursory phenomena reflecting the nucleation process of the main event.

Keywords: statistical seismology; numerical modeling; probabilistic forecasting; time-series analysis

1. Introduction

The Epidemic Type Aftershock Sequence (ETAS) model is widely regarded as the gold standard for seismic predictions and validating hypotheses related to seismic clustering [1–5]. In this model, the increase in seismic activity immediately after the mainshock is attributed to a "bottom-up" triggering process [6]. Essentially, any earthquake can generate a certain number of aftershocks, which are typically of smaller magnitudes, but there is also a non-negligible chance that it could trigger a larger magnitude earthquake. In the latter case, the seismic rate increases before the occurrence of the mainshock, and earthquakes responsible for this increase are referred to as foreshocks. While foreshocks originating from a cascade of triggered events, as in the ETAS model, follow the same patterns as aftershocks and are therefore not particularly informative for predicting large earthquakes, there is a possibility that foreshocks are due to the cumulative effects of tectonic loading processes on the fault that will host the mainshock [7,8]. This effect is commonly referred to as "top-down" loading, and through this mechanism, foreshocks can act as passive tracers of the preparatory process for the impending mainshock. Therefore, in principle, foreshocks can be used to improve predictions about mainshock occurrence.

The debate about the origin of foreshocks and their prognostic value is still ongoing. However, much effort has been made on both sides to extract useful information from instrumental catalogs. Supporting the bottom-up triggering process, several research lines have indicated that the Epidemic Type Aftershock Sequence (ETAS) model could explain the most relevant statistical features of foreshocks [9–12]. Marzocchi and Zhuang [13] have also shown that foreshock activity observed in the Italian catalog is consistent with what is expected by the ETAS model, and they attribute the variability of the statistical features

Citation: Petrillo, G.; Lippiello, E. Incorporating Foreshocks in an Epidemic-like Description of Seismic Occurrence in Italy. *Appl. Sci.* **2023**, *13*, 4891. https://doi.org/10.3390/app13084891

Academic Editor: Jianbo Gao

Received: 27 February 2023
Revised: 14 March 2023
Accepted: 16 March 2023
Published: 13 April 2023

of foreshocks to the limited sample size. On the other hand, supporting the top-down loading process, Brodsky [14] documented a lack of foreshocks in the ETAS synthetic catalog. Other studies [15–20] have confirmed that the number of foreshocks predicted by the ETAS model is less than the number observed in the instrumental catalog. In order to solve the insufficiencies in foreshock predictions in the ETAS catalog, Petrillo and Lippiello [21] introduced the ETAFS model, integrating both the aftershock and foreshock phenomena. By incorporating this innovative framework, the ETAFS model promises a new description of seismicity patterns and provides us with new insights into earthquake prediction and hazard assessment. This model is in good agreement with the number of foreshocks and aftershocks observed in the Southern California instrumental catalog. This perspective is supported by recent mechanical models of seismic faults, such as those proposed in ref. [22–27]. These models are a generalization of the original spring-block model [28,29] and present seismic patterns with the occurrence of the largest earthquakes frequently preceded by smaller foreshocks. This paper extends the work of [21] to the Italian seismic catalog. We improve the ETAS model by first accounting for the aftershock incompleteness present in the experimental catalog, which is caused by the overlapping of the coda waves [30,31]. We then add the ingredient of foreshocks, which represents a conceptual change for the model. By preserving the point process nature of the model, we replace the occurrence probability of a single earthquake with the occurrence probability of a cluster of earthquakes, which is composed of an earthquake anticipated by its own foreshocks. As for aftershocks, the number of events in each cluster depends on the magnitude of the final earthquake, and their space-time occurrence depends on the space-time of the final earthquake in the cluster. After defining the ETAFS model, we conduct rigorous statistical tests for the aftershock and foreshock numbers for both the ETAS and ETAFS models.

The paper is structured as follows. In Section 2, we introduce the definitions of mainshock, aftershock, and foreshock, and discuss declustering techniques. In Section 3, we present the various models under consideration. Section 4 explains the statistical validation method and the null hypothesis. We present our results in Section 5, followed by the conclusions in the final section.

2. General Definitions and Declustering

Aftershock, Foreshocks and Declustering Techniques

There are several ways to decluster a numerical earthquake catalog, resulting in different definitions of aftershocks, foreshocks, and mainshocks. In this paper, we consider the Baiesi–Paczusky–Zaliapin–Ben-Zion (BPZB) declustering method proposed in [32–35], which relies on a metric to quantify the correlation between events. The nearest neighbor of an event is defined by the metric $\eta_{ij} = t_{ij} r_{ij}^d 10^{-b(m_i - m_c)}$, where t_{ij} is the time difference between events i and j, r_{ij} is their epicentral distance, d is the fractal dimensionality of epicenters, b is the exponent of the Gutenberg–Richter (GR) law, m_i is the magnitude of the first event, and m_c is the completeness magnitude. Two earthquakes form a cluster if their distance η_{ij} is below a certain threshold η_c. To determine η_c, we define the magnitude-normalized time and space components as follows

$$\tau_{ij} = t_{ij} \times 10^{-bm_i/2} \qquad s_{ij} = r_{ij} \times 10^{-bm_i/2} \tag{1}$$

and for any pair of events we plot $log(s_{ij})$ vs. $log(\tau_{ij})$ in Figure 1. We use $d = 1.6$ for the fractal dimensionality, and $b = 0.95$ obtained from the magnitude distribution (Figure 2). Similar results are obtained for similar choices of b and d. In Figure 1, two distinct populations can be observed: the foreshock/aftershock clustering, consisting of events with $s_{ij}\tau_{ij} < \eta_c$, and the stationary Poisson seismicity, consisting of events i with $s_{ij}\tau_{ij} > \eta_c$ for any j. The red line in Figure 1 represents the clear boundary between these two populations, which is achieved with $\eta_c = 10^6$. For each event i in the Poisson population, we identify a seismic sequence that includes event i and all k clustered events with $\eta_{ki} < \eta_c$. The largest event in the sequence is defined as the mainshock, while events occurring before the mainshock are classified as foreshocks and those occurring after are classified as aftershocks.

To avoid obscuring the observation of foreshocks, we apply an additional filter proposed by [19] to the declustering procedure. Specifically, if an event i is identified as a mainshock after the BPZB procedure, but it occurs close enough in time and space to a previous large earthquake ($m \geq 5$), then it is considered an aftershock of the earlier event rather than a background event. More precisely, if the probability that the i-event is triggered by the $m \geq 5$ earthquake is higher than the background probability μ, the entire cluster is excluded from the study.

After identifying the complete cluster of seismic events, we group all mainshocks into magnitude classes $m \in [m_M, m_M + 1)$, and for each aftershock and foreshock, we define a temporal distance Δt_M and an epicentral distance Δr_M from the corresponding mainshock. We then define $n_A(T, R, m_L, m_M)$ and $n_F(T, R, m_L, m_M)$ as the total number of aftershocks (and foreshocks) with $\Delta t_M < T$, $\Delta r_M < R$, and magnitude $m \in [m_L, m_L + 0.5)$ linked to a mainshock with magnitude $m \in [m_M, m_M + 1)$.

In this study, we use $T = 10$ days, $m_M = 4, 5$, $m_L = 2, 2.5, 3, 3.5$, and $R = L(m_M) = 0.01 \times 10^{0.5 m_M}$.

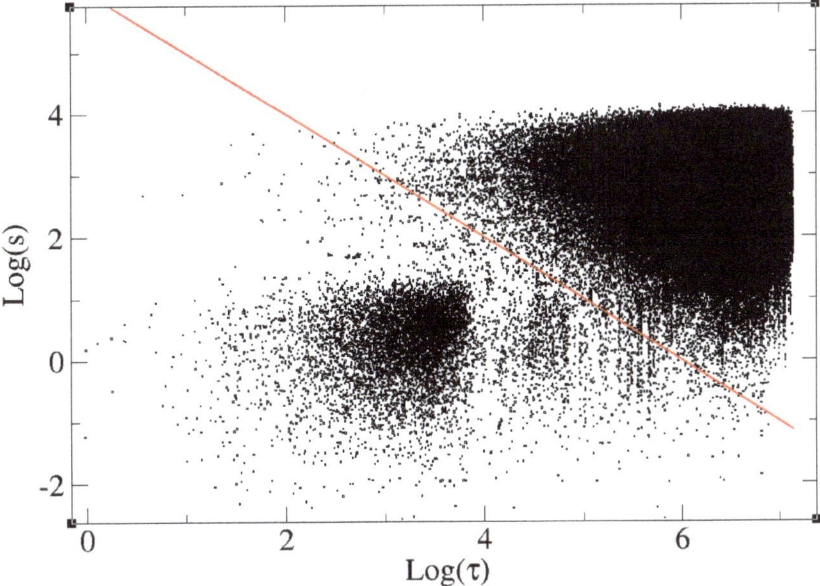

Figure 1. Bimodal distribution of time and space components of the nearest-neighbor for the observed seismicity in Italy. Solid red line corresponds to $log_{10}(s) + log_{10}(\tau) = 6$. The fractal dimensionality is fixed at $d = 1.6$.

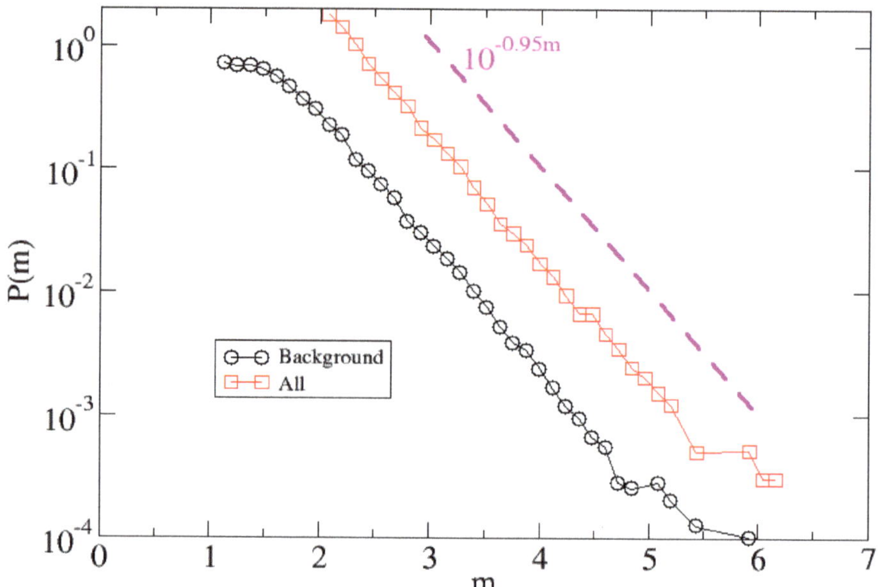

Figure 2. Instrumental magnitude distribution $P(m)$ for the Italian seismicity. Black circles represents only the background activity whereas red squares the whole catalog.

3. The Models

3.1. The Fixed-α ETAS Model

In the Epidemic Type Aftershock Sequence (ETAS) the occurrence rate λ of an event with magnitude $m > m_0$, at the position (x, y), at a time t is given by

$$\Lambda(m, x, y, t) = \mu(x, y) + \sum_{j:t_j<t} K10^{\alpha(m_j-m_0)} \frac{(p-1)c^{p-1}}{(t-t_j+c)^p} \frac{q-1}{\pi} (\delta(m_j))^{q-1}[(x-x_j)^2 + (y-y_j)^2 + \delta(m_j)]^{-q} \tag{2}$$

with $\delta(m_j) = d10^{\gamma(m_j-m_0)}$. In the fixed-α ETAS model, $\alpha = b$, a choice which reduces the incompleteness of the seismic catalog [19,36] and take into account spatial anisotropy [37,38] of the seismic events.

3.2. The Incomplete ETAS Model: The ETASI Model

A crucial prerequisite for an accurate estimation of the ETAS parameters is the completeness of the seismic catalog. Unfortunately, due to the strong temporal correlation between earthquakes, there is a significant lack of events, particularly in the immediate aftermath of high-magnitude mainshocks [39–43]. This is caused by the overlapping of coda waves, which is even more pronounced when focusing on larger earthquakes where aftershocks are more challenging to detect and report in experimental catalogs [30,44–48]. To address this issue, we introduce the concept of short-term aftershock incompleteness into the ETAS model, resulting in the ETAS Incomplete (ETASI) model [30]. The core idea is that the probability of detecting an event at time $t - t_i$ after an m_i earthquake depends on the difference between its magnitude m and the detection threshold $m_i^{th}(t - t_i)$. Specifically, we adopt the functional form proposed by [39,40].

$$m_i^{th}(t - t_i) = m_i - w\log(t - t_i) - \delta_0 \tag{3}$$

where w and δ_0 are two parameters obtained to reach the best agreement with the instrumental catalog. The functional form for m_i^{th} (Equation (3)) is the most compatible with experimental data, and its logarithmic decay can be explained by the behavior of the seismic waveform envelope $\mu(t)$ after a mainshock [44,49]. This envelope is always greater than a minimum value $\mu_c(t)$, which decays logarithmically. Lippiello et al. [44] have linked the existence of $\mu_c(t)$ to the overlap between aftershock coda waves, and demonstrated that its decay incorporates parameters related to the Omori–Utsu law governing the decay of aftershocks [50]. Therefore, the expected number of aftershocks in the immediate aftermath of a mainshock can be estimated [49]. We implement the expression Equation (3) in the ETAS model by means of the function $\Phi(m|m_j^{th}(t-t_j),\sigma)$ which represents the cumulative distribution of the normal distribution. In particular, Φ is a decreasing function of m which is roughly equal 1 when $m > m_j^{th}(t-t_j)+\sigma$, whereas $\Phi \simeq 0$ when $m < \lambda - \sigma$.

The function Φ can be implemented in the ETAS model Equation (2) as

$$\Lambda(m,x,y,t) = \Lambda(x,y,t) \times \prod_{j:t_j<t} \Phi(m|m_j^{th}(t-t_j),\sigma) \tag{4}$$

The implementation of the ETASI model is straightforward, as it involves generating a synthetic ETAS catalog and then applying the Φ function to remove a portion of the events. The parameters of the ETASI model include all those of the ETAS, as well as two additional parameters that account for the incompleteness function. However, we do not restrict our analysis to fitting only the last two parameters (w and δ_0); rather, we perform a global optimization of all 10 parameters in the model.

3.3. The Top-Down Loading ETAS Model—The ETAFS Model

By construction, the previously defined models assume bottom-up triggering as the hypothesis for foreshocks. In other words, a foreshock is considered a normal earthquake that triggers an offspring with a magnitude greater than itself. However, it is important to also consider the possibility of top-down loading as an explanation for foreshocks. The Epidemic Type Aftershocks and Foreshocks Sequence ETAFS model is a generalization of the ETAS model that incorporates the mechanism of loading by aseismic slip to account for top-down triggering. In the ETAFS model, aftershocks are triggered with the same probability rate as the ETAS model, but each earthquake can also be preceded by foreshock activity. Mathematically, this can be expressed as:

$$\Lambda_{ETAFS}(m,x,y,t) = \Lambda_{ETAS} + \sum_k 10^{-bm} Q_f(d_{ik}, t_k - t, m_k) \tag{5}$$

where d_{ik} is the distance between the event i and k and the sum extends over all events with magnitudes m_k, occurred at time t_k, in the position (x_k, y_k) triggered according to the ETAS probability (Equation (2)), whereas $Q_f(d_{ik}, t_k - t, m_k)$ is the rate of foreshocks potentially occurring at times $t < t_k$. There is no specific constraint on the functional form of Q_f. In order to reduce the number of model parameters we assume [21] that the spatio-temporal organization of the event in the cluster is similar to the aftershock one, setting

$$Q_f(d_{ik}, t_k - t, m_k) = K_f 10^{\alpha_f(m_k-m_0)} \frac{(p_f-1)c_f^{p_f-1}}{(t_k-t+c_f)^p} \frac{q_f-1}{\pi}(\delta(m_k))^{q_f-1}\left(d_{ik}^2 + \delta(m_k)\right)^{-q_f} \tag{6}$$

where $\delta(m_k) = d_f 10^{\gamma_f(m_k-m_0)}$, $\gamma_f = \gamma$, $d_f = d$, and $q_f = q$. As in the ETAS model, here we extract the number of events belonging to k-th cluster from a Poissonian distribution with average $K_f 10^{\alpha_f(m_k-m_0)}$. Moreover we implement the inverse Omori Law with the same p as in the aftershock occurrence and $c_f = c$. In principle, different functional forms could be adopted to achieve similar results. Finally, we apply the same aftershock removal procedure as done for the ETAS model in order to take into account the hiding of events caused by the overlapping of coda waves. In mathematical terms

$$\Lambda_{ETAFS}^{inc}(m,x,y,t) = \Lambda_{ETAFS}(m,x,y,t) \times \prod_{j:t_j<t} \Phi(m|m_j^{th}(t-t_j),\sigma) \tag{7}$$

For the simulations of the ETAFS model, we use exactly the same parameters as the ETASI model to generate a complete ETAS catalog. Starting from each simulated ETAS earthquake, we then use the kernel Q_f to generate foreshocks. The magnitude of each foreshock is extracted from the Gutenberg–Richter law, but with the constraint that the magnitude of each event belonging to the cluster must be smaller than the final magnitude m_k. We finally apply the filtering procedure by means of the function Φ, according to Equation (7), to take into account incompleteness.

The apparent problem with the ETAFS model is that the spatio-temporal organization of the cluster's events depends directly on the characteristics of the incoming event, the mainshock. In practice, it seems that this model violates the temporal causality principle since the occurrence probability at a certain time depends on the future. However, from the point of view of the point process-like model, the ETAFS remains well defined if one considers each single ETAS event as a cluster of events. In particular, in a formulation practically similar to the ETAS model, the ETAFS model can be viewed as a point process, where each single point has an internal structure represented by an earthquake and its anticipating foreshocks. With this approach, Equation (2) gives the occurrence probability of the last event of the cluster, and all events that belong to the same cluster are deterministically correlated with each other. This reflects the idea that events belonging to the same cluster are the manifestation of the same underlying process and contain information on the incoming event.

We generate synthetic catalogs with the same algorithm used in [21] and the parameter values used in the three models are given in Table 1.

Table 1. Parameters for the three models presented in this study. We consider the lower magnitude threshold $m_0 = 2$.

Model	K_0	α	b	p	c	d	γ	q	ω	δ_0	K_f	α_f
ETAS	0.07	0.95	0.95	1.2	0.024	0.006	1.958	1.3	-	-	-	-
ETASI	0.1	0.93	0.95	1.2	0.01	0.006	1.958	1.3	0.3	1	-	-
ETAFS	0.1	0.93	0.95	1.2	0.01	0.006	1.958	1.3	0.3	1	0.05	0.5

4. Data Catalog, Methods and Null Models

The purpose of the statistical test that will be carried out in this paper is to evaluate whether the three models defined in the previous section are able to describe the experimental data.

In the null-hypothesis test we compare the number of aftershocks and foreshocks for each mainshock. In practice we consider the ISIDE Italian Catalog (from 2005/04/16 to 2021/04/30). We use the BPZB selection procedure to identify aftershocks and foreshocks and we evaluate the quantities $n_A(T, R, m_L, m_M)$ and $n_F(T, R, m_L, m_M)$ for a wide range of parameter $m_L = [2, 2.5, 3, 3.5]$, $m_M = [4, 5, 6]$, and for $T = 1, 3, 10$ days, always considering $R = L(m_M)$ km. We compute these quantities for each model we defined before, namely ETAS, ETASI and ETAFS and we consider each of them as null model for the hypothesis test. We choose model parameters such as the models lead to synthetic catalogs of equal duration of the ISIDE catalog and roughly the same number of events with $M > 2.5$. For each model and for a defined set of model parameters, we generate 1000 catalogues that we concatenate and we identify seismic sequences by means of a BPZB procedure. After the identification of the sequences, we divide the whole catalog in N_s subset where each one contains the same number of the earthquakes of instrumental Italian catalogue. Now we compute the quantity $n_{A,j}^X$ and $n_{F,j}^X$, which represents the number of aftershocks and foreshocks, respectively, in the j-th subset produced by the X model. The same quantity is computed for the instrumental Italian catalogue. The superscript, X, indicates a specific numerical model with the three possible entries $X = [ETAS, ETASI, ETAFS]$, according to the corresponding model. Conversely, we no longer use the superscript X for the number of aftershocks and foreshocks in the instrumental Italian catalogue.

We evaluate the difference $\psi_{j,A} = \frac{n^X_{A,j} - n_A}{n^X_{A,j} + n_A}$ and $\psi_{j,F} = \frac{n^X_{F,j} - n_F}{n^X_{F,j} + n_F}$. This allows us to quantify the difference in the number of aftershocks and foreshocks, respectively, between the j-th realization of the synthetic catalogue of the X model and the instrumental ISIDE catalogue. In particular we evaluate these quantities for a different sets of parameters (T, R, m_L, m_M). and compute two histograms $H(\psi_{j,A})$ and $H(\psi_{j,F})$ defined as the fraction of the N_s subsets with a value $\psi_A = \psi \pm d\psi$, or $\psi_F = \psi \pm d\psi$, where $d\psi = 0.005$. If the histogram $H(\psi)$ is very peaked around $\psi = 0$, the prediction of the model is in good agreement with the experimental catalogue, i.e., the number of aftershocks (foreshocks) in the model is similar the instrumental one. The two limits $\psi = -1$ and $\psi = 1$ represents the worst cases. In particular, with $\psi = -1$, the synthetic catalogue predicts a number of events equal to 0. The opposite behaviour is observed with $\psi = 1$, which represents the limit for infinite events predicted in the simulated catalogue. For the foreshock statistical evaluation, it is useful to define the quantities $I^X_+ = \sum_{\psi > 0} H(\psi_{j,F}) d\psi$, representing the area under the histogram for positive ψ values.

The overall procedure is shown in detail in Figure 3.

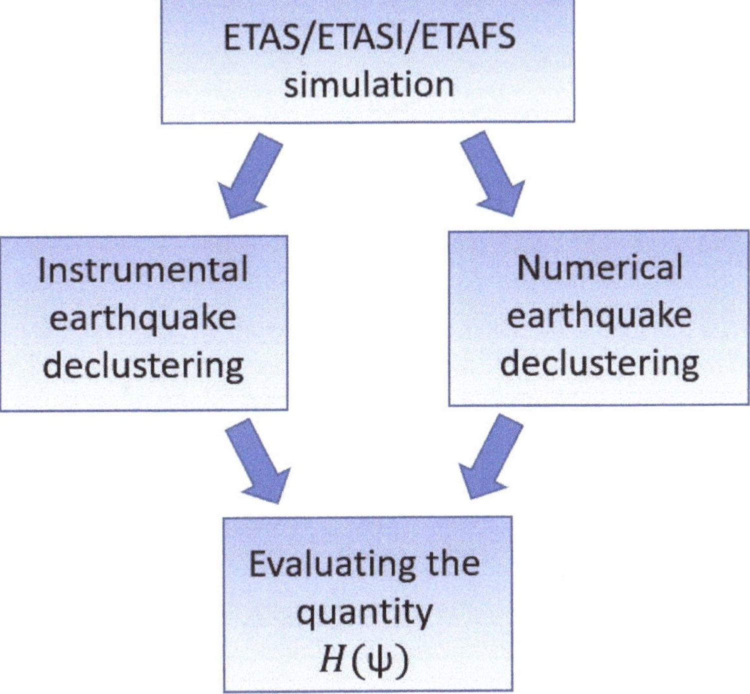

Figure 3. Flowchart of the method employed in the study.

5. Results

5.1. Aftershock Comparison

In this subsection we present results of the comparison for the aftershock number observed in the synthetic catalogs of the ETAS and ETAFS models with those selected in the instrumental catalog. The number of aftershocks in the ETAFS and the ETASI model roughly coincides, therefore we limit ourselves to present results for the ETAFS model.

In Figure 4 the histogram $H(\psi)$ is plotted for the ETAS model and the ETAFS model, considering different values of m_M, m_L and different time windows T. The best agreement with the instrumental catalog is obtained when $H(\psi)$ is a very peaked distribution around $\psi \simeq 0$. This corresponds to the situation when the majority of synthetic catalogs present a number of aftershocks similar to the one of the instrumental catalog. This condition is clearly satisfied for $m_M \leq 5$ for both the ETAS and the ETAFS model. Only for $m_M = 5$ and $m_L = 2$, we observe that the maximum of $H(\psi)$ in the ETAS model is located at $\psi \simeq 0.25$ indicating an excess of aftershocks in the ETAS catalog compared to the instrumental one. This shift on the right of the peak of $H(\psi)$ can be attributed to the short term aftershock incompleteness, as confirmed by the fact that the peak of $H(\psi)$ in the ETAFS model is close to $\psi = 0$. This shift on the right is still present for $m_m = 5$ and $m_L = 2.5$ but it is less relevant and it disappears by increasing m_L, as expected since short-term aftershock incompleteness is less relevant the larger the aftershock magnitude is. The situation for $m_M = 6$ is less clear since the distribution is much broader, both for the ETAS and the ETAFS models, probably because there are only 3 $m > 6$ mainshocks in the instrumental catalog. Nevertheless, we notice that for $m_M = 6$ and $m_L \leq 3$ the distribution $H(\psi)$ for the ETAS model is significantly asymmetric with a clear excess of aftershocks in the ETAS catalog compared to the instrumental catalog. This can be again attributed to short term aftershock incompleteness, which is more relevant for larger mainshocks, and it is confirmed by the observation that $H(\psi)$ for the ETAFS model is more symmetric.

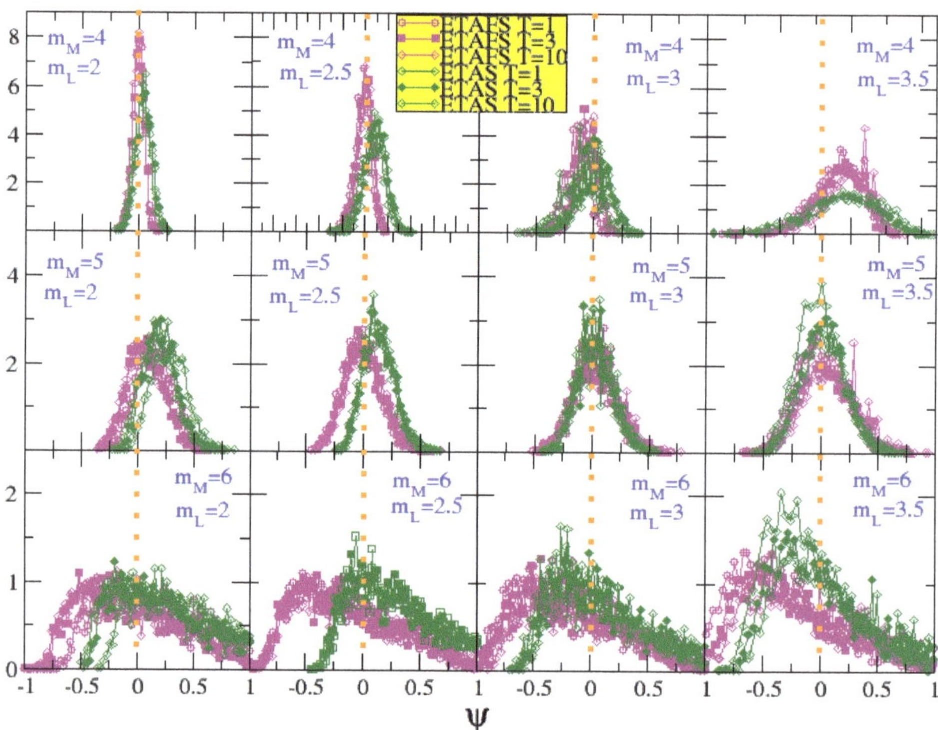

Figure 4. The fraction of numerical subsets $H(\psi)$ of the ETAFS (ETASI) and ETAS simulated catalogue, with a number of aftershocks $\dfrac{n_{A,j}^X(T,R,m_L,m_M)-n_A(T,R,m_L,m_M)}{n_{A,j}^X(T,R,m_L,m_M)+n_A(T,R,m_L,m_M)} = \psi \pm 0.005$. Results are for different values of the time window T. Moving horizontally, from left to right, $m_L = 2, 2.5, 3, 3.5$ whereas vertically, from top to bottom, $m_M = 4, 5, 6$. The orange dashed line indicate the optimal description of the seismicity $\psi = 0$.

5.2. Foreshock Comparison

In Figure 5 we plot the histograms $H(\psi)$ of $\psi_{j,F}$ for the ETAS and ETAFS catalogues. We will not present results for the ETASI model which are comparable to those of the ETAS model, since the two models only differ in the implementation of short term incompleteness, which is not relevant for foreshocks. As for the aftershocks, we consider different windows of time T, foreshock minimum magnitude m_L and mainshock threshold $m_M = 4, 5$. We will present results for mainshocks with $m > 6$ separately. This figure shows that the ETAS catalogue presents significant less foreshocks than the instrumental Italian one. Indeed $H(\psi)$ is in all cases peaked around $\psi \simeq -0.5$, which indicates that the number of foreshocks in the ETAS catalogue is, on average, about $1/3$ the number found in the Italian catalogue. Moreover, we note that the probability of producing a numerical catalog that has the same number of foreshocks as the observed catalog is practically zero. Indeed, no ETAS synthetic catalog presents a number of foreshocks equal or larger than the one observed in the instrumental catalog when $m_M = 4$ and, at the same time, when $m_M = 5$ less than 1% of synthetic catalogs satisfies this condition. This conclusion can be drawn for all the foreshocks magnitude thresholds m_L, and for each time window T considered. This interpretation is supported from the measurement of the quantity I_+^{ETAS} reported in Table 2. Indeed, for all values of m_M, m_L and T the value of I_+^{ETAS} is nearly zero. We conclude that it is very unlikely or even impossible that a synthetic catalog for the ETAS model presents a number of foreshocks equal or larger than the one observed in the Italian seismic catalogue. The situation slightly changes when one considers the ETASI model, with I_+^{ETASI} (Table 2) still indicating that the hypothesis that the ETASI model predicts a number of foreshocks equal to or larger than those actually observed can be rejected with a high confidence.

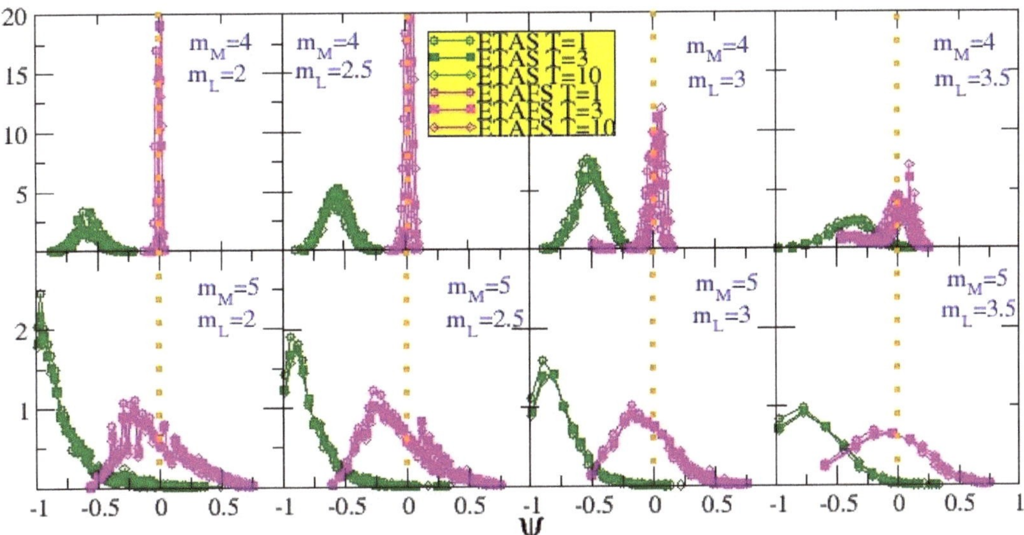

Figure 5. The fraction of numerical subsets $H(\psi)$ of the ETAS and ETAFS simulated catalogue, with a number of foreshocks $\frac{n_{F,j}^X(T,R,m_L,m_M) - n_F(T,R,m_L,m_M)}{n_{F,j}^X(T,R,m_L,m_M) + n_F(T,R,m_L,m_M)} = \psi \pm 0.005$. Results are for different values of time window T. Moving horizontally from left to right, $m_L = 2, 2.5, 3$, and 3.5, whereas from top to bottom, $m_M = 4$ and 5. The orange dashed line indicates the optimum description of the seismicity $\psi = 0$.

On the contrary, the numerical catalog simulated by means of the ETAFS model captures the seismicity of the foreshocks very well; in fact, all the $H(\psi)$ histograms exhibit a peak that is very close to $\psi = 0$. We observe this behavior for all minimum magnitude m_L and temporal domains T. It is worth noting that a dramatic change in $H(\psi)$ is obtained by adding just a small percentage of earthquakes (about 10%) to the ETASI catalog. This small percentage of earthquakes are foreshocks triggered according to the second term in Equation (5), indicating that their addition is strictly necessary to obtain a good agreement with the observed seismicity in the Italian catalogue. This is confirmed quantitatively in

Table 2, where all I_+^{ETAFS} values are significantly larger than 0 and smaller than 1, presenting fluctuations around the optimal value ($I_+^{ETAFS} = 0.5$).

Table 2. The positive area below the curve $H(x)$ for ETAS and ETASI model for the foreshocks.

m_M	m_L	$T(d)$	I_+^{ETAS}	I_+^{ETASI}	I_+^{ETAFS}
4	2	1	0	0	0.6
4	2	3	0	0.01	0.73
4	2	10	0	0.02	0.76
4	2.5	1	0	0	0.50
4	2.5	3	0	0	0.8
4	2.5	10	0	0.01	0.85
4	3	1	0	0.01	0.43
4	3	3	0	0.03	0.80
4	3	10	0	0.05	0.92
4	3.5	1	0	0.10	0.30
4	3.5	3	0.02	0.10	0.51
4	3.5	10	0.03	0.15	0.58
5	2	1	0	0.03	0.32
5	2	3	0.01	0.05	0.39
5	2	10	0.02	0.07	0.42
5	2.5	1	0	0.01	0.29
5	2.5	3	0	0.03	0.37
5	2.5	10	0	0.05	0.40
5	3	1	0	0.01	0.28
5	3	3	0	0.03	0.34
5	3	10	0	0.05	0.37
5	3.5	1	0.01	0.16	0.22
5	3.5	3	0.03	0.26	0.27
5	3.5	10	0.04	0.33	0.29

Foreshocks for Mainshocks with $m_M \geq 6$

The ISIDE catalog presents only three $m_M \geq 6$ earthquakes, Table 3, which are here analysed separately. The L'Aquila earthquake was a Mw6 magnitude event that occurred on 6 April 2009 at 01:32 UTC in the Abruzzo region of central Italy. After BPZB declustering, we identified 540 foreshocks without considering any time or magnitude constraints. The Amatrice and Norcia events, on the other hand, were a Mw6 and Mw6.5 magnitude earthquake that occurred on 24 August 2016 and 30 October 2016, respectively. Unlike the L'Aquila earthquake, these events did not present any foreshocks. To test the effectiveness of the models, we simulated 1000 events for each of the three earthquakes using the ETAS and ETAFS models, using the magnitude of the actual mainshocks as the parent magnitude. We then calculated the distribution of the number of foreshocks for each case. We found that the ETAFS model produced significantly more foreshocks than the ETAS model. However, neither model was able to accurately describe the experimental data, as the L'Aquila earthquake had a much higher number of foreshocks than even the ETAFS model predicted. For the Amatrice and Norcia events, the ETAS model seemed to be the best fit as it did not predict any foreshocks. However, due to the limited statistics available for Mw6 magnitude earthquakes, the results are not statistically significant. Indeed, in our study, we do not consider Mw7 earthquakes because in the ISIDE catalog and in the time window considered there are no earthquakes with a magnitude greater than 7. The method has been verified for such earthquakes for the Southern California catalog [21].

Table 3. The Mw6 mainshock registered in the ISIDE seismic catalogue.

Name	Mw	Date
L'Aquila	6.0	6 April 2009
Amatrice	6.0	24 August 2016
Norcia	6.5	30 October 2016

6. Conclusions

In this study, we analyzed the Italian seismic catalog (ISIDE) and compared different types of ETAS models. The first is the standard ETAS model with the only constraint that $\alpha = b$, as proposed in [19]. The second and third variants, ETASI [30] and ETAFS [21], explicitly account for the incompleteness of the experimental catalog and a higher probability of foreshock occurrence, respectively. The ETAFS model combines the concept of bottom-up triggering for the aftershock side and top-down loading for the occurrence of foreshocks. By simulating the Italian seismicity with the ETAS model, we show that the fixed-α ETAS model accurately reproduces the number of aftershocks in the ISIDE catalog when the magnitude difference between mainshock and aftershocks is relatively small. However, an excess of aftershocks is observed in the ETAS model compared to the ISIDE catalog when this difference increases. We attribute this effect to the short-term aftershock incompleteness, which becomes much more important as the magnitude difference between mainshock and aftershocks increases. To address this issue, we demonstrate that introducing the incompleteness ingredient in the ETAS model (ETASI model) makes it possible to better describe the occurrence of aftershocks by removing events that were not recorded by seismic stations.

The key observation, however, is that both the ETAS model and the ETASI model fail to capture the number of foreshocks present in the ISIDE catalog. Our study clearly shows an excess of foreshocks in the ISIDE catalog for mainshocks with magnitude $m < 6$ compared to what is predicted by the ETAS and ETASI models. In contrast, the ETAFS model, which introduces a preparatory phase accompanied by foreshocks in a point process description, accurately describes the seismicity reported by the ISIDE catalog for both foreshocks and aftershocks. These patterns are consistent with those found for the Southern California seismic catalog in ref. [21], suggesting that they are a stable feature of seismic occurrences. The situation for the three mainshocks with $m > 6$ in the ISIDE catalog is different. Indeed, two of them, the Amatrice and the Norcia earthquakes, do not present foreshocks, whereas the L'Aquila earthquake presents a number of foreshocks significantly larger than those predicted by the ETAFS model.

We finally remark that the ETAFS model assumes that the magnitude of the mainshock is encoded in the spatial organization of the foreshocks. The spatial kernel of Q_f is of the order of the area fractured by the mainshock. Therefore, while the prediction of a mainshock with a bottom-up triggering ETAS model is purely random, in the case of a top-down loading model, it would be possible to use the organization of foreshock epicenters as forecasting information. It is important to note that we did not assume a direct dependency between the magnitude of foreshocks and the magnitude of the mainshock, since the introduction of such dependence is still being studied [51] and is beyond the scope of this article.

Author Contributions: G.P. and E.L. contributed to the research, numerical results and writing of the manuscript. All authors have read and agreed to the published version of the manuscript.

Funding: This research received no external funding.

Data Availability Statement: The complete seismic catalog is available at http://terremoti.ingv.it/en/iside, accessed on 30 April 2021.

Acknowledgments: G.P. acknowledges support by MEXT Project for Seismology TowArd Research innovation with Data of Earthquake (STAR-E Project), Grant Number: JPJ010217. E.L. acknowledges support from project PRIN201798CZLJ and from VALERE project of the University of Campania "L. Vanvitelli". G.P. and E.L. also thank Jiancang Zhuang for helpful discussions.

Conflicts of Interest: The authors declare no conflict of interest.

References

1. Ogata, Y. Statistical Models for Earthquake Occurrences and Residual Analysis for Point Processes. *J. Am. Stat. Assoc.* **1988**, *83*, 9–27. [CrossRef]
2. Ogata, Y. Space-time point-process models for earthquake occurrences. *Ann. Inst. Stat. Math.* **1998**, *50*, 379–402. [CrossRef]
3. Petrillo, G.; Zhuang, J. Bayesian Earthquake Forecasting approach based on the Epidemic Type Aftershock Sequence model. *Res. Sq.* **2022**. [CrossRef]
4. Zhuang, J. Next-day earthquake forecasts for the Japan region generated by the ETAS model. *Earth Planets Space* **2011**, *63*, 207–216. [CrossRef]
5. Zhuang, J. Long-term earthquake forecasts based on the epidemic-type aftershock sequence (ETAS) model for short-term clustering. *Res. Geophys.* **2012**, *2*, e8. [CrossRef]
6. Mignan, A. The debate on the prognostic value of earthquake foreshocks: A meta-analysis. *Sci. Rep.* **2014**, *4*, 4099. [CrossRef]
7. Mignan, A. Seismicity precursors to large earthquakes unified in a stress accumulation framework. *Geophys. Res. Lett.* **2012**, *39*, 21308. [CrossRef]
8. Das, S.; Scholz, C. Theory of time-dependent rupture in the Earth. *J. Geophys. Res. Solid Earth* **1981**, *86*, 6039–6051. [CrossRef]
9. Helmstetter, A.; Sornette, D. Foreshocks explained by cascades of triggered seismicity. *J. Geophys. Res. Solid Earth* **2003**, *108*, 457. [CrossRef]
10. Helmstetter, A.; Sornette, D.; Grasso, J. Mainshocks are aftershocks of conditional foreshocks: How do foreshock statistical properties emerge from aftershock laws. *J. Geophys. Res. Solid Earth* **2003**, *108*, 2046. [CrossRef]
11. Felzer, K.R.; Abercrombie, R.E.; Ekstrm, G. A Common Origin for Aftershocks, Foreshocks, and Multiplets. *Bull. Seismol. Soc. Am.* **2004**, *94*, 88–98. [CrossRef]
12. Hardebeck, J.; Felzer, K.; Michael, A. Improved tests reveal that the accelerating moment release hypothesis is statistically insignificant. *J. Geophys. Res.* **2008**, *113*, B08310.
13. Marzocchi, W.; Zhuang, J. Statistics between mainshocks and foreshocks in Italy and Southern California. *Geophys. Res. Lett.* **2011**, *38*, L09310. [CrossRef]
14. Brodsky, E.E. The spatial density of foreshocks. *Geophys. Res. Lett.* **2011**, *38*, L10305. [CrossRef]
15. Lippiello, E.; Marzocchi, W.; de Arcangelis, L.; Godano, C. Spatial organization of foreshocks as a tool to forecast large earthquakes. *Sci. Rep.* **2012**, *2*, 1–6. [CrossRef]
16. Shearer, P.M. Self-similar earthquake triggering, Båth's law, and foreshock/aftershock magnitudes: Simulations, theory, and results for southern California. *J. Geophys. Res. Solid Earth* **2012**, *117*, B06310. [CrossRef]
17. Ogata, Y.; Katsura, K. Comparing foreshock characteristics and foreshock forecasting in observed and simulated earthquake catalogs. *J. Geophys. Res. Solid Earth* **2014**, *119*, 8457–8477. [CrossRef]
18. de Arcangelis, L.; Godano, C.; Grasso, J.R.; Lippiello, E. Statistical physics approach to earthquake occurrence and forecasting. *Phys. Rep.* **2016**, *628*, 1–91.
19. Seif, S.; Zechar, J.D.; Mignan, A.; Nandan, S.; Wiemer, S. Foreshocks and Their Potential Deviation from General Seismicity. *Bull. Seismol. Soc. Am.* **2018**, *109*, 1–18. [CrossRef]
20. Trugman, D.T.; Ross, Z.E. Pervasive Foreshock Activity Across Southern California. *Geophys. Res. Lett.* **2019**, *46*, 8772–8781. [CrossRef]
21. Petrillo, G.; Lippiello, E. Testing of the foreshock hypothesis within an epidemic like description of seismicity. *Geophys. J. Int.* **2020**, *225*, 1236–1257. [CrossRef]
22. Landes, F.P. Viscoelastic interfaces driven in disordered media. In *Viscoelastic Interfaces Driven in Disordered Media*; Springer Theses: New York, NY, USA, 2016 ; pp. 113–166.
23. Jagla, E.A.; Landes, F.P.; Rosso, A. Viscoelastic effects in avalanche dynamics: A key to earthquake statistics. *Phys. Rev. Lett.* **2014**, *112*, 174301. [PubMed]
24. Lippiello, E.; Petrillo, G.; Landes, F.; Rosso, A. Fault Heterogeneity and the Connection between Aftershocks and Afterslip. *Bull. Seismol. Soc. Am.* **2019**, *109*, 1156–1163.
25. Petrillo, G.; Landes, F.P.; Lippiello, E.; Rosso, A. The influence of the brittle-ductile transition zone on aftershock and foreshock occurrence. *Nat. Commun.* **2020**, *11*, 3010. [PubMed]
26. Lippiello, E.; Petrillo, G.; Landes, F.; Rosso, A. The Genesis of Aftershocks in Spring Slider Models. *Stat. Methods Model. Seism.* **2021**, *1*, 131–151.
27. Petrillo, G.; Rosso, A.; Lippiello, E. Testing of the Seismic Gap Hypothesis in a Model With Realistic Earthquake Statistics. *J. Geophys. Res. Solid Earth* **2022**, *127*, e2021JB023542. [CrossRef]
28. Burridge, R.; Knopoff, L. Model and theoretical seismicity. *Bullettin Seismol. Soc. Am.* **1967**, 341–371. [CrossRef]

29. Olami, Z.; Feder, H.J.S.; Christensen, K. Self-organized criticality in a continuous, nonconservative cellular automaton modeling earthquakes. *Phys. Rev. Lett.* **1992**, *68*, 1244–1247.
30. de Arcangelis, L.; Godano, C.; Lippiello, E. The Overlap of Aftershock Coda Waves and Short-Term Postseismic Forecasting. *J. Geophys. Res. Solid Earth* **2018**, *123*, 5661–5674. [CrossRef]
31. Hainzl, S. ETAS-Approach Accounting for Short-Term Incompleteness of Earthquake Catalogs. *Bull. Seismol. Soc. Am.* **2022**, *112*, 494–507.
32. Baiesi, M.; Paczuski, M. Scale-free networks of earthquakes and aftershocks. *Phys. Rev. E* **2004**, *69*, 066106. [CrossRef]
33. Baiesi, M.; Paczuski, M. Complex networks of earthquakes and aftershocks. *Nonlinear Process. Geophys.* **2005**, *12*. [CrossRef]
34. Zaliapin, I.; Gabrielov, A.; Keilis-Borok, V.; Wong, H. Clustering Analysis of Seismicity and Aftershock Identification. *Phys. Rev. Lett.* **2008**, *101*, 018501. [PubMed]
35. Zaliapin, I.; Ben-Zion, Y. Earthquake clusters in southern California I: Identification and stability. *J. Geophys. Res. Solid Earth* **2013**, *118*, 2847–2864. [CrossRef]
36. Seif, S.; Mignan, A.; Zechar, J.D.; Werner, M.J.; Wiemer, S. Estimating ETAS: The effects of truncation, missing data, and model assumptions. *J. Geophys. Res. Solid Earth* **2017**, *122*, 449–469. [CrossRef]
37. Helmstetter, A.; Kagan, Y.Y.; Jackson, D.D. Importance of small earthquakes for stress transfers and earthquake triggering. *J. Geophys. Res. Solid Earth* **2005**, *110*, B05S08. [CrossRef]
38. Hainzl, S.; Christophersen, A.; Enescu, B. Impact of Earthquake Rupture Extensions on Parameter Estimations of Point-Process Models. *Bull. Seismol. Soc. Am.* **2008**, *98*, 2066–2072. [CrossRef]
39. Kagan, Y.Y. Short-Term Properties of Earthquake Catalogs and Models of Earthquake Source. *Bull. Seismol. Soc. Am.* **2004**, *94*, 1207–1228. [CrossRef]
40. Helmstetter, A.; Kagan, Y.Y.; Jackson, D.D. Comparison of Short-Term and Time-Independent Earthquake Forecast Models for Southern California. *Bull. Seismol. Soc. Am.* **2006**, *96*, 90–106. [CrossRef]
41. Enescu, B.; Mori, J.; Miyazawa, M. Quantifying early aftershock activity of the 2004 mid-Niigata Prefecture earthquake (Mw6.6). *J. Geophys. Res. Solid Earth* **2007**, *112*. [CrossRef]
42. Peng, Z.; Vidale, J.E.; Ishii, M.; Helmstetter, A. Seismicity rate immediately before and after main shock rupture from high-frequency waveforms in Japan. *J. Geophys. Res. Solid Earth* **2007**, *112*, B03306. [CrossRef]
43. Peng, Z.; Zhao, P. Migration of early aftershocks following the 2004 Parkfield earthquake. *Nat. Geosci.* **2009**, *2*, 877–881. [CrossRef]
44. Lippiello, E.; Cirillo, A.; Godano, G.; Papadimitriou, E.; Karakostas, V. Real-time forecast of aftershocks from a single seismic station signal. *Geophys. Res. Lett.* **2016**, *43*, 6252–6258. [CrossRef]
45. Hainzl, S. Apparent triggering function of aftershocks resulting from rate-dependent incompleteness of earthquake catalogs. *J. Geophys. Res. Solid Earth* **2016**, *121*, 6499–6509. [CrossRef]
46. Hainzl, S. Rate-Dependent Incompleteness of Earthquake Catalogs. *Seismol. Res. Lett.* **2016**, *87*, 337–344.
47. Lippiello, E.; Giacco, F.; Marzocchi, W.; Godano, G.; Arcangelis, L.D. Statistical features of foreshocks in instrumental and ETAS catalogs. *Pure Appl. Geophys.* **2017**, *174*, 1679–1697. [CrossRef]
48. Lippiello, E.; Godano, C.; de Arcangelis, L. The Relevance of Foreshocks in Earthquake Triggering: A Statistical Study. *Entropy* **2019**, *21*, 173. [CrossRef]
49. Lippiello, E.; Petrillo, G.; Godano, C.; Papadimitriou, E.; Karakostas, V. Forecasting of the first hour aftershocks by means of the perceived magnitude. *Nat. Commun.* **2019**, *10*, 2953. [CrossRef]
50. Utsu, T.; Ogata, Y.; Matsuura, R. The centenary of the Omori formula for a decay law of aftershock activity. *J. Phys. Earth* **1995**, *43*, 1–33.
51. Petrillo, G.; Zhuang, J. The debate on the earthquake magnitude correlations: A meta-analysis. *Sci. Rep.* **2022**, *12*, 20683.

Article

Preliminary Study on the Generating Mechanism of the Atmospheric Vertical Electric Field before Earthquakes

Ruilong Han [1,2], **Minghui Cai** [1,2,*], **Tao Chen** [1], **Tao Yang** [1], **Liangliang Xu** [1], **Qing Xia** [1], **Xinyu Jia** [1] **and Jianwei Han** [1,2]

[1] State Key Laboratory of Space Weather, National Space Science Center, Chinese Academy of Sciences, Beijing 100190, China; hanruilong@nssc.ac.cn (R.H.); tchen@nssc.ac.cn (T.C.); yangtao@nssc.ac.cn (T.Y.); xuliangliang@nssc.ac.cn (L.X.); xiaqing@nssc.ac.cn (Q.X.); jiaxinyu@nssc.ac.cn (X.J.); hanjw@nssc.ac.cn (J.H.)
[2] College of Astronomy and Space Science, University of Chinese Academy of Sciences, Beijing 100049, China
* Correspondence: caiminghui@nssc.ac.cn

Abstract: Precursor signals for earthquakes, such as radon anomalies, thermal anomalies, and water level changes, have been studied in earthquake prediction over several centuries. The atmospheric vertical electric field anomaly has been observed in recent years as a new and valuable signal for short-term earthquake prediction. In this paper, a physical mechanism of the atmospheric vertical electric field anomaly before the earthquake was proposed, based on which the Wenchuan earthquake verified the correctness of the model. Using Monte Carlo simulations, the variation of the radon concentration with height before the earthquake was used to simulate and calculate the ionization rates of radioactive radon decay products at different heights. We derived the atmospheric vertical electric field from -593 to -285 V/m from the surface to 10 m before the earthquake by solving the system of convection-diffusion partial equations for positive and negative particles. Moreover, negative atmospheric electric field anomalies were observed in both Wenjiang and Pixian before the Wenchuan earthquake on 12 May, with peaks of -600 V/m in Pixian and -200 V/m in Wenjiang. The atmospheric electric field data obtained from the simulation were shown to be in excellent concordance with the observed data of the Wenchuan earthquake. The physical mechanism can provide theoretical support for the atmospheric electric field anomaly as an earthquake precursor.

Keywords: atmospheric electric field; earthquake precursors; radon concentration

Citation: Han, R.; Cai, M.; Chen, T.; Yang, T.; Xu, L.; Xia Q.; Jia, X.; Han, J. Preliminary Study on the Generating Mechanism of the Atmospheric Vertical Electric Field before Earthquakes. *Appl. Sci.* **2022**, *12*, 6896. https://doi.org/10.3390/app12146896

Academic Editors: Eleftheria E. Papadimitriou and Alexey Zavyalov

Received: 7 June 2022
Accepted: 6 July 2022
Published: 7 July 2022

Publisher's Note: MDPI stays neutral with regard to jurisdictional claims in published maps and institutional affiliations.

1. Introduction

Earthquakes(EQs) are one of the most serious natural disasters. There have been hundreds of Ms $\geqslant$ 7 EQs worldwide from 2000 to now [1]. EQs cause direct damage to the buildings in the residential areas in the center of the EQ and cause different levels of damage to buildings of different materials and structures [2,3]. Secondary disasters such as tsunamis and landslides triggered by EQs cause enormous human casualties, construction damages, and property losses [4–7]. Increasingly vital research on pre-EQ disaster phenomena was conducted by researchers around the world to avoid significant losses caused by EQ disasters. For instance, Namgaladze, Karpov, and Knyazeva have found that low-frequency electromagnetic radiation is emitted in the preparation period for EQs in the ionosphere [8]. The abnormal phenomenon has been observed by ground-based stations and satellite detectors [8,9]. Moreover, the movement of the crust during the preparation period of many EQs also caused anomalous increases in radon concentrations, including water radon and gas radon [10–14]. In addition, there were also other observable anomalies that the researchers studied before the EQ, such as thermal anomalies and ground light phenomena [15,16]. However, the pre-EQ anomalies mentioned above generally occur on a time scale of a few days to a few months during the preparation period of an EQ, and it is impossible to predict the occurrence time and location of an EQ accurately. Furthermore, the measurement of radon concentration and low-frequency electromagnetic waves are associated with high uncertainties.

In recent decades, EQ forecasters have repeatedly detected negative abnormal signals in the atmospheric vertical electric field in front of the EQ. During fair weather and excluding weather conditions such as haze, thunder, and dust, this signal has occurred at the hourly level ahead of an EQ, and numerous observations have demonstrated the accuracy of this anomalous signal [17–19]. Although the atmospheric electric field anomaly was proven to be an important precursor before EQs, the cause of its formation was not clearly understood. In previous work [20], researchers claimed that atmospheric vertical electric field anomalies were caused by radon ionization of the air; however, the authors failed to propose a specific mechanism of interaction from radon concentration to atmospheric electric field. A mechanism for the formation of anomalous electrode effects by near-surface air ionization was proposed by Boyarchuk, Lomonosov, and Pulinets in 1997 [21]. They obtained an anomalous reduction effect of the atmospheric electric field within 1 m of the surface. Existing studies have confirmed that radon ionization of air contributes to the atmospheric vertical electric field anomaly, and some explorations of radon ionization atmospheric processes have also been conducted. However, up to now, there is no quantitative study on the signal of atmospheric vertical electric field anomaly caused by radon concentration.

In this paper, the physical mechanism of the formation of negative atmospheric vertical electric field anomalies during the EQ preparation period is proposed by studying the pre-EQ radon gas concentration data. This paper aims to obtain detailed information on the variation of the atmospheric electric field with height obtained by solving the proposed physical model using radon gas and atmospheric electric field data before the Wenchuan EQ. The first part of this paper briefly analyzes the formation mechanism of the atmospheric vertical electric field before the earthquake and describes in detail the formation process of anomalous atmospheric electric fields in a progressive way, including the variation of radon concentration with height, simulating the ionization rate of radon stable decay daughter at different heights and solving the convective diffusion equation of positive and negative particles to obtain the atmospheric electric field. In the second part, the 2008 Wenchuan EQ in China is used as an example for simulation and calculation, and the correctness of the model is verified by taking the radon concentration before the Wenchuan EQ as input. In the model of this paper, a Monte Carlo simulation of the radon ionization rate at different heights is proposed for the first time, and the model is verified by using EQ cases. The electrode effect proposed by Boyarchuk, Lomonosov, and Pulinets only considered the effect of radon ionization in the atmosphere within 10 cm of the surface, and it was a small-scale model [21]. Moreover, Boyarchuk, Lomonosov, and Pulinets only pointed out that the electrode effect can cause atmospheric electric field anomalies, and they did not validate it for a particular EQ. This mechanism provides a theoretical explanation for the pre-EQ atmospheric vertical electric field anomaly as an EQ precursor.

2. Methodology

In the build-up to a strong EQ, the crushing and tearing of the Earth's crust occur during the crushing process. Cracks in the Earth's crust have been extended to the surface, resulting in numerous microcracks at the surface [22]. In this context, radon, a radioactive gas in the Earth's crust, diffuses into the atmosphere along with the cracks. As shown in Figure 1, the negative signal of the atmospheric vertical electric field due to the radon ionization process in the atmosphere is demonstrated. Radon diffuses from the Earth's crust to the atmosphere through microcracks and then undergoes alpha decay, beta decay, and gamma decay in the air. A high amount of positive and negative ion pairs are produced by ionizing alpha particles, electrons, and gamma-rays in the air. The positive and negative ion pairs diffuse and migrate in the air. Then eventually, the distribution of positive and negative particles reverses the normal downward positive atmospheric vertical electric field in fair weather.

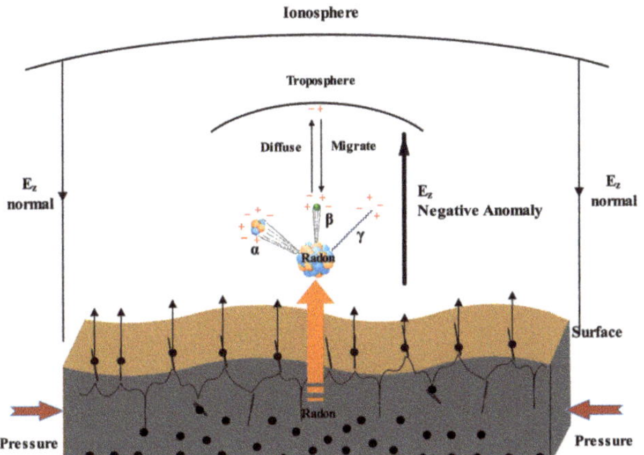

Figure 1. Schematic diagram of radon ionization atmosphere.

2.1. Radon Concentration

2.1.1. Vertical Distribution

In the troposphere over the mainland, the radioisotope of primary interest for small ion production is ^{222}Rn, with a half-life of 3.8 days. Radioactive radon is transported upward with atmospheric movement, resulting in radon concentration variation with height. We reviewed the ^{222}Rn in fair weather relative to altitude profile radioactive decay product distribution. For the fair weather mentioned above, Harrison and Nicoll (2018) defined fair weather in terms of the following detailed descriptions: (1) the visual range should be above 2 to 5 km, and the relative humidity should be less than 95%, (2) there are no low-level clouds and negligible cumulonimbus clouds, and (3) the wind speed at 2 m should be less than 7 m/s [23]. The average vertical distribution of ^{222}Rn in different parts of the continent in fair weather was demonstrated by the experimental results [24]. As shown in Figure 2, the experimental results and best-fit curves for radon gas concentrations at different altitudes are presented. The best-fit trend line is logarithmically described by the following Equation:

$$y = -2.351 \ln(x) + 13.377 \tag{1}$$

where x (dpm/m^3) is the Radon concentration and y (km) is the height. The fitting accuracy is $R^2 = 0.70453$.

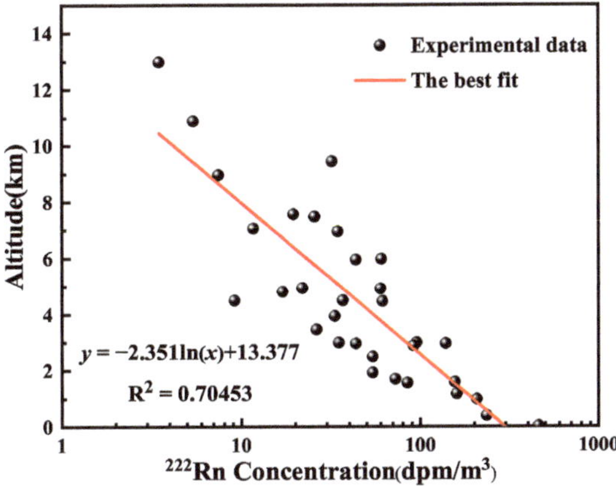

Figure 2. Radon concentration distribution at various altitudes.

2.1.2. Decay Products

Radioactive radon can undergo alpha decay, beta decay, and gamma decay in the air, producing many alpha particles, electrons, and gamma rays. A quantitative relationship between radon concentration and the corresponding decay products is first required to obtain the radon ionization rate at various heights. The decay paths of ^{222}Rn and its daughter are depicted in Figure 3. We simulated the decay of 100,000 radon in a vacuum using Monte Carlo simulation software and recorded the number of stable decay products. As shown in Table 1, the alpha to beta particles ratio in the decay products is 4:5.2.

Table 1. Stabilization products of radon decay.

Particle Type	Number	Average Energy	Range in Air
^{222}Rn	100,000	0	0
alpha	400,000	6.123 MeV	4.94 cm
e	520,866	263.4 keV	63.16 cm

The radon concentration is the number of alpha particles resulting from decay per cubic meter per minute. As a result, we deduced the variation in the concentration of radon's main decay products, alpha and beta particles, with the height from the radon distribution.

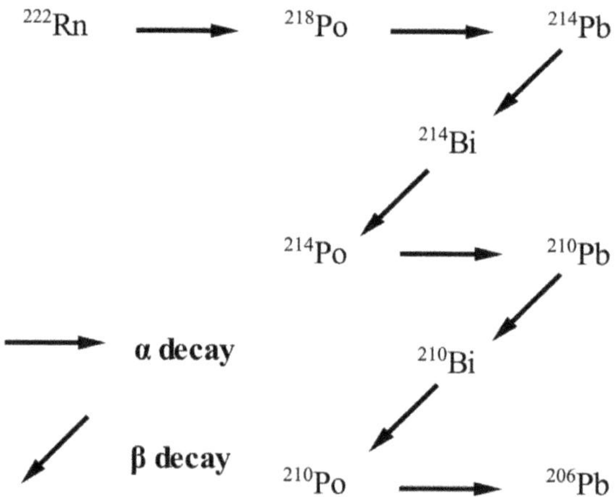

Figure 3. Decay paths of radon and its daughters.

2.2. Ionization Rate at Different Heights

To obtain the ionization rates caused by radon at different heights, we used Monte Carlo simulation software GEANT4 (Geometry And Tracking) for simulation [25]. The ionization rates of alpha and beta particles in the air were simulated from 0 to 100 m. A series of isotropic surface sources of alpha and beta particles, corresponding to the concentration of the corresponding particles at different heights, were set up separately. Following that, the ionization rates at different heights were calculated by combining the average ionization energy of the particles in the air with the energy deposition of detectors set at different heights.

2.2.1. Geometric Models and Particle Source

As shown in Figure 4, a geometric model was built that involved 38 identical detectors, recording energy deposition at different altitudes. Each detector is a thin slice with a radius of 5 m and a height of 1 mm. We configured one detector per 1 cm interval for 0–9 cm, one detector per 10 cm interval for 10–90 cm, one detector per 1 m interval for 1–9 m, and one detector per 10 m interval for 10–100 m. Since the atmospheric density varies very little within 0–100 m, we assumed sea level air for the world and the detector material. Sea level air is comprised of 0.0000124 C, 0.755267 N, 0.231781 O and 0.012827 Ar with a density of 1.20479×10^{-3} g/cm^3.

Whereas alpha particles have a range of 4.94 cm in air, all isotropic sources were arranged ±5 cm around the detector; if this distance was exceeded, alpha particles could never access the detector. Moreover, surface sources were established at 1 cm intervals to measure the alpha emitted within a volume of 5 m radius and 1 cm height at that location. For instance, the detector located at 1 m was surrounded by an alpha source of 5 m radius at 0.95, 0.96, 0.97, 0.98, 0.99, 1, 1.01, 1.02, 1.03, 1.04, and 1.05 m. The range for beta particles is 63.16 cm. Due to the longer range of the beta particles, we did not utilize all 38 detectors as we did for alpha particles. Only 10–100 m detectors were used, and a beta isotropic surface source was designed with a radius of 5 m and spacing of 10 cm. The beta surface sources around the ten detectors were positioned at ±60 cm. As with alpha particles, surface sources were placed at 10 cm intervals to measure the beta emitted within a volume of 5 m radius by 10 cm height at that location. The number of particles emitted per second from the source at different heights was fixed from the distribution of alpha and beta particles with height.

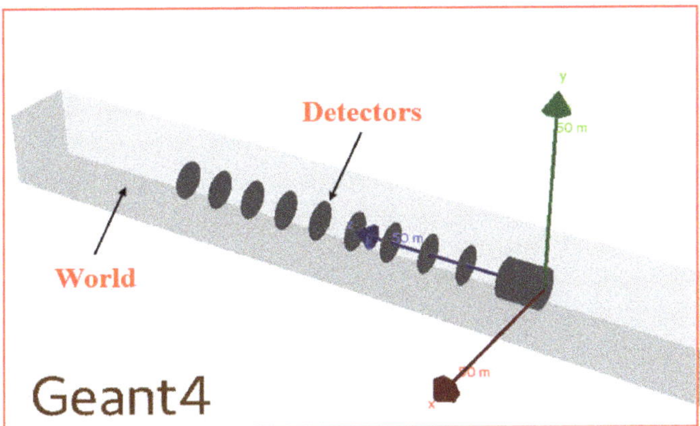

Figure 4. Geant4 geometric model.

2.2.2. Physical Lists and Data Statistics

The stable products of ^{222}Rn decay are alpha with an average energy of about 6.123 MeV and beta with an average energy of about 263.4 keV. Alpha and beta particles entering the air undergo complex physical processes that produce many secondary charged particles. Secondary particles are tracked to the lowest possible energy to improve the accuracy of the simulated process. Different physical processes occur when particles of different types and energies interact with air. This paper considered the following physical processes: low-energy electromagnetic processes, hadron processes, decay processes, and photonuclear processes. A secondary particle generation threshold of 250 eV was imposed within the entire model.

Deposition of energy at different height locations was recorded by distinct detectors in the model. First, the total energy deposition for each detector was calculated by multiplying the average energy deposition by the number of simulated particles identified and the duration. Jesse discussed the generation of ions in the air by ionizing radiation [26]. He reported the mean energy required to generate an ionization in the air for alpha and beta particles, 35 eV for alpha particles and 33.8 eV for beta particles. Next, the number of positive and negative particle pairs deposited within each detector was obtained by dividing the total energy deposited by the average ionization energy of alpha and beta particles. Finally, the ionization rates of detectors at various heights were evaluated.

The following case is illustrated with an example of a detector. In this paper, we assumed that the average energy deposition of the detector is E eV, the number of simulated events is N pcs, the ionization duration event is t s, the total energy deposition is N_{tot} eV, the average ionization energy of particles in the air is W eV, the number of positive and negative particle pairs in the detector is N_i ion parts, and the ionization rate is Q cm^3·s. The technique outlined above was presented as Equation (2), and then Equation (3) was obtained.

$$E \times N = N_{\text{tot}}; \quad \frac{N_{\text{tot}}}{W} = N_i; \quad \frac{N_i \times 10^{-6}}{\pi \times 5^2 \times t} = Q \tag{2}$$

$$\frac{E \times N}{W} \times \frac{10^{-6}}{\pi \times 5^2 \times t} = Q \tag{3}$$

2.3. Physical Lists and Data Statistics

Positive and negative particles are ionized by radon decay particles at different heights. However, the different mobility of differently charged ions leads to the inability of positive and negative particles to fully compensate for each other. Then, the reverse atmospheric vertical electric field is produced in the case of strong turbulent conditions diffusion. For simulating the electrical state of the turbulent ground layer, a steady-state kinetic equation was applied in this paper.

$$-\frac{\partial}{\partial z}\left[D_t \frac{\partial n_+}{\partial z}\right] + b_+ \frac{\partial}{\partial z}(En_+) = Q - \alpha n_+ n_-$$
$$-\frac{\partial}{\partial z}\left[D_t \frac{\partial n_-}{\partial z}\right] - b_- \frac{\partial}{\partial z}(En_-) = Q - \alpha n_+ n_- \tag{4}$$
$$\frac{\partial E}{\partial z} = 4\pi e(n_+ - n_-)$$

Here, $n\pm$ is the concentration of positive and negative ions; $D_t(z) = (Kz + \gamma)/(z + \beta)$ is the turbulent diffusion coefficient; $b\pm$ is the mobility of positive and negative ions; Q is the ionization rate, which is mainly caused by radon ionization of air near the ground; α is the ion recombination coefficient; e is the electron charge; and ε_0 is the vacuum dielectric constant.

The fluctuations in positive and negative ion concentrations and the atmospheric vertical electric field with height are provided by solving Equation (4). It proves that the formation of the atmospheric vertical anomaly is caused by the ionization of radioactive radon decay products into the atmosphere.

3. Wenchuan EQ Analysis and Discussion

3.1. Radon Concentration before the Wenchuan EQ

The Wenchuan Ms8.0 EQ in 2008 in China was used as an example for our study to verify the validity of the mechanism. The distances between the epicenter of the Wenchuan EQ and the detection stations at Pixian, Wenjiang, and Guzan are 49, 52, and 202 km, respectively. As shown in Figure 5a, before the Wenchuan EQ, Guzan shows a significant increase in the daily median radon concentration from May 7 to the moment of the EQ [19]. As shown in Figure 5b, the visibility and low cloud cover at the Wenjiang and Pixian stations on the day of the EQ. As shown in Figure 5c, the wind velocity at Wenjiang and Pixian stations throughout the day on 12 May. Therefore, it can be determined that it was fair weather in the epicenter area of the Wenchuan EQ on 12 May. The atmospheric vertical electric field is disturbed only a little by meteorological effects.

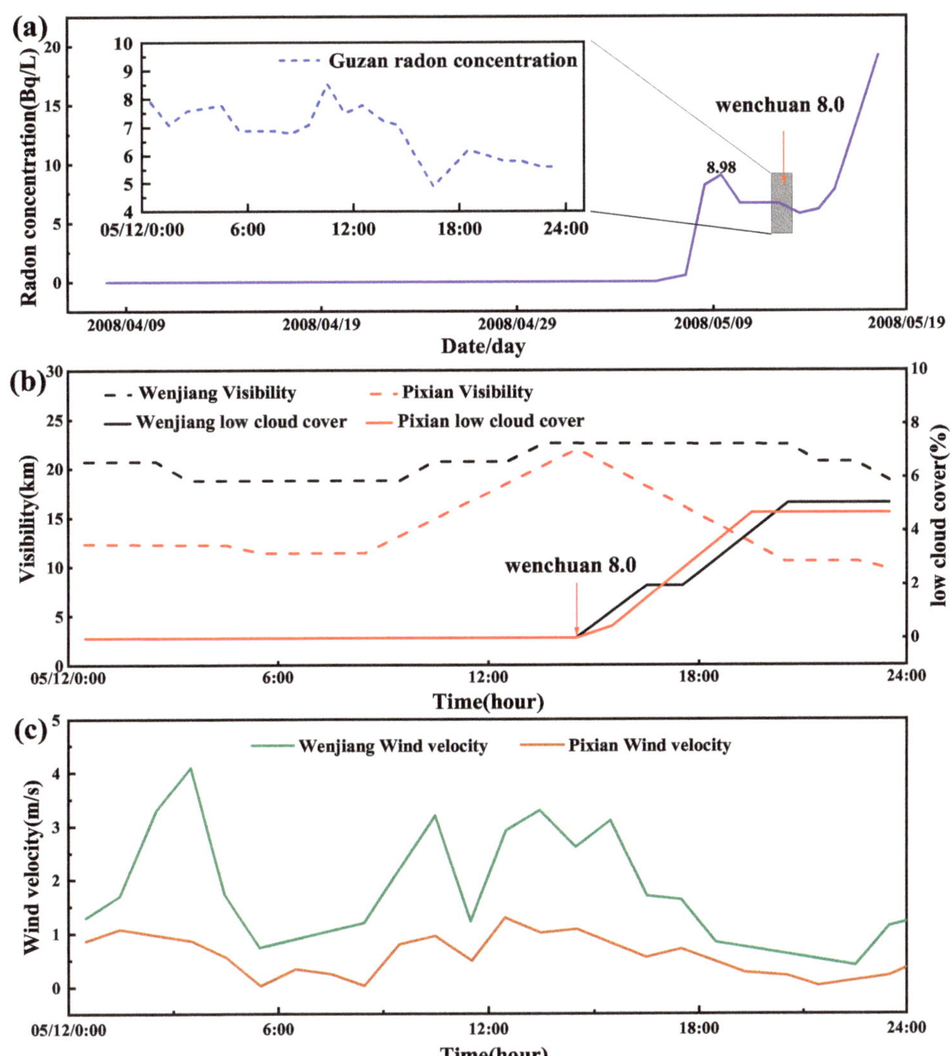

Figure 5. (**a**) Variation of radon concentration at Guzan station before the Wenchuan EQ. (**b**) Visibility and low cloud cover at Wenjiang and Pixian stations before the Wenchuan EQ. (**c**) Wind velocity at Wenjiang and Pixian stations before the Wenchuan EQ.

It can be seen from Figure 5a that the radon concentration before the Wenchuan EQ peaked at 8.98 Bq/L on May 10. For receiving the variation of radon concentration with height before the Wenchuan EQ, it was considered that the peak radon concentration before the Wenchuan EQ was the ground, y = 0 km, radon concentration, i.e., (8.98 Bq/L, 0 km). The vertical distribution of radon concentration varies with the changes in meteorological parameters [27]. For the simulation in this paper, we considered the constant migration and turbulent diffusion coefficients. Thus the slope of the natural logarithm of radon concentration with height was considered to be the same as the curve of Equation (1). The intercept of the curve is determined from the radon concentration data before Wenchuan earthquake (8.98 Bq/L, 0 km). The radon concentration variation with height was obtained as:

$$y = -2.351 \ln(x) + 31.026 \tag{5}$$

where x (dpm/m^3) is the Radon concentration and y (km) is the height.

The concentration of alpha particles at different heights was obtained from Equation (5), and the variation of beta particles with height was obtained from the alpha to beta ratio in Section 2.1.2. Since we configured the Monte Carlo simulation alpha particle source equivalent to the number of alpha particles generated in a cylinder with a radius of 5 m and a height of 1 cm in 1 h, it was possible to obtain the number of alpha particles simulated at different heights from the alpha particle concentration at different heights.

3.2. Results and Discussion

3.2.1. Monte Carlo Simulation of Ionization Processes

Based on the variation of alpha and beta ion concentrations with height, the Monte Carlo simulations were used to simulate their ionization rates at different heights. For the Wenchuan EQ alpha and beta particle concentrations given in Section 3.1, the variation of the ionization rate with height was derived from the simulation method in the methodology, as shown in Tables 2 and 3 and Figure 5. Tables 2 and 3 show that the heights and simulation number are the simulation input conditions, and the average energy deposition is the output from the GEANT4 calculation. It was noteworthy that the total energy deposition is the case of one-hour accumulated radon concentration. Next, the ionization rate variation was obtained by Equation (3).

Table 2. Alpha particle simulation parameters and results.

Altitude/m	Average Energy Deposition/keV	Simulation Number	Total Energy Deposition/eV	Ion Pairs	Ionization Rate/cm$^3\cdot$s
0.1	60.8249	2.7924×10^8	1.6985×10^{13}	4.8527×10^{11}	1716.3
1	60.8241	2.7913×10^8	1.6978×10^{13}	4.8508×10^{11}	1715.62
2	60.824	2.7901×10^8	1.6971×10^{13}	4.8487×10^{11}	1714.89
3	60.8242	2.7889×10^8	1.6963×10^{13}	4.8467×10^{11}	1714.17
4	60.8238	2.7877×10^8	1.6956×10^{13}	4.8446×10^{11}	1713.43
5	60.824	2.7866×10^8	1.6949×10^{13}	4.8426×10^{11}	1712.7
6	60.8241	2.7854×10^8	1.6942×10^{13}	4.8405×10^{11}	1711.98
7	60.8241	2.7842×10^8	1.6935×10^{13}	4.8384×10^{11}	1711.25
8	60.8239	2.7830×10^8	1.6927×10^{13}	4.8364×10^{11}	1710.52
9	60.8239	2.7818×10^8	1.6920×10^{13}	4.8343×10^{11}	1709.79
10	60.8237	2.7806×10^8	1.6913×10^{13}	4.8322×10^{11}	1709.06
20	60.823	2.7688×10^8	1.6841×10^{13}	4.8117×10^{11}	1701.78
30	60.8226	2.7571×10^8	1.6769×10^{13}	4.7912×10^{11}	1694.55
40	60.8215	2.7454×10^8	1.6698×10^{13}	4.7708×10^{11}	1687.32
50	60.8206	2.7337×10^8	1.6627×10^{13}	4.7505×10^{11}	1680.14
60	60.8196	2.7221×10^8	1.6556×10^{13}	4.7302×10^{11}	1672.98
70	60.8209	2.7106×10^8	1.6486×10^{13}	4.7103×10^{11}	1665.91
80	60.8216	2.6991×10^8	1.6416×10^{13}	4.6903×10^{11}	1658.86
90	60.8207	2.6876×10^8	1.6346×10^{13}	4.6703×10^{11}	1651.8
100	60.8228	2.6762×10^8	1.6277×10^{13}	4.6507×10^{11}	1644.84

Table 3. Beta particle simulation parameters and results.

Altitude/m	Average Energy Deposition/keV	Simulation Number	Total Energy Deposition/eV	Ion Pairs	Ionization Rate/cm^3·s
10	462.855	4.27×10^9	9.89×10^{11}	2.93×10^{10}	103.45
20	462.958	4.25×10^9	9.85×10^{11}	2.91×10^{10}	103.04
30	462.936	4.24×10^9	9.80×10^{11}	2.90×10^{10}	102.59
40	462.963	4.22×10^9	9.76×10^{11}	2.89×10^{10}	102.17
50	462.893	4.20×10^9	9.72×10^{11}	2.88×10^{10}	101.72
60	462.878	4.18×10^9	9.68×10^{11}	2.86×10^{10}	101.28
70	462.954	4.16×10^9	9.64×10^{11}	2.85×10^{10}	100.87
80	462.938	4.15×10^9	9.60×10^{11}	2.84×10^{10}	100.44
90	462.892	4.13×10^9	9.56×10^{11}	2.83×10^{10}	100
100	462.978	4.11×10^9	9.52×10^{11}	2.82×10^{10}	99.59

As shown in Figure 6, the black dots are the results obtained from the simulation, and the red curve is the best-fit result. An exponential fit was used as the fitting function to match the actual ionization rate variation more closely with height. As a result, the ionization rate fitted alpha and beta ionization rate curves were Equations (6) and (7).

$$Q_\alpha = e^{21.263 - 4.25 \times 10^{-4} z} \tag{6}$$

$$Q_\beta = e^{18.459 - 4.245 \times 10^{-4} z} \tag{7}$$

here the ionization rate fit is close to perfect, where the Q_α fitting accuracy $R^2 = 1$ and the β fitting accuracy $R^2 = 0.99995$. Therefore, the variation of the ionization rate with height was $Q_\alpha + Q_\beta$ caused by radon.

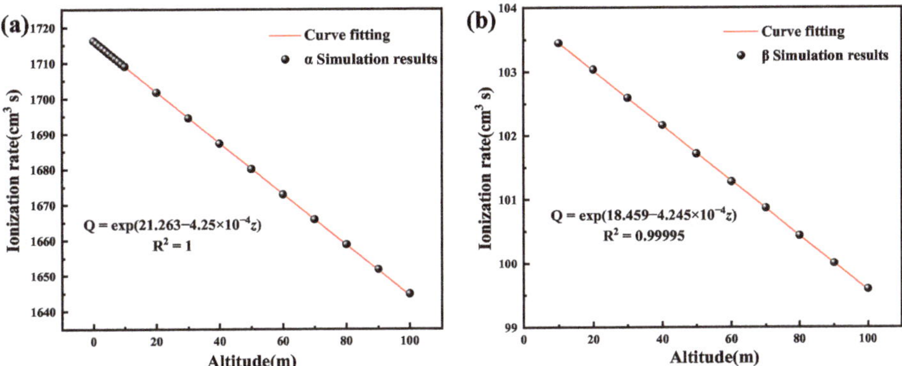

Figure 6. (**a**) Alpha particle simulation data with the best fit. (**b**) Alpha particle simulation data with the best fit.

3.2.2. Simulation Results and Discussion

From the previous work, the variation of the ionization rate with height due to radon was given as $Q = Q_\alpha + Q_\beta$. In this model, the mobility $b_\pm$ of positive and negative ions is set to 2.5×10^{-4} and 2.2×10^{-4} m^2·(V·s)$^{-1}$ [28], respectively, and the ion recombination coefficient α is set to 1.6×10^{-12}. For the turbulent diffusion coefficient $D_t(z) = (Kz + \gamma)/(z + \beta)$, where β is set to 10 m, γ is set to 5×10^{-5} m^3·s^{-1} and the strong turbulence coefficient K is set to 5 m^2·s^{-1} [21].

For the set of partial differential Equation (4), the boundary conditions near the Earth's surface were considered as:

$$n_+(z=0) = n_-(z=0) = 450\,\text{cm}^{-3}, n_1(10) = n_2(10) = (Q(10)/\alpha)^{1/2} \tag{8}$$

$$E(10) = \frac{j_0}{\lambda_{10}}, \lambda_{10} = e \times (b_+ n_+(10) + b_- n_-(10)) \tag{9}$$

In Equation (8), j_0 is the current density at $z = 0$. For the Wenchuan EQ, Kuo et al. developed a model for the piezoelectric effect in rocks and have calculated a current density of I_{rock} = 500 nA·m^{-2} for an EQ fault area of A_{rock} = 200 × 30 km^2 [29]. The total current input to the atmosphere equals the current output from rocks.

$$I_{air} = A_{air} \times J_0 = I_{rock} = A_{rock} \times J_{rock} \tag{10}$$

The electromagnetic field parameters of a large area of China were abnormal before the Wenchuan EQ. Abnormal electromagnetic signals were observed at the Gao Bei station in Hebei Province, 1440 km from the epicenter, on May 9, three days before the main EQ. Electromagnetic anomalies were also observed at the Qingxian station in Hebei province, 1439 km from the epicenter, and at Ningjin county in Hebei province, 1241 km from the epicenter. However, there was no electromagnetic signal anomaly observed in Sanhe County, Hebei Province, 1541 km from the epicenter [30]. Thus, it was assumed that the area of the pre-EQ air anomaly region A = 1440 × 1440 km^2 calculated by Equation (9) yields $J_0 \approx 1.5$ nA.

The partial differential Equation (4) was solved mathematically by the boundary conditions given for the Wenchuan EQ. As shown in Figure 7, we solve the variation of the positive and negative particle concentrations and the atmospheric vertical electric field with height. From the data in Figure 7, the separation of positive and negative particles can be seen.

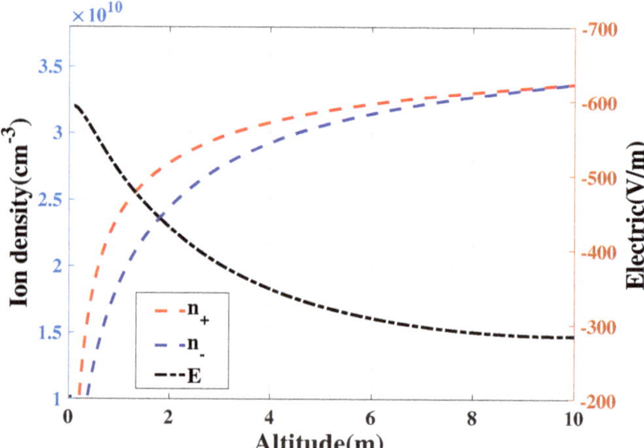

Figure 7. Variation of positive and negative particle concentration and atmospheric vertical electric field with height.

As shown in Figure 7, we calculate that the atmospheric vertical electric field from the surface to 10 m is −593 to −285 V/m. Negative atmospheric electric field anomalies were observed at both Wenjiang and Pixian on 12 May, with a peak of −600 V/m at Pixian and −200 V/m at Wenjiang [19]. The data of atmospheric electric field sounding at Wenjiang station were consistent with the data obtained from the model calculation in this paper. Moreover, there were some deviations in the data of the Pixian station. The results were consistent with those calculated by Boyarchuk, Lomonosov, and Pulinets, which showed an anomaly of reduction in the atmospheric electric field [21]. The difference was that the calculations used as input conditions were the atmospheric electric field on fair weather and the radon ionization source within 10 cm of the ground. An analogous process was observed in the case of measuring the electric field in the underground nuclear weapons test area: the electric field drastically decreased at the time of the explosion [31]. In this case, the near-surface layer was ionized by the fission fragments emerging from the soil. This experiment was similar to the mechanism of this paper and was a direct demonstration of this paper. Since we could not obtain the detailed radon concentration detection data before the Wenchuan EQ, the radon concentration used in the calculation was based on the gradient of the natural logarithm of radon concentration with height in fair weather. The variation of radon concentration with height was obtained using the radon concentration at the ground level y = 0 km before the Wenchuan EQ. Moreover, the parameters in the equation of motion were according to the average parameters on fair weather, which might have some errors with the actual situation. Therefore, the parameters used in the simulation deviated from the actual parameters. If the actual meteorological parameters before the EQ and the variation of radon concentration with height were available in the simulation, the accuracy of the simulation would be improved. In general, these results still indicate that the mechanism of our initial exploration matches well with the measured values and indicates a good validity of our model.

4. Conclusions

A research method based on a combination of Monte Carlo simulations and the numerical solution was proposed to study the mechanism of the formation of the atmospheric vertical electric field of the short-term precursor signals of EQs. An innovative Monte Carlo simulation method was established to calculate the ionization rate at different heights from data on the variation of radon concentration with height before an EQ. Next, the ionization rate data were combined with the partial differential equations describing the motion of positive and negative particles to numerically solve for the variation of positive and negative particles and radon concentration with height. Finally, the model was validated with the 2008 Wenchuan EQ in China, which had both radon concentration data and atmospheric electric field data before the EQ. In addition, the ionization of alpha and beta particles at different heights, which were the main decay stabilization products of radon, was considered in the model. Alpha particles were the best ionized of the radon decay products, and their fitted ionization rates decayed exponentially with height. The results showed that the atmospheric electric field results obtained from the simulations compounded well with the atmospheric electric field data detected before the Wenchuan EQ. The atmospheric electric field measured by the Wenchuan EQ is direct evidence of the correctness of the model. Moreover, our atmospheric electric field simulations' results were consistent with those obtained by Boyarchuk, Lomonosov, and Pulinets. In this paper, Holzer's experiments on the atmospheric electric field effects of nuclear explosions in 1972 fundamentally proved the feasibility of the atmospheric electric field anomalies generated by radon ionization. Therefore, the mechanism model can effectively explain the mechanism of atmospheric electric field anomalies before EQ.

Author Contributions: Conceptualization, R.H., M.C., T.C. and J.H.; methodology, R.H., M.C. and T.C.; software, T.Y., L.X. and X.J.; validation, R.H., M.C. and J.H.; formal analysis, R.H. and Q.X.; investigation, R.H., M.C. and T.C.; resources, M.C., T.C. and J.H.; data curation, T.Y.; writing—original draft preparation, R.H.; writing—review and editing, R.H. and M.C.; visualization, R.H.; supervision, T.C. and J.H.; project administration, M.C. and J.H.; funding acquisition, M.C. All authors have read and agreed to the published version of the manuscript.

Funding: This work was supported by the National Space Science Center director fund (Grant No.E0PD41A11S); "Climbing Plan" of National Space Science Center, Chinese Academy of Sciences(Grant No.E1PD30047S); and Strategic Priority Research Program of Chinese Academy of Sciences (Grant No.XDA17010301).

Institutional Review Board Statement: Not applicable.

Informed Consent Statement: Not applicable.

Data Availability Statement: The data used to support the findings of this study are available from the corresponding author upon request.

Acknowledgments: We acknowledge for the data resources from "National Space Science Data Center, National Science & Technology Infrastructure of China (https://www.nssdc.ac.cn)".

Conflicts of Interest: The authors declare no conflict of interest.

Reference

1. China Earthquake Network Center. Available online: https://news.ceic.ac.cn/ (accessed on 25 February 2022).
2. Li, S.Q.; Liu, H.B. Vulnerability prediction model of typical structures considering empirical seismic damage observation data. *Bull. Earthq. Eng.* **2022**. [CrossRef]
3. Bilgin, H.; Shkodrani, N.; Hysenlliu, M.; Ozmen, H.B.; Isik, E.; Harirchian, E. Damage and performance evaluation of masonry buildings constructed in 1970s during the 2019 Albania earthquakes. *Eng. Fail. Anal.* **2022**, *131*, 105824. [CrossRef]
4. Shibusawa, H. A dynamic spatial CGE approach to assess economic effects of a large earthquake in China. *Prog. Disaster Sci.* **2020**, *6*, 100081. [CrossRef]
5. Rose, A.; Benavides, J.; Chang, S.E.; Szczesniak, P.; Lim, D. The regional economic impact of an earthquake: Direct and indirect effects of electricity lifeline disruptions. *J. Reg. Sci.* **1997**, *37*, 437–458. [CrossRef]
6. Barone, G.; Mocetti, S. Natural disasters, growth and institutions: A tale of two earthquakes. *J. Urban Econ.* **2014**, *84*, 52–66. [CrossRef]
7. Yeh, H.; Sato, S.; Tajima, Y. The 11 March 2011 East Japan earthquake and tsunami: Tsunami effects on coastal infrastructure and buildings. *Pure Appl. Geophys.* **2013**, *170*, 1019–1031. [CrossRef]
8. Namgaladze, A.; Karpov, M.; Knyazeva, M. Seismogenic disturbances of the ionosphere during high geomagnetic activity. *Atmosphere* **2019**, *10*, 359. [CrossRef]
9. Ye, Q.; Fan, Y.; Du, X.; Cui, T.; Zhou, K.; Singh, R.P. Diurnal characteristics of geoelectric fields and their changes associated with the Alxa Zuoqi MS5. 8 earthquake on 15 April 2015 (Inner Mongolia). *Earthq. Sci* **2018**, *31*, 35–43. [CrossRef]
10. Holub, R.; Brady, B. The effect of stress on radon emanation from rock. *J. Geophys. Res. Solid Earth* **1981**, *86*, 1776–1784. [CrossRef]
11. Walia, V.; Virk, H.S.; Bajwa, B.S. Radon precursory signals for some earthquakes of magnitude> 5 occurred in NW Himalaya: An overview. *Pure Appl. Geophys.* **2006**, *163*, 711–721. [CrossRef]
12. Kim, J.; Lee, J.; Petitta, M.; Kim, H.; Kaown, D.; Park, I.W.; Lee, S.; Lee, K.K. Groundwater system responses to the 2016 ML 5.8 Gyeongju earthquake, South Korea. *J. Hydrol.* **2019**, *576*, 150–163. [CrossRef]
13. Papachristodoulou, C.; Stamoulis, K.; Ioannides, K. Temporal variation of soil gas radon associated with seismic activity: A case study in NW Greece. *Pure Appl. Geophys.* **2020**, *177*, 821–836. [CrossRef]
14. Nevinsky, I.; Tsvetkova, T.; Dogru, M.; Aksoy, E.; Inceoz, M.; Baykara, O.; Kulahci, F.; Melikadze, G.; Akkurt, I.; Kulali, F.; et al. Results of the simultaneous measurements of radon around the Black Sea for seismological applications. *J. Environ. Radioact.* **2018**, *192*, 48–66. [CrossRef]
15. Derr, J.S.; St-Laurent, F.; Freund, F.; Thériault, R. Earthquake lights. *Encycl. Earth Sci. Ser. Encycl. Solid Earth Geophys.* **2011**, *5*, 165–167.
16. Pulinets, S.; Ouzounov, D.; Karelin, A.; Boyarchuk, K.; Pokhmelnykh, L. The physical nature of thermal anomalies observed before strong earthquakes. *Phys. Chem. Earth, Parts A/B/C* **2006**, *31*, 143–153. [CrossRef]
17. Pulinets, S.; Alekseev, V.; Legen'ka, A.; Khegai, V. Radon and metallic aerosols emanation before strong earthquakes and their role in atmosphere and ionosphere modification. *Adv. Space Res.* **1997**, *20*, 2173–2176. [CrossRef]
18. Choudhury, A.; Guha, A.; De, B.K.; Roy, R. A statistical study on precursory effects of earthquakes observed through the atmospheric vertical electric field in northeast India. *Ann. Geophys.* **2013**, *56*, R0331.
19. Jin, X.; Zhang, L.; Bu, J.; Qiu, G.; Ma, L.; Liu, C.; Li, Y. Discussion on anomaly of atmospheric electrostatic field in Wenchuan Ms8. 0 earthquake. *J. Electrost.* **2020**, *104*, 103423. [CrossRef]
20. Pulinets, S.; Ouzounov, D.; Ciraolo, L.; Singh, R.; Cervone, G.; Leyva, A.; Dunajecka, M.; Karelin, A.; Boyarchuk, K.; Kotsarenko, A. Thermal, atmospheric and ionospheric anomalies around the time of the Colima M7. 8 earthquake of 21 January 2003. *Ann. Geophys.* **2006**, *24*, 835–849. [CrossRef]
21. Boyarchuk, K.; Lomonosov, A.; Pulinets, S. Electrode effect as an earthquake precursor. *Bull. Russ. Acad. Sci. Phys.* **1997**, *61*, 175–179.
22. Main, I.G.; Bell, A.F.; Meredith, P.G.; Geiger, S.; Touati, S. The dilatancy–diffusion hypothesis and earthquake predictability. *Geol. Soc. Lond. Spec. Publ.* **2012**, *367*, 215–230. [CrossRef]
23. Harrison, R.; Nicoll, K. Fair weather criteria for atmospheric electricity measurements. *J. Atmos. Sol.-Terr. Phys.* **2018**, *179*, 239–250. [CrossRef]
24. Martell, E.A. Enhanced ion production in convective storms by transpired radon isotopes and their decay products. *J. Geophys. Res. Atmos.* **1985**, *90*, 5909–5916. [CrossRef]

25. Geant4: A Toolkit for the Simulation. Available online: https://geant4.web.cern.ch/ (accessed on 25 February 2022).
26. Jesse, W.P. Precision measurements of W for polonium alpha particles in various gases. *Radiat. Res.* **1968**, *33*, 229–237. [CrossRef]
27. Vargas, A.; Arnold, D.; Adame, J.A.; Grossi, C.; Hernández-Ceballos, M.A.; Bolivar, J.P. Analysis of the vertical radon structure at the Spanish "El Arenosillo" tower station. *J. Environ. Radioact.* **2015**, *139*, 1–17. [CrossRef]
28. Riousset, J.A.; Pasko, V.P.; Bourdon, A. Air-density-dependent model for analysis of air heating associated with streamers, leaders, and transient luminous events. *J. Geophys. Res. Space Phys.* **2010**, *115*. [CrossRef]
29. Kuo, C.; Huba, J.; Joyce, G.; Lee, L. Ionosphere plasma bubbles and density variations induced by pre-earthquake rock currents and associated surface charges. *J. Geophys. Res. Space Phys.* **2011**, *116*. [CrossRef]
30. Li, M.; Lu, J.; Parrot, M.; Tan, H.; Chang, Y.; Zhang, X.; Wang, Y. Review of unprecedented ULF electromagnetic anomalous emissions possibly related to the Wenchuan M S= 8.0 earthquake, on 12 May 2008. *Nat. Hazards Earth Syst. Sci.* **2013**, *13*, 279–286. [CrossRef]
31. Holzer, R.E. Atmospheric electrical effects of nuclear explosions. *J. Geophys. Res.* **1972**, *77*, 5845–5855. [CrossRef]

Article

Variability in the Statistical Properties of Continuous Seismic Records on a Network of Stations and Strong Earthquakes: A Case Study from the Kamchatka Peninsula, 2011–2021

Galina Kopylova [1],*, Victoriya Kasimova [1], Alexey Lyubushin [2] and Svetlana Boldina [1]

[1] Kamchatka Branch of the Geophysical Survey of the Russian Academy of Sciences (KB GS RAS), 683023 Petropavlovsk-Kamchatsky, Russia

[2] Schmidt Institute of Physics of the Earth, Russian Academy of Sciences, 123995 Moscow, Russia

* Correspondence: gala@emsd.ru

Abstract: A study of spatiotemporal variability and synchronization effects in continuous seismic records (seismic noise) on a network of 21 broadband seismic stations on the Kamchatka Peninsula was carried out in connection with the occurrence of strong earthquakes, $M = 7.2$–8.3. Data of 1-min registrations of the vertical movements velocity Earth's surface were used for constructing time series of daily values of the generalized Hurst exponent α^*, singularity spectrum support width $\Delta\alpha$, wavelet-based spectral exponent β, and minimum normalized entropy of squared orthogonal wavelet coefficients En for all stations during the observation period 2011–2021. Averaged maps and time-frequency diagrams of the spectral measure of four noise parameters' coherent behavior were constructed using data from the entire network of stations and by groups of stations taking into account network configuration, volcanic activity and coastal sea waves. Based on the distribution maps of noise parameters, it was found that strong earthquakes arose near extensive areas of the minimum values of α^*, $\Delta\alpha$, β, and the En maximum values advance manifestation during several years. The time-frequency diagrams revealed increased amplitudes of the spectral measure of the coherent behavior of the 4-dimensional time series (synchronization effects) before three earthquakes with $M_w = 7.5$–8.3 over months to about one year according to observations from the entire network of stations, as well as according to data obtained at groups of continental and non-volcanic stations. A less-pronounced manifestation of coherence effects diagrams plotted from data obtained at coastal and volcanic groups of stations and is apparently associated with the noisiness in seismic records caused by coastal waves and signals of modern volcanic activity. The considered synchronization effects correspond to the author's conceptual model of seismic noise behavior in preparation of strong earthquakes and data from other regions and can also be useful for medium-term estimates of the place and time of seismic events with $M_w \geq 7.5$ in the Kamchatka.

Keywords: seismic noise; time series; singularity spectrum; wavelet analysis; spectral measure of coherence; Kamchatka Peninsula; earthquake prediction

Citation: Kopylova, G.; Kasimova, V.; Lyubushin, A.; Boldina, S. Variability in the Statistical Properties of Continuous Seismic Records on a Network of Stations and Strong Earthquakes: A Case Study from the Kamchatka Peninsula, 2011–2021. *Appl. Sci.* **2022**, *12*, 8658. https://doi.org/10.3390/app12178658

Academic Editors: Eleftheria E. Papadimitriou and Alexey Zavyalov

Received: 28 July 2022
Accepted: 26 August 2022
Published: 29 August 2022

Publisher's Note: MDPI stays neutral with regard to jurisdictional claims in published maps and institutional affiliations.

1. Introduction

Continuous records of microseismic oscillations recorded by broadband seismometers at a network of stations may contain information about geodynamic processes in the Earth's interior, despite the main part of the energy of such oscillations being due to atmospheric and oceanic influences [1–3]. The conditions for the propagation of natural and man-made signals in the geological environment and its transmission properties can vary over time, including under the influence of earthquake preparation processes [4]. Statistical characteristics of microseisms can reflect changes in properties of the geological environment, so the study of their spatiotemporal structure is a promising direction in search for signs of strong earthquake preparation in seismically active regions.

New opportunities for studying the properties of continuous microseismic oscillations (below "seismic noise") are provided by the method developed by A.A. Lyubushin [5–17]. This method implements ideas about the general patterns of complex dynamic systems' behavior when they approach a critical state associated with irreversible changes. In particular, an increase in the synchronization (coherence) of changes in the properties of such systems is considered as one of the signs of the approach to a catastrophic state [12,18].

Synchronization effects can manifest themselves both in the behavior of natural objects and in experiments. For example, it was shown in [19] that weak mechanical periodic forcing applied to a model spring-slider system imitating the behavior of a seismogenic fault can lead to synchronization of slip events, documented as acoustic emission bursts with the frequency of a weak forcing.

Strong earthquakes are associated with abrupt changes in seismicity over vast areas and can be considered as "catastrophes" preceded by manifestations of various precursors, including an increase in the correlation radius of random fluctuations of microseismic oscillations. A physical description of the growth mechanism of microseismic synchronization is impossible due to the complexity of geological environmental structures and a large number of external influences and processes, many of which cannot be traced during the observation period. Therefore, the use of synchronous behavior in the statistical measures of a certain set of seismic noise parameters at a network of stations makes it possible to formally solve the problem of diagnosing preparation for a strong earthquake.

The possibility of the method for identifying areas of strong earthquakes in the Northwestern part of the Pacific seismic belt, North America, and the whole world with a lead time of months–a few years was demonstrated in [5,13,14,16]. In particular, regular spatiotemporal changes in seismic noise parameters before the 2003 and 2011 earthquakes with M_w = 8.3 and 9.0 were found according to data of the Japanese F-net network. Such changes showed the evolution of the statistical structure of microseismic oscillations to white noise [8].

This paper presents the results of applying the method of A.A. Lyubushin to the data of continuous recording of seismic signals at the network of stations in the Kamchatka Peninsula to search for signs of preparation for strong earthquakes in the variations of seismic noise taking into account the specific conditions of the region such as modern volcanic activity and coastal sea disturbances. To do this, the maps of the spatial distribution of four noise statistics were created and increased values of spectral coherence in the time series of noise parameters for 2011–2021 were analyzed in comparison with major earthquakes that have occurred. Diagrams of the spectral measure of coherence for four noise parameter time series were constructed using data from the entire network of stations and separately from data of northern, central, and southern groups of stations, as well as separately for stations allocated taking into account their remoteness from active volcanoes and separately for continental and coastal stations (Figure 1, Table 1). Using this approach, the features of the spatiotemporal distribution and coherent behavior of seismic noise statistics were found in connection with the preparation of local earthquakes with magnitudes of 7–8.

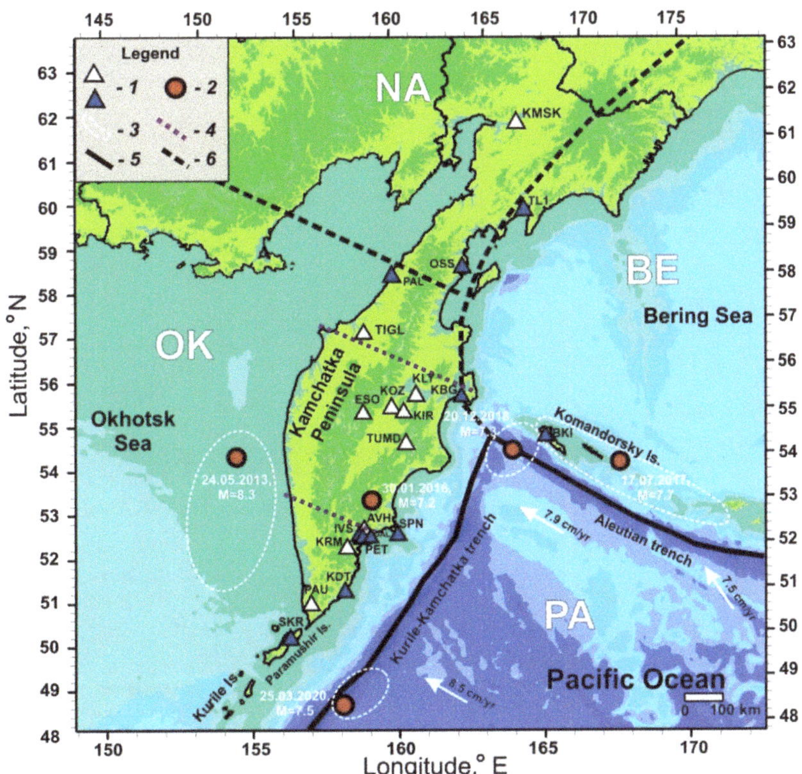

Figure 1. Map of the Kamchatka Peninsula showing the location of seismic stations (Table 1), earthquake epicenters with M_w = 7.2–8.3, and tectonic plate boundaries: 1—seismic stations with codes indicated: blue designates coastal stations, and white designates continental stations; 2—earthquake epicenters; 3—areas of earthquake sources according to [20–24]; 4—boundaries of the northern, central, and southern groups of stations (from top to bottom); 5—northwestern and northeastern boundaries of the considered area of the Pacific Oceanic Plate (PA); 6—boundary of the North American continental plate (NA) with PA and small lithospheric plates Beringia (BE) and Okhotsk (OK) [25,26]. White arrows indicate the direction of PA movement; numbers—the speed of TO movement [27].

Table 1. Seismic stations, their equipment [28], and the belonging of stations to selected groups.

Seismic Station	Station Code	Coordinates			Equipment	Frequency Range/ Registration Frequency, Hz	Belonging to Dedicated Groups of Stations		
		N °	E °	h asl, m					
Avacha	AVH	53.264	158.740	942	CMG-6TD	0.033–40/100	southern	continental	volcanic
Bering	BKI	55.194	165.984	12	CMG-3TB	0.0083–40/100	central	coastal	non-volcanic
Dal'niy	DAL	53.031	158.754	57	CMG-6TD	0.033–40/100	southern	coastal	non-volcanic
Institut	IVS	53.066	158.608	160	CMG-3TB	0.0083–40/100	southern	coastal	non-volcanic
Kamenskaya	KMSK	62.467	166.206	40	CMG-6TD	0.033–40/100	northern	continental	non-volcanic
Karymshina	KRM	52.828	158.131	100	CMG-3TB	0.033–40/100	southern	continental	non-volcanic
Kirisheva	KIR	55.953	160.342	1470	CMG-6TD	0.033–40/100	central	continental	volcanic
Klyuchi	KLY	56.317	160.857	35	KS2000	0.01–40/100	central	continental	volcanic
Kozyrevsk	KOZ	56.058	159.872	60	CMG-6TD	0.033–40/100	central	continental	volcanic

Table 1. *Cont.*

Seismic Station	Station Code	Coordinates			Equipment	Frequency Range/ Registration Frequency, Hz	Belonging to Dedicated Groups of Stations		
		N °	E °	h asl, m					
Krutoberegovo	KBG	56.258	162.713	30	CMG-3TB	0.0083–40/100	central	coastal	non-volcanic
Ossora	OSS	59.262	163.072	35	CMG-6TD	0.033–40/100	northern	coastal	non-volcanic
Palana	PAL	59.094	159.968	70	STS-2	0.0083–40/100	northern	coastal	non-volcanic
Pauzhetka	PAU	51.468	156.815	130	CMG-6TD	0.033–40/100	southern	continental	non-volcanic
Petropavlovsk	PET	53.023	158.650	100	STS-1	0.0027–10/20	southern	coastal	non-volcanic
Severo-Kuril'sk	SKR	50.670	156.116	30	CMG-3TB	0.0083–40/100	southern	coastal	volcanic
Tigil	TIGL	57.765	158.671	115	CMG-6TD	0.033–40/100	northern	continental	non-volcanic
Tilichiki	TL1	60.446	166.145	25	CMG-3TB	0.0083–40/100	northern	coastal	non-volcanic
Tumrok	TUMD	55.203	160.399	478	CMG-6TD	0.033–40/100	central	continental	volcanic
Khodutka	KDT	51.809	158.077	22	CMG-6TD	0.033–40/100	southern	coastal	non-volcanic
Shipunskiy	SPN	55.106	160.011	95	CMG-6TD	0.033–40/100	southern	coastal	non-volcanic
Esso	ESO	55.932	158.695	490	CMG-6TD	0.033–40/100	central	continental	non-volcanic

2. Region, Data, Method

2.1. Network of Stations

The Kamchatka Peninsula is one of the most seismically active regions of the Earth due to its location at the junction of the Pacific oceanic plate and the continental North American and Eurasian tectonic plates. Here, strong earthquakes occur within the Kamchatka fragment of the Kuril–Kamchatka seismic focal zone and the western fragment of the Aleutian seismic focal zone as a result of Pacific oceanic plate movement in the northwest direction (Figure 1). Since 2011, a network of digital broadband seismic stations (Table 1) has been operating here. This provided the technical conditions for studying the variations in microseismic oscillations and evaluating the potential of such data to search for precursors of strong earthquakes. The geometric dimensions of the network are characterized by the maximum distances between stations, 1450 km in the N–S direction (between SKR and KMSK stations) and 560 km in the W–E direction (between PAL and BKI stations). The average distance between two adjacent stations is 120 km (median 110 km) with a range of values from 10 to 240 km.

2.2. Stations Identification

2.2.1. Northern, Central and Southern Groups of Station

Stations located north of the Pacific Plate northern boundary (Figure 1) were identified as the northern group. The territory of the central and southern groups of stations is located within the Kuril–Kamchatka island arc. The region of the southern group belongs to the Kuril segment of the island arc, and the region of the central group belongs to the Kamchatka segment. The boundary between the southern and central groups of stations runs along the Nachikinskaya zone of transverse dislocations separating the Kamchatka and Kuril segments of the island arc [26,29].

2.2.2. Volcanic Stations

In the eastern part of the Kamchatka Peninsula there is an area of modern volcanism including active volcanoes. Broadband seismic stations are located in the vicinity of some volcanoes (Figure 2). During volcanic eruptions, volcanic and seismic signals can be recorded in seismic records at distances up to a few tens of kilometers from the centers of eruption [30,31]. Data on volcano activity from the Institute of Volcanology and Seismology of the Russian Academy of Sciences and the Kamchatka Branch of the Geophysical Survey of the Russian Academy of Sciences (http://geoportal.kscnet.ru/volcanoes/ (accessed on 31 December 2021); http://www.emsd.ru/~ssl/monitoring/main.htm (accessed on 31 December 2021)) were used when identifying a group of volcanic stations (Table 1).

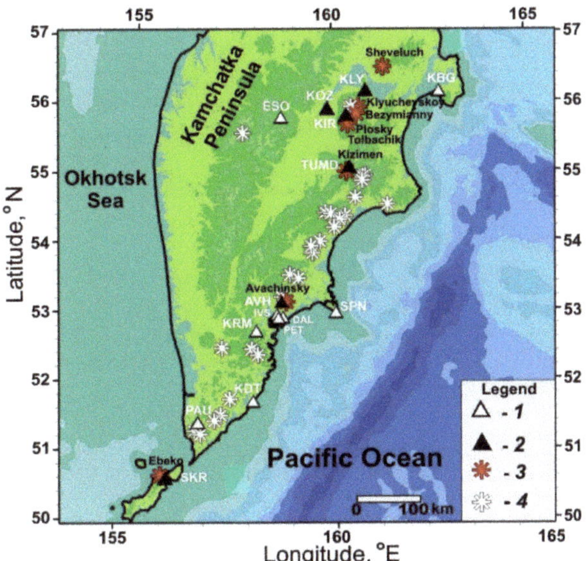

Figure 2. Location map of seismic stations and active volcanoes: 1—non-volcanic station, 2—volcanic station, 3—active volcano near which a seismic station is located, 4—active volcano at rest or near which there is no seismic station, 2011–2021.

In 2011–2021 the Sheveluch, Klyuchevskoy, Bezymyanny, Plosky Tolbachik, Kizimen, Avachinsky, and Ebeko volcanoes were in the stages of eruptions and increased fumarole and seismic activity (Figure 2). The high activity of northern volcanoes Sheveluch, Klyuchevskoy, Bezymyanny, Plosky, and Tolbachik was manifested in the records at seismic stations KIR, KLY, KOZ, and at TUMD station during the Kizimen volcano activation in 2011–2013. Weak signals from gas–ash emissions appeared in the records at SKR station during the Ebeko volcano eruption in 2016–2021 [32–34]. Brief episodes of increased activity from the Avachinsky volcano appeared in records at AVH station.

2.2.3. Coastal and Continental Stations

Stations located near seacoasts were allocated to the group of coastal stations. Other stations were allocated to the group of continental stations (Figure 1, Table 1). It should be noted that the diurnal and semidiurnal tidal harmonics are more pronounced in the spectra of hourly time series of noise from coastal stations, compared to stations located at a distance from seacoasts. In records from continental stations, mainly diurnal maxima are distinguished. Whereas diurnal and semidiurnal maxima, as well as maxima on the periods of powerful lunar waves M_2 (period 12.42 h) and O_1 (period 25.82 h), are clearly distinguished in the spectra of records from coastal stations [35,36]. Such features of seismic noise records showed a high degree of signal noisiness at coastal stations due to sea tidal waves.

The distributions of stations by groups, taking into account the number of stations affected by sea waves (coastal stations, 11 or 52% in total) and volcanic activity (volcanic stations, 6 or 29% in total), are presented in Table 2.

Table 2. Distribution of stations into groups, taking into account the number of stations affected by sea waves (coastal stations) and volcanic activity (volcanic stations).

Station Groups Influence Factors *	All Stations, N= 21 (100%)	Northern, N = 5 (100%)	Central, N = 7 (100%)	Southern, N = 9 (100%)	Non-Volcanic, N = 15 (100%)
coastal, n (%)	11 (52%)	3 (60%)	2 (29%)	6 (67%)	10 (67%)
volcanic, n (%)	6 (29%)	–	4 (57%)	2 (22%)	–

* n denotes the number of stations and the % of the total number of stations in group N is indicated in brackets.

2.3. Strong Earthquakes

Comparison of network linear dimensions (1450 × 560 km, average distance 120 km between neighboring stations) with the maximum linear sizes of earthquake sources L, km according to the formula $lgL = 0.440\, M_w - 1.289$ [37] and the calculated radii of level deformations 10^{-8} (R, km) at preparation of earthquakes with M_w: $R = 10^{0.43 \cdot Mw}$ [38], makes it possible to roughly estimate the network sensitivity to the preparation of earthquakes with magnitudes of at least 7–9. For such earthquakes, the maximum sizes of sources are about 60–450 km according to [37]. The radii of deformation sensitivity of 10^{-8}, equivalent to the areas of earthquake preparation with magnitudes 7–9 by [38], are 1–7 thousand km. Therefore, we believe that the configuration of the operating network (Figure 1) can cover completely or partially the areas of earthquake preparation in magnitude range 7–9. However, a more reasonable estimate of the network sensitivity for earthquake magnitudes can be obtained from experimental data. As will be shown in this study, the method used, based on the existing network of stations, may be sensitive to the preparation for earthquakes with a magnitude of at least 7.5.

Data on the five most powerful earthquakes of 2013–2020 with $M_w = 7.2$–8.3 that occurred during the observation period are presented in Table 3. The location of these earthquakes' epicenters and sources according to the aftershocks of the first day is presented in Figure 1.

2.4. Seismic Noise Parameters

To analyze spatiotemporal variations of seismic noise, time series of the daily values of four statistical parameters were used: generalized Hurst exponent α^*, singularity spectrum support width $\Delta\alpha$, wavelet-based spectral exponent β, and minimum normalized entropy of squared orthogonal wavelet coefficients En calculated for all stations during the 2011–2021 observation period [35,36,39]. The statistical parameters of seismic noise were estimated according to daily data of 1-min signal registration on BHZ channels for all stations (Figure 1, Table 1).

To calculate the noise parameters, an updated archive of 1-min continuous recordings was created at 21 stations of the KB GS RAS formed from daily fragments of records with a frequency of 100 and 20 Hz by averaging them in a window of one minute (1440 min values in each day) (http://www.ceme.gsras.ru/new/infres/, accessed on 31 December 2021). Below is a brief description of the four seismic noise statistics used in this study.

Multi-fractal parameters $\Delta\alpha$ and α^*. Consider some random fluctuation $x(t)$ on a time interval $[t - \delta/2,\, t + \delta/2]$ of length δ centered at the time point t. Consider the range $\mu(t, \delta)$ of random fluctuations on this interval, that is, the difference between the maximum and minimum values:

$$\mu(t, \delta) = max\; x(s) - min\; x(s), \tag{1}$$

when $t - \delta/2 \leq s \leq t + \delta/2$.

If we force $\delta \to 0$, then the value $\mu(t, \delta)$ will also tend to zero, but the speed of this decrease is important here. If the speed is determined by the law $\delta^{h(t)}$: $\mu(t, \delta) \sim \delta^{h(t)}$, or if there is a limit $h(t) = \lim\limits_{\delta \to 0} \frac{log(\mu(t,\delta))}{log(\delta)}$, then the $h(t)$ is called the Hölder-Lipschitz exponent.

If the value $h(t)$ does not depend on the moment of time t: $h(t) = const = H$, then the random fluctuation $x(t)$ is called mono-fractal, and the value H is called the Hurst exponent. If the Hölder-Lipschitz exponents $h(t)$ differ for different moments of time t, then the random fluctuation is called a multi-fractal, and the notion of a singularity spectrum $F(\alpha)$ can be defined for it [40].

To do this, we select a set $C(\alpha)$ of such moments of time t that have the same value α as the Hölder-Lipschitz exponent: $h(t) = \alpha$. For some values α the set $C(\alpha)$ is not empty, that is, there are some minimum α_{min} and maximum α_{max}, such that only for $\alpha_{min} < \alpha < \alpha_{max}$ does the set $C(\alpha)$ contain some elements.

The multi-fractal spectrum of a singularity $F(\alpha)$ is the fractal dimension of a set of points $C(\alpha)$.

The parameter $\Delta\alpha = \alpha_{max} - \alpha_{min}$, called the singularity spectrum support width, is an important multi-fractal characteristic. In addition, of considerable interest is the argument α^* that provides the maximum of the singularity spectrum: $F(\alpha^*) = max\, F(\alpha)$, when $\alpha_{max} \geq \alpha \geq \alpha_{min}$, called the generalized Hurst exponent. To get the estimates of the multi-fractal characteristics of signals, we used a method based on the analysis of fluctuations after the removal of scale-dependent trends [41] by polynomials of the 8th order.

2.4.1. Minimum Normalized Entropy of Squared Orthogonal Wavelet Coefficients En

When processing signals using orthogonal wavelets, the choice of a basis is determined using the criterion for the minimum entropy of the distribution of wavelet coefficients [42]. Let $c_j^{(k)}$ are the wavelet coefficients of the analyzed signal $x(t)$, $t = 1, ..., L$ are a discrete indexes numbering successive values of the time series.

The superscript k is the number of the level of detail of the orthogonal wavelet decomposition; the subscript j numbers the sequence of time interval centers in the vicinity of which the signal convolution, with finite elements as its basis, is calculated.

We used 17 orthogonal Daubechies wavelets: 10 ordinary bases with minimal support and a number of vanishing moments from 1 to 10, and 7 Daubechies symlets [42] with a number of vanishing moments from 4 to 10.

For each basis, the normalized entropy of the distribution of the squared coefficients was calculated and a basis was found that ensures the minimum entropy:

$$En = -\sum_{k=1}^{m}\sum_{j=1}^{M_k} p_j^{(k)}\cdot\ln\left(p_j^{(k)}\right)/ln(N_r) \rightarrow min, \text{ when } p_j^{(k)} = \left|c_j^{(k)}\right|^2 / \sum_{l,i}\left|c_i^{(l)}\right|^2.$$

Here m is the number of levels of detail taken into consideration; M_k is the number of wavelet coefficients at the level of detail with number k.

The number of levels m depends on the length L of the analyzed samples.

For example, if $L = 2^n$, then $m = n$, $M_k = 2^{(n-k)}$. If the length L is not equal to a power of two, then the signal $x(t)$ is padded with zeros to the minimum length N, which is greater than or equal to L: $N = 2^n \geq L$.

In this case, among the number $2^{(n-k)}$ of all wavelet coefficients at level k, only $L\cdot2^{-k}$ coefficients correspond to the decomposition of the real signal, while the remaining coefficients are equal to zero due to the addition of zeros to the signal $x(t)$.

Thus, $M_k = L\cdot2^{-k}$, and only "real" coefficients are used to calculate the entropy. The number N_r is equal to the number of "real" coefficients, that is, $N_r = \sum_{k=1}^{m} M_k$. By construction, $0 \leq En \leq 1$.

When estimating the wavelet spectral exponent β, the orthogonal wavelet decomposition of the signal fragments in the current daily time window [5] is used, which was previously used in [43] to analyze the noise component of signals from a network of 1203 GPS stations in Japan before the Tohoku mega earthquake on March 11, 2011, and in [36,39] when processing noise data from the Kamchatka Peninsula.

For each station, the time series of the four seismic noise parameters were obtained with a time step of 1 day for the time period 2011–2021. When calculating the statistical parameters of the noise, the low-frequency polygenic components in the daily continuous records of the seismic signal with a frequency of 1 min were preliminarily filtered by the 8th order polynomial at each station. Such filtering and the subsequent transition to the first differences of the averaged 1-min data provided the suppression and compensation of tidal, atmospheric, other natural effects, and anthropogenic activity in the original continuous records.

2.4.2. Visualization of the Seismic Noise Parameters Distribution

For all four statistical parameters, daily GRD files were created, representing tables of their values at the nodes of a regular grid of 50×50 nodes in size, covering the area in the latitude range of 50–64° N and in the longitude range of 155–168° E for the entire observation period. The distribution of each noise statistic over the territory, obtained by interpolating the median values of the parameters from the three stations closest to each node of the grid, was reflected on digital maps created using a geographic information system [44].

When averaging daily maps over all days within a given time interval, averaged maps were obtained that characterize the features of noise parameters' changes over space for the corresponding time intervals. An analysis of the set of such maps for the same time interval and their variability in time in comparison with maps over a long-term period makes it possible to trace the features of the seismic noise parameters' distribution for the territory under consideration as a whole and in the areas of earthquake sources.

When interpreting maps, we analyzed areas located at distances not exceeding the average minimum distance between network stations (120 km). The shaded area of the maps is limited by a line corresponding to the envelope of circles with centers in the regions of outlying stations and a radius of 120 km (Figure 3a). Such a limitation allowed us to uniformly adjust the coloring area on the maps (Figure 3b). On digital maps, only this selected area was colored and analyzed (Figure 4).

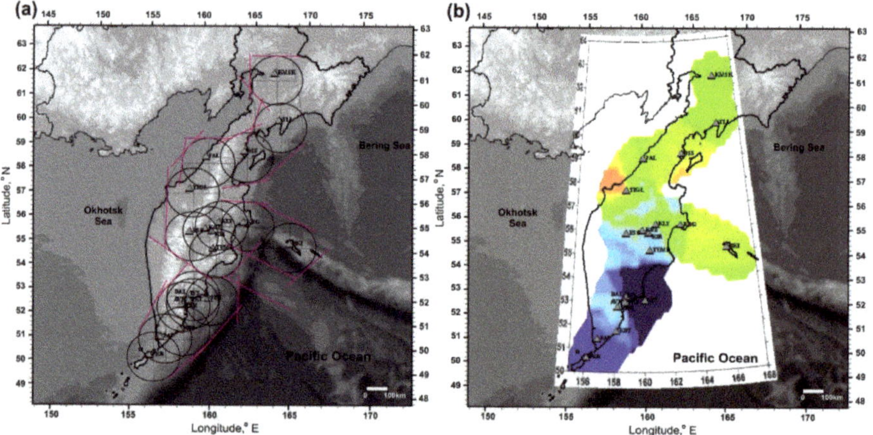

Figure 3. Determination of the coloring area on maps showing the distribution of noise parameters. (**a**) circular areas with a radius of 120 km around seismic stations; (**b**) corresponding colored area (see text for explanation).

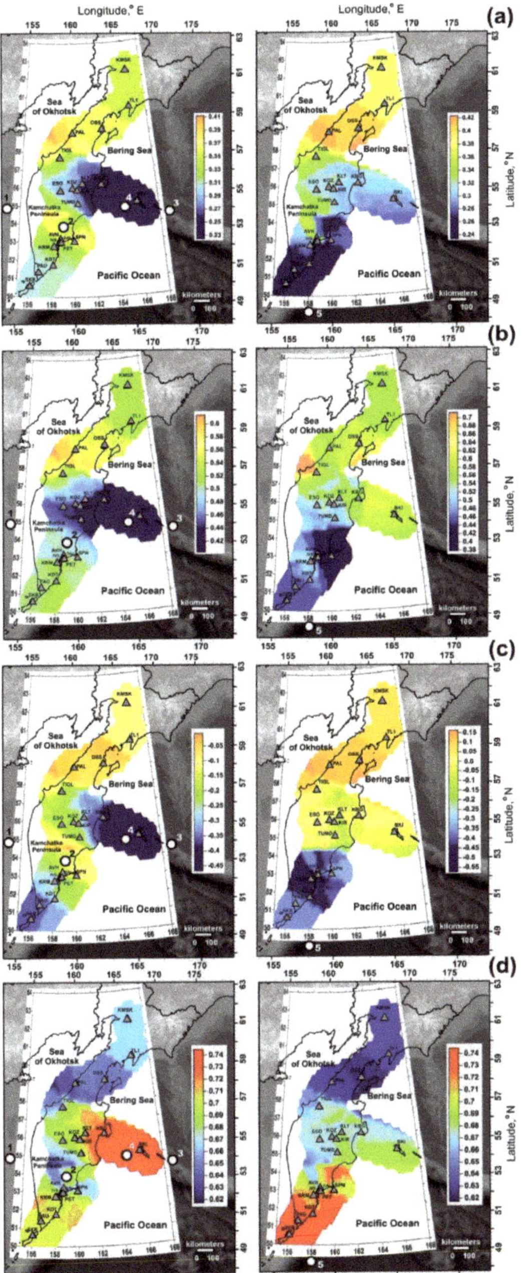

Figure 4. Maps of seismic noise parameters' distribution for 2011–2018 (left) and for 2019–2021 (right). (**a**) Generalized Hurst exponent α*; (**b**) singularity spectrum support width Δα; (**c**) wavelet-based spectral exponent β; and (**d**) normalized entropy of the squared orthogonal wavelet coefficients *En*. The white circles show the earthquake epicenters (Table 3) that occurred over the corresponding time periods. Rectangles with a coordinate grid show the area of noise parameters' calculation. The coloring corresponding to the color scales was carried out for the area at a distance of no more than 120 km from the edge stations of the network (Figure 3).

Table 3. Earthquakes with M_w = 7.2–8.3 (http://earthquake.usgs.gov/earthquakes (accessed on 31 December 2021)).

No	Date dd.mm.yyyy Name	Time hh:mm:ss	Coordinates, deg. N° E°		H, km	M_w	$M_0,$ N·m·10^{20}
1	24 May 2013 Sea of Okhotsk	05:44:48	54.89	153.22	598	8.3	38.4
2	30 January 2016 Zhupanovskoe	03:25:12	53.98	158.55	177	7.2	0.8
3	17 July 2017 Near Islands Aleutian	23:34:13	54.44	168.86	10	7.7	5.2
4	20 December 2018 Uglovoye Podnyatiye	17:01:55	55.10	164.70	17	7.3	1.0
5	25 March 2020 North Kuril	02:49:21	48.96	158.70	58	7.5	2.0

2.5. Spectral Measure of Coherent Behavior of Seismic Noise Parameters

To diagnose synchronization effects in changes to the 4-dimensional time series of noise parameters, we used the value of the spectral measure of their coherent behavior $\nu(\tau, \omega)$, estimated in a sliding time window. Previously, the spectral measure of coherent behavior was used in [5,11,12] for diagnosing synchronization effects in changes in geophysical, geochemical, hydrological, meteorological and other multidimensional time series.

The spectral measure of coherence of a 4-dimensional series of seismic noise parameters is constructed as the modulus of the product of the by-component canonical coherences of a multidimensional series $\nu(\tau, \omega) = \prod_{j=1}^{m} |\mu_j(\tau, \omega)|$, where $m = 4$ is the total number of analyzed time series that make up a multidimensional series (dimension of a multivariate time series); ω–frequency, day^{-1}; τ is the time coordinate of the right end of the sliding time window, consisting of a given number of samples of the time series; $\mu_j(\tau, \omega)$ is the canonical coherence of the j-th scalar time series, which characterizes the degree of connection of this series with all other series that make up the multidimensional series.

The value $|\mu_j(\tau, \omega)|^2$ is a generalization of the quadratic coherence spectrum between two signals, where the first signal represents the j-th scalar time series, and the second signal is a vector that reflects the overall changes of the remaining three series.

The quantity $\mu_j(\tau, \omega)$ satisfies the inequalities $0 \leq |\mu_j(\tau, \omega)| \leq 1$, from which it follows that the closer the value $|\mu_j(\tau, \omega)|$ to unity, the more linearly related are the variations at the frequency ω in the time window with the coordinate τ of the j-th time series, with similar variations in the other three time series.

Thus, the value $0 \leq \nu(\tau, \omega) \leq 1$, by virtue of its construction, describes the effect of the cumulative coherent (synchronous, collective) behavior in the time-frequency domain of all seismic noise parameters' time series that form a 4-dimensional time series.

Since the values of $\nu(\tau, \omega)$ vary in the interval [0, 1], the closer the corresponding value is to unity, the stronger relationship between the variations of the 4-dimensional time series components at the frequency ω for the time window with the coordinate τ.

It should be noted that comparison of the $\nu(\tau, \omega)$ absolute values is possible only for the same number m of simultaneously processed time series, because, by virtue of the formula for $\nu(\tau, \omega)$, as m grows, the value of ν decreases, as the product m values less than one. In this paper, the number of simultaneously analyzed time series is $m = 4$.

To estimate the spectral matrix of 4-dimensional time series, a 5th order vector autoregressive model was used (AR = 5) [9]. Taking into account the problem of identifying prognostic effects preceding earthquakes, the $\nu(\tau, \omega)$-calculated values in all diagrams are referred to the right edge of the current time window.

Examples of the $\nu(\tau, \omega)$ distribution in the time-frequency domain during 2011–2021 calculated in a sliding time window 365 days long with a step of 3 days are shown in Figures 5 and 6.

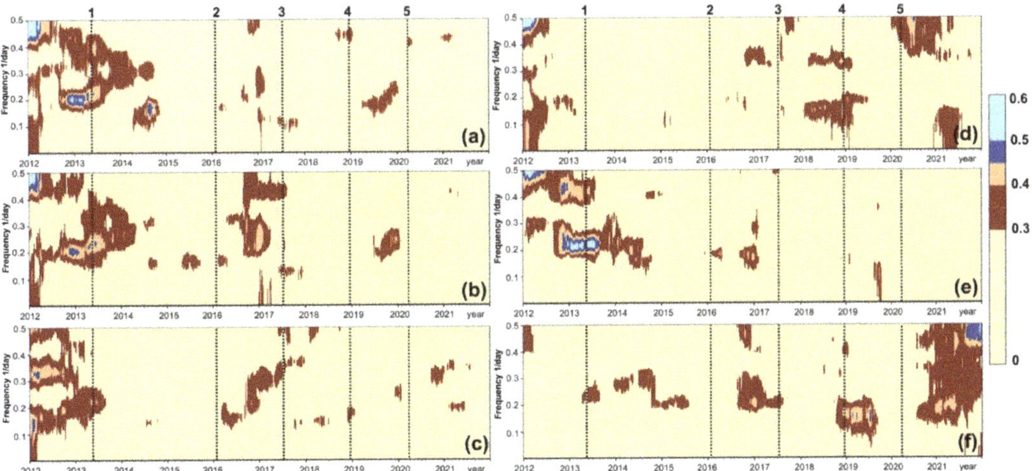

Figure 5. Time-frequency diagrams of the spectral measure of coherence of 4-dimensional time series of seismic noise parameters $\nu(\tau, \omega)$ in comparison with earthquakes (Table 3) according to data from (**a**) the entire network of stations, (**b**) non-volcanic, and (**c**) continental stations, as well as for the (**d**) northern, (**e**) central, and (**f**) southern groups of stations, 2011–2021. Synchronization effects of seismic noise parameters are distinguished by the values $\nu(\tau, \omega) \geq 0.3$.

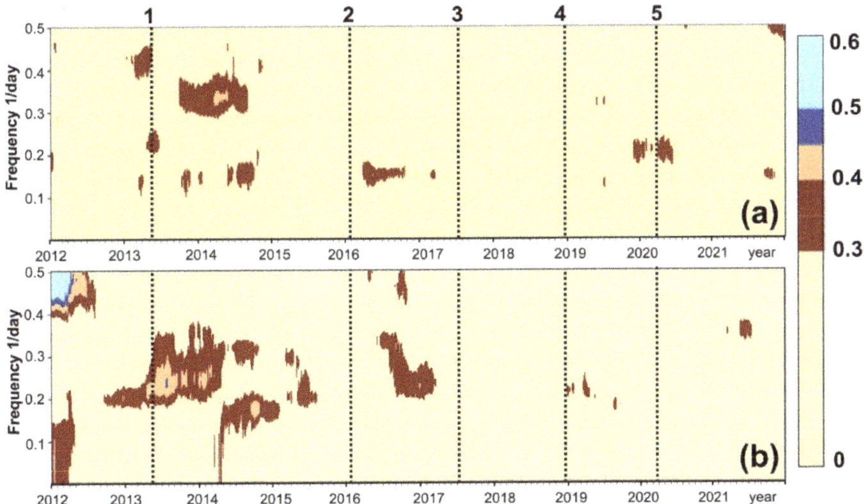

Figure 6. Time-frequency diagrams of the spectral measure of coherence of 4-dimensional time series of seismic noise parameters $\nu(\tau, \omega)$ in comparison with occurred earthquakes (Table 3) constructed from (**a**) the group of volcanic stations and (**b**) the group of coastal stations, 2011–2021.

The identification of synchronization effects in the diagrams was carried out taking into account the range of values $\nu(\tau, \omega)$ in case of random manifestation. For this, a 4-dimensional time series was generated consisting of four independent realizations of Gaussian white noise, each with a length of 365×10^4 samples. For this multidimensional series, a time-frequency diagram of the evolution of the spectral measure of coherence was constructed in successive non-overlapping time windows 365 samples long, which provides 10,000 independent estimates of $\nu(\tau, \omega)$. The resulting time-frequency diagram was a chaotic pattern, for which the average value of random bursts of the coherence measure is 0.008, the median is 0.006, and the maximum value is 0.15. The length of the time window corresponded to the same length of 365 samples as in construction of the diagrams in Figures 5 and 6.

The double maximum value of the random fluctuations of the measure of spectral coherence in the 4-dimensional time series $\nu(\tau, \omega) = 0.30$ was used as the upper limit of the random occurrence of the values $\nu(\tau, \omega)$. To visually highlight the effects of synchronization on the diagrams, the regions of spectral coherence manifestation with the values $\nu(\tau, \omega) \geq 0.30$ are colored (Figures 5 and 6).

Based on these diagrams, the time intervals and frequency bands of the anomalous coherent behavior of the discussed noise statistics' time series were identified and then correlated to the timeline of the earthquakes in Table 3.

2.6. Conceptual Model Used in Data Interpretation

A conceptual model of noise behavior at a network of stations in a seismically active region was proposed in [5–17,35] using seismic records and theoretical modeling. It was shown that the high values of the multi-fractal parameters $\Delta\alpha$, α^* and the β value, as well as the low values of the minimum normalized entropy of the squared orthogonal wavelet coefficients En, are due to an increase in the number of outliers in the original seismic records. For example, an increase in the number of outliers in the time series of a continuous seismic signal can occur when seismicity is activated during the aftershock stages of strong earthquakes. On the other hand, the consolidation of individual elements of the geological environment and the weakening of near-surface movements may manifest itself in a decrease in the number of high-amplitude outliers in seismic records and will be reflected in high values of entropy En and low values of $\Delta\alpha$, α^*, and β.

In seismically active areas, the formation of a large, consolidated block contributes to the accumulation of energy in it and, consequently, increases the danger of a strong earthquake. Such large, consolidated blocks can be long-existing areas of seismic quietness distinguished by a decrease in the number of weak earthquakes or "seismic gaps" [45,46]. Formation of a large, consolidated block can also be accompanied by a decrease in the diversity of the transfer and resonance properties of the geological environment and loss of the multi-fractality of noise time series and, accordingly, a decrease in $\Delta\alpha$ u α^* parameters [5–7].

Such a model of noise parameters' behavior during the preparation for strong earthquakes was used in interpreting the results of processing data from a network of stations in Kamchatka [35,36,39], as well as in the present work. We also note that this simple model of the behavior of seismic noise parameters during the preparation of a strong earthquake is in good agreement with the data on the decrease in the multifractal parameters α^* and $\Delta\alpha$ in the regions of future earthquake sources with $M_w = 8.3$ and 9.0 in Japan [8].

In accordance with the general patterns of complex systems' behavior before catastrophic changes, we also considered an increase in coherence in the variations of the four-dimensional series of the statistical parameters of seismic noise as a possible sign of strong earthquake preparation, diagnosed in time-frequency diagrams [12].

3. Data Analysis

3.1. Variability of Seismic Noise Parameters' Spatiotemporal Distribution

Main features of the noise parameters' spatiotemporal distribution for the considered timeframe, 2011–2021, are shown on maps in Figure 4. On these maps, the areas of danger of strong earthquakes are distinguished by the minimum α^*, $\Delta\alpha$, and β values and maximum En values in accordance with the conceptual model used.

During the first seven years in 2011–2018, the danger area was located in the northern part of Kamchatka Peninsula, at the junction of the Kurile–Kamchatka and Aleutian island arcs (Figure 4, maps on the left). All four earthquakes in 2013–2018 with M_w = 7.2–8.3 occurred in the latitude range 54–58° N (Table 3), identified in previous authors' publications [35,36] as "dangerous" for the emergence of strong earthquakes with $M_w \geq 7.5$. In this case, the magnitudes of two events out of four, the Sea of Okhotsk (No. 1 in Table 3) and Near Islands Aleutian (No. 3 in Table 3), corresponded to the magnitude range of expected events. The Sea of Okhotsk deep-focus earthquake on 24 May 2013, with M_w = 8.3 was the strongest seismic event in the region of the Kamchatka Peninsula during detailed seismological observations since 1961 [20]. Its seismic moment exceeded by 7–48 times the seismic moments of all other considered earthquakes (Table 3).

In 2017–2018, seismicity intensified in the indicated danger area over a section about 750 km long in the zone of contact of the Pacific oceanic plate with the Beringia and Okhotsk small continental plates and the Commander block. The epicenters and focal areas of the two strongest earthquakes with magnitudes 7.7 and 7.3 (No. 3, 4 in Table 3) are shown in Figure 1. The Near Islands Aleutian earthquake (No. 3 in Table 3), with a rupture length of 500 km [23], occurred on July 17, 2017, also in the "dangerous" range of latitudes 54–58° N.

Over the next three years, 2019–2021, significant changes took place in the spatial distribution of seismic noise parameters (Figure 4, maps on the right). The danger area moved to southern part of the region in the latitude range of 50–54° N [39]. On 25 March 2020, the North Kuril earthquake with M_w = 7.5 (No. 5 in Table 3) occurred to the south of the indicated area.

According to the maps of the spatial distribution of noise parameters at the end of 2021, the position of dangerous areas remained in the southern part of the region, which may indicate the possibility of strong earthquakes here.

Thus, a certain correspondence between the distinguished areas of strong earthquake danger and the occurred seismic events gives reason to believe that the processes of preparation for strong earthquakes are reflected in the regular behavior of seismic noise parameters.

3.2. Synchronization Signals in Noise Parameter Changes

Synchronization signals allocated by increased values of the spectral coherence of noise parameters $\nu(\tau, \omega) \geq 0.3$ are shown in the time-frequency diagrams in Figures 5 and 6 for the time interval 2011–2021, including the moments of all five strong earthquakes (Table 3). Such diagrams were constructed based both on the data of the entire network of stations and on individual groups of stations.

According to data from the entire network of stations as well as from non-volcanic and continental stations (Figure 5a–c), the most pronounced synchronization signals appeared during the period from six months to one year before three earthquakes with $M_w \geq 7.5$ (No. 1, 3, 5 in Table 3). The maximum amplitudes, $\nu(\tau, \omega) \geq 0.45$, were recorded before the strongest Sea of Okhotsk earthquake (No. 1 in Table 3, M_w = 8.3) for half a year. Before the two considered earthquakes with $M_w < 7.5$, similar signals of spectral coherence growth either did not appear (Zhupanovskoe, No. 2 in Table 3, M_w = 7.2), or were much less pronounced (Uglovoye Podnyatiye, No. 4 in Table 3, M_w = 7.3).

Increased values of spectral coherence also appeared during the aftershock stages of all considered earthquakes. After the Sea of Okhotsk earthquake, the duration of postseismic synchronization was about one year and no more than 1–2 months after other earthquakes.

The diagrams in Figure 5d,f show the $\nu(\tau, \omega) \geq 0.3$ distributions according to the data (from top to bottom) for the northern, central, and southern groups of stations. According to the data from the northern stations (Figure 5d), increased values of spectral coherence manifested themselves during the 7–9 months before the Near Islands Aleutian (No. 3 in Table 3) and Uglovoye Podnyatiye (No. 4 in Table 3) earthquakes, as well as at their aftershock stages. Both of these two earthquakes occurred in the northern part of the region. Other manifestations of increased $\nu(\tau, \omega)$ values are difficult to associate with strong earthquakes. Perhaps they reflect regional movements associated with geodynamic activity on the periphery of strong earthquake preparation areas, or they are random due to local features of seismic noise at the northern group of stations.

According to the data from the central group of stations (Figure 5e), increased $\nu(\tau, \omega)$ values with amplitudes up to 0.45–0.60 were most pronounced 6–9 months before the Sea of Okhotsk earthquake.

Figure 5f shows the spectral coherence distribution diagram for the southern stations. In this diagram, noise synchronization before the Sea of Okhotsk and North Kuril earthquakes with $M_w \geq 7.5$ (No. 1, 5 in Table 3) is very weak. This may be due to the fact that 67% of the stations in the southern group (six stations out of nine, Table 2) are coastal, and seismic records from them are noisy due to sea waves. In addition, the Ebeko volcano, located 6 km from the SKR station, has been erupting since 2019, and volcanic microseisms could mask the North Kuril earthquake preparation. However, it should be noted that Figure 5f shows an increase in the synchronization of seismic noise parameters during all 11 years of observations. The most pronounced increase in synchronization has manifested itself over the past year and a half, from mid-2020 to the end of 2021. This may indicate an increased danger of strong earthquakes in the southern part of the region under consideration. This assumption is consistent with the spatial distribution of noise parameters on maps for 2019–2021 (Figure 4, maps on the right), showing the increased danger of strong earthquakes in the southern part of the region.

On the diagrams constructed from coastal and volcanic stations, the distribution of $\nu(\tau, \omega) \geq 0.3$ values is mainly mosaic and non-systematic (Figure 6a,b). While the effects of increasing noise synchronization before earthquakes with $M_w \geq 7.5$ on the diagrams based on data from all network stations, as well as data from non-volcanic and continental stations (Figure 5a–c), are quite pronounced in the frequency range 0.15–0.35 day^{-1} (periods 3–7 days).

4. Discussion and Conclusions

The methodological basis of the used approach to data analysis is the general property of synchronization of the behavior of the constituent parts of complex systems as they approach critical states [18]. The goal of all methods used is to search for synchronization effects by estimating the coherence of seismic noise in different areas of a seismically active region. In this paper, a phenomenological approach is applied to the study of complex multicomponent systems, which include Earth's crust, based on the general property of increasing the radius and the degree of strong coherence of random fluctuations in the parameters of a complex system as it approaches a sharp change in its properties as a result of its own dynamics. As a result of studying long-term continuous records of low-frequency seismic noises on a network of broadband seismic stations covering the study area, it was possible to establish the characteristic time points for changing trends in the coherence of seismic noise properties. The described approach to the analysis of continuous seismic noise records has a long history of application in various regions of the planet, both at the regional and global levels, which is reflected in publications [5–17].

Using the presented methodological approach to processing data from continuous recording of microseismic oscillations at the network of stations on the Kamchatka Peninsula and the authors' conceptual model of the seismic noise behavior based on empirical data and general ideas about the behavior of complex dynamic systems in critical conditions, a study was made of spatiotemporal variations in noise parameters for 2011–2021 in connection with five earthquakes with $M_w = 7.2$–8.3.

According to the distribution maps of the noise statistical parameters, the manifestation of decreased α^*, $\Delta\alpha$, and β values and increased En values in the areas of future strong earthquake sources during months to years was found (Figure 4). In 2011–2018, the earthquake hazard area was located in the northern part of the region in the latitude range 54–58° N, in which four strong earthquakes occurred (Table 3). Since 2019, there have been changes in the spatial distribution of noise parameters, and the danger area in 2019–2021 was located to the southern part of the region (latitude range 50–54° N). The source of the North Kuril earthquake with $M_w = 7.5$ occurred on 25 March 2020, near this area.

Thus, the conceptual model of the relationship between the changes in noise parameters and the preparation of strong earthquakes has found convincing confirmation in spatiotemporal variations in seismic noise parameters and occurred seismic events with magnitudes 7.2–8.3 for the Kamchatka region in the data for 2011–2021. Using this model, according to the observations for 2011–2016 in the author's publications made an early prediction of the area (54–58° N) of future subsequent earthquakes, including the Near Islands Aleutian earthquake on 17 July 2017, with $M_w = 7.7$.

According to observations at of the end of 2021 and in accordance with the model, the danger area for strong earthquakes in the southern part of the region remains.

Before the Sea of Okhotsk, Near Islands Aleutian and North Kuril earthquakes with $M_w \geq 7.5$ (Table 3), a noise parameter synchronization effect was found by increased values of the spectral measure of coherent behavior of noise parameters' time series constructed from the data from the entire network and for groups of stations least affected by volcanic activity and sea waves. A property of this type of synchronization is an increase in the measure of spectral coherence $\nu(\tau, \omega) \geq 0.3$ during the time period from several months to a few years before seismic events.

It can be assumed that for the Kamchatka Peninsula, the preparation and realization of strong earthquakes with $M_w \geq 7.5$ manifests itself on the time-frequency diagrams of the evolution of spectral coherence as a time-compact increase in the spectral coherence at frequencies of 0.15–0.35 day^{-1}. Meanwhile, bursts of increase of spectral coherence of presumably volcanic and marine genesis can manifest in a wider frequency range.

The revealed type of synchronization of the seismic noise behavior on the network of stations can be of prognostic value to the issue of predicting earthquakes with $M_w \geq 7.5$ in the region of the Kamchatka Peninsula.

The results of long-term studies on the relationship between seismic noise variations and strong earthquakes in the Kamchatka region allow one to consider the presented method for processing and interpreting data from continuous recordings of signals from a network of broadband stations as a way to dynamically assess the danger of strong earthquakes. Despite the obvious prognostic shortcomings of this method, such as the large size of the dangerous area and the wide uncertainty in the time frame of expected events' occurrence, it can be used in the aggregate of seismic forecast data to predict the strongest seismic events, accompanied by catastrophic consequences for the population and infrastructure of the region.

Since 2021, the authors have been implementing this method with the issuance of quarterly forecast conclusions about the danger of strong earthquakes to the Russian Expert Council for Earthquake Prediction and its Kamchatka branch [39].

An important element of the considered method for study spatio-temporal variations of seismic noise to search for signs of preparation for strong earthquakes is the effective suppression of the natural and technogenic components present in the original seismic records. When estimating the statistical parameters of noise, it is necessary to carry out preliminary filtering of low-frequency polygenic components in the records of a continuous seismic signal at each station. In the present study, for such filtering of minute data, an 8th order polynomial was used, and after applying it to daily data fragments, a transition to the first differences was carried out.

The results of the work also showed that, in areas of modern volcanism, in order to search for earthquake preparation signals, it is necessary to take into account the activity of active volcanoes near broadband seismic stations, giving preference to data obtained at remote stations from the centers of volcanic activity.

Further advancement of the presented method suggests a more detailed study of local noise in records of microseismic oscillations on the individual stations with the development of effective methods for the compensation of natural and anthropogenic components and an increase in the number of broadband seismic stations for the more reliable diagnostics of earthquake preparation signals, especially in the northern part of the Kamchatka Peninsula.

Author Contributions: Conceptualization, G.K. and A.L.; Data curation, G.K. and S.B.; Formal analysis, V.K.; Funding acquisition, A.L. and G.K.; Investigation, G.K. and V.K.; Methodology, G.K. and A.L.; Project administration, G.K.; Resources, S.B.; Software, V.K. and A.L.; Supervision, G.K.; Validation, G.K. and A.L.; Visualization, V.K. and S.B.; Writing—original draft, G.K., V.K. and A.L.; Writing—review & editing, G.K. and A.L. All authors listed have made a substantial, direct, and intellectual contribution to the work. All authors have read and agreed to the published version of the manuscript.

Funding: The work was supported by the Ministry of Science and Higher Education of the Russian Federation (075-00576-21) and carried out within the framework of the state assignment of the Institute of Physics of the Earth of the Russian Academy of Sciences. The data used in the work were obtained from the large-scale research facilities "Seismic infrasound array for monitoring Arctic cryolitozone and continuous seismic monitoring of the Russian Federation, neighbouring territories and the world" (https://ckp-rf.ru/usu/507436/, http://www.gsras.ru/unu).

Institutional Review Board Statement: Not applicable.

Informed Consent Statement: Not applicable.

Data Availability Statement: The original contributions presented in the study are included in the article; further inquiries regarding the initial data and research software can be directed to the corresponding authors G.N. Kopylova and A.A. Lyubushin.

Conflicts of Interest: The authors declare that the research was conducted in the absence of any commercial or financial relationships that could be construed as a potential conflict of interest.

References

1. Tanimoto, T. The oceanic excitation hypothesis for the continuous oscillations of the Earth. *Geophys. J. Int.* **2005**, *160*, 276–288. [CrossRef]
2. Tanimoto, T.; Um, J.; Nishida, K.; Kobayashi, N. Earth's continuous oscillations observed on seismically quiet days. *Geophys. Res. Lett.* **1998**, *25*, 1553–1556. [CrossRef]
3. Stehly, L.; Campillo, M.; Shapiro, N.M. A study of the seismic noise from its long-range correlation properties. *J. Geophys. Res.* **2006**, *111*, 10306. [CrossRef]
4. Berger, J.; Davis, P.; Ekstrom, G. Ambient Earth Noise: A survey of the Global Seismographic Network. *J. Geophys. Res.* **2004**, *109*, 11307. [CrossRef]
5. Lyubushin, A.A. Microseismic noise in the low frequency range (periods of 1-300 min): Properties and possible prognostic features. *Izvestiya. Phys. Solid Earth* **2008**, *44*, 275–290. [CrossRef]
6. Lyubushin, A.A. The statistics of the time segments of low-frequency microseisms: Trends and synchronization. *Izvestiya. Phys. Solid Earth* **2010**, *46*, 544–554. [CrossRef]
7. Lyubushin, A.A. The great Japanese earthquake. *Priroda* **2012**, *8*, 23–33. (In Russian)
8. Lyubushin, A.A. Mapping the properties of low-frequency microseisms for seismic hazard assessment. *Izvestiya. Phys. Solid Earth* **2013**, *49*, 9–18. [CrossRef]
9. Lyubushin, A.A. Analysis of coherence in global seismic noise for 1997–2012. *Izvestiya. Phys. Solid Earth* **2014**, *50*, 325–333. [CrossRef]
10. Lyubushin, A.A. Coherence between the fields of low-frequency seismic noise in Japan and California. *Izvestiya. Phys. Solid Earth* **2016**, *52*, 810–820. [CrossRef]
11. Lyubushin, A. Synchronization of Geophysical Fields Fluctuations. In *Complexity of Seismic Time Series: Measurement and Applications*; Chelidze, T., Telesca, L., Vallianatos, F., Eds.; Elsevier: Amsterdam, The Netherlands, 2018; pp. 161–197. [CrossRef]
12. Lyubushin, A. Field of coherence of GPS-measured earth tremors. *GPS Solut.* **2019**, *23*, 120. [CrossRef]
13. Lyubushin, A. Trends of Global Seismic Noise Properties in Connection to Irregularity of Earth's Rotation. *Pure Appl. Geophys.* **2020**, *177*, 621–636. [CrossRef]
14. Lyubushin, A. Connection of Seismic Noise Properties in Japan and California with Irregularity of Earth's Rotation. *Pure Appl. Geophys.* **2020**, *177*, 4677–4689. [CrossRef]
15. Lyubushin, A. Global Seismic Noise Entropy. *Front. Earth Sci.* **2020**, *8*, 611663. [CrossRef]
16. Lyubushin, A.A. Seismic Noise Wavelet-Based Entropy in Southern California. *J. Seismol.* **2021**, *25*, 25–39. [CrossRef]
17. Lyubushin, A.A. Low-Frequency Seismic Noise Properties in the Japanese Islands. *Entropy* **2021**, *23*, 474. [CrossRef]
18. Nicolis, G.; Prigogine, I. *Exploring Complexity, an Introduction*; W.H. Freedman and Co.: New York, NY, USA, 1989.

19. Chelidze, T.; Matcharashvili, T.; Lursmanashvili, O.; Varamashvili, N.; Zhukova, N.; Mepariddze, E. Triggering and Synchronization of Stick-Slip: Experiments on Spring-Slider System. In *Synchronization and Triggering: From Fracture to Earthquake Processes. GeoPlanet: Earth and Planetary Sciences*; Rubeis, V., Czechowski, Z., Teisseyre, R., Eds.; Springer: Berlin/Heidelberg, Germany, 2010; pp. 123–164. [CrossRef]
20. Chebrov, V.N.; Kugayenko, Y.A.; Vikulina, S.A.; Kravchenko, N.M.; Matveenko, E.A.; Mityushkina, S.V.; Lander, A.V. Deep M_w = 8.3 sea of Okhotsk earthquake on May 24, 2013—The largest event ever recorded near Kamchatka Peninsula for the whole period of regional seismic network operation. *Vestn. KRAUNTS Seriya Nauk. O Zemle* **2013**, *21*, 17–24. (In Russian)
21. Chebrov, V.N.; Kugayenko, Y.A.; Abubakirov, I.R.; Droznina, S.Y.; Ivanova, E.I.; Matveenko, E.A.; Mityushkina, S.V.; Ototyuk, D.A.; Pavlov, V.M.; Raevskaya, A.A.; et al. The 30 January 2016 earthquake with K_s = 15.7, M_w = 7.2, I = 6 in the Zhupanovsky region (Kamchatka). *Vestn. KRAUNTS Seriya Nauk. O Zemle* **2016**, *29*, 5–16. (In Russian)
22. Chebrov, D.V.; Kugayenko, Y.A.; Lander, A.V.; Abubakirov, I.R.; Voropayev, P.V.; Gusev, A.A.; Droznin, D.V.; Droznina, S.Y.; Ivanova, E.I.; Kravchenko, N.M.; et al. The 29 March 2017 earthquake with Ks = 15.0, M_w = 6.6, I = 6 in Ozernoy Gulf (Kamchatka). *Vestn. KRAUNTS Seriya Nauk. O Zemle* **2017**, *35*, 7–21. (In Russian)
23. Chebrov, D.V.; Kugaenko, Y.A.; Abubakirov, I.R.; Gusev, A.A.; Droznina, S.Y.; Mityushkina, S.V.; Ototyuk, D.A.; Pavlov, V.M.; Titkov, N.N. Near Islands Aleutian earthquake with M_w = 7.8 on July 17, 2017: I. Extended rupture along the Commander block of the Aleutian Island arc from observations in Kamchatka. *Izv. Phys. Solid Earth* **2019**, *55*, 576–599. [CrossRef]
24. Chebrov, D.V.; Kugayenko, Y.A.; Lander, A.V.; Abubakirov, I.R.; Droznina, S.Y.; Mityushkina, S.V.; Pavlov, V.M.; Saltykov, V.A.; Serafimova, Y.K.; Titkov, N.N. The Uglovoye Podnyatiye Earthquake on 20 December 2018 (M_w = 7.3) in the Junction Zone between Kamchatka and Aleutian Oceanic Trenches. *Vestn. KRAUNTs Seriya Nauk. O Zemle* **2020**, *45*, 100–117. (In Russian) [CrossRef]
25. Gordeev, E.I.; Pinegina, T.K.; Kozhurin, A.I.; Lander, A.V. Beringia: Seismic hazard and fundamental problems of geotectonics. *Izv. Phys. Solid Earth* **2015**, *51*, 512–521. [CrossRef]
26. Kozhurin, A.I. Active Faulting in the Kamchatsky Peninsula, Kamchatka-Aleutian Junction. In *Geophysical Monograph Series Volcanism and Subduction: The Kamchatka Region*; Eichelberger, J., Gordeev, E., Kasahara, M., Lees, J., Eds.; American Geophysical Union: Washington, DC, USA, 2007; Volume 172, pp. 107–116.
27. Argus, D.F.; Gordon, R.G. No-net-rotation model of current plate velocities incorporating plate motion model NUVEL-1. *Geophys. Res. Lett.* **1991**, *18*, 2039–2042. [CrossRef]
28. Chebrova, A.Y.; Chemarev, A.S.; Matveenko, E.A.; Chebrov, D.V. Seismological data information system in Kamchatka branch of GS RAS: Organization principles, main elements and key functions. *Geophys. Res.* **2020**, *21*, 66–91. [CrossRef]
29. Geology of the USSR. *Kamchatka, the Kuril and Commander Islands. Geological Description*; Nedra: Moscow, Russia, 1964; Volume 31, 733p. (In Russian)
30. Gorelchik, V.I.; Garbuzova, V.T.; Storcheus, A.V. Deep-seated volcanic processes beneath Klyuchevskoi volcano as inferred from seismological data. *J. Volcanol. Seismol.* **2004**, *6*, 21–34.
31. Thompson, G. Seismic Monitoring of Volcanoes. In *Encyclopedia of Earthquake Engineering*; Beer, M., Kougioumtzoglou, I.A., Patelli, E., Au, I.K., Eds.; Springer: Berlin/Heidelberg, Germany, 2015. [CrossRef]
32. Kotenko, T.A.; Sandimirova, E.I.; Kotenko, L.V. The 2016–2017 eruptions of Ebeko Volcano (Kuriles Islands). *Vestn. KRAUNTs Seriya Nauk. O Zemle* **2018**, *37*, 32–42. (In Russian)
33. Belousov, A.; Belousova, M.; Auer, A.; Walter, T.; Kotenko, T.; Belousova, M.; Auer, A.; Walter, T.; Kotenko, T. Mechanism of the historical and the ongoing Vulcanian eruptions of Ebeko volcano, Northern Kuriles. *Bull. Volcanol.* **2021**, *83*, 4. [CrossRef]
34. Walter, T.; Belousov, A.; Belousova, M.; Kotenko, T.; Auer, A. The 2019 Eruption Dynamics and Morphology at Ebeko Volcano Monitored by Unoccupied Aircraft Systems (UAS) and Field Stations. *Remote Sens.* **2020**, *12*, 1961. [CrossRef]
35. Lyubushin, A.A.; Kopylova, G.N.; Kasimova, V.A.; Taranova, L.N. The Properties of Fields of Low Frequency Noise from the Network of Broadband Seismic Stations in Kamchatka. *Bull. Kamchatka Reg. Assoc. Educ. Sci. Cent. Earth Sci.* **2015**, *26*, 20–36. (In Russian)
36. Kasimova, V.A.; Kopylova, G.N.; Lyubushin, A.A. Variations in the parameters of background seismic noise during the preparation stages of strong earthquakes in the Kamchatka region. *Izv. Phys. Solid Earth* **2018**, *54*, 269–283. [CrossRef]
37. Riznichenko, Y.V. The source dimensions of the crustal earthquakes and the seismic moment. In *Issledovaniya po Fizike Zemletryasenii (Studies in Earthquake Physics)*; Nauka: Moscow, Russia, 1976; pp. 9–27. (In Russian)
38. Dobrovolsky, I.P.; Zubkov, S.I.; Miachkin, V.I. Estimation of the size of earthquake preparation zones. *Pageoph* **1979**, *117*, 1025–1044. [CrossRef]
39. Kopylova, G.N.; Lyubushin, A.A.; Taranova, L.N. New prognostic technology for analysis of low-frequency seismic noise variations (on the example of the Russian Far East). *Russ. J. Seismol.* **2021**, *3*, 75–91. (In Russian) [CrossRef]
40. Feder, J. *Fractals*; Plenum Press: New York, NY, USA, 1988; 254p.
41. Kantelhard, J.W.; Zschiegner, S.A.; Konscienly-Bunde, E.; Havlin, S.; Bunde, A.; Stanley, H.E. Multifractal detrended fluctuation analysis of nonstationary time series. *Phys. A Stat. Mech. Appl.* **2002**, *316*, 87–114. [CrossRef]
42. Mallat, S. *A Wavelet Tour of Signal Processing*; Academic Press: San Diego, CA, USA, 1998; 671p.
43. Lyubushin, A.; Yakovlev, P. Properties of GPS noise at Japan islands before and after Tohoku mega-earthquake. *Springer Plus Open Access J.* **2014**, *3*, 364. [CrossRef]

44. Kopylova, G.N.; Ivanov, V.Y.; Kasimova, V.A. The implementation of information system elements for interpreting integrated geophysical observations in Kamchatka. *Russ. J. Earth Sci.* **2009**, *11*, ES1006. [CrossRef]
45. Fedotov, S.A. *Dolgosrochnyy Seysmicheskiy Prognoz Dlya Kurilo-Kamchatskoy Dugi (Long-Term Seismic Forecast for Kurilo-Kamchatka Arcs)*; Nauka Publ.: Moscow, Russia, 2005; 302p. (in Russian)
46. Mogi, K. *Earthquake Prediction*; Mir: Moscow, Russia, 1988; 382p. (In Russian)

 *applied sciences*

 MDPI

Article

Analyzing the Correlations and the Statistical Distribution of Moderate to Large Earthquakes Interevent Times in Greece

Christos Kourouklas [1,*], George Tsaklidis [2], Eleftheria Papadimitriou [1] and Vasileios Karakostas [1]

[1] Geophysics Departments, School of Geology, Aristotle University of Thessaloniki, 541 24 Thessaloniki, Greece; ritsa@geo.auth.gr (E.P.); vkarak@geo.auth.gr (V.K.)

[2] Statistics and Operational Research Department, School of Mathematics, Aristotle University of Thessaloniki, 541 24 Thessaloniki, Greece; tsaklidi@math.auth.gr

* Correspondence: ckouroukl@geo.auth.gr

Abstract: Seismic temporal properties constitute a fundamental component in developing probabilistic models for earthquake occurrence in a specific area. Earthquake occurrence is neither periodic nor completely random but often accrues into bursts in both short- and long-term time scales, and involves a complex summation of triggered and independent events (ΔT). This behavior underlines the impact of the correlations on many potential applications such as the stochastic point process for the earthquake interevent times. In this respect, we intend firstly to determine the appropriate magnitude thresholds, M_{thr}, indicating the temporal crossover between correlated and statistically independent earthquakes in each 1 of the 10 distinctive sub-areas of the Aegean region. The second goal is the investigation of the statistical distribution that optimally fits the interevent times' data for earthquakes with $M \geq M_{thr}$ after evaluating the Gamma, Weibull, Lognormal and Exponential distributions performance. Results concerning the correlations analysis evidenced that the temporal crossover of the earthquake interevent time data ranges from $M_{thr} \geq 4.7$ up to $M_{thr} \geq 5.1$ among the 10 sub-areas. The distribution fitting and comparison reveals that the Gamma distribution outperforms the other three distributions for all the data sets. This finding indicates a burst or clustering behavior in the earthquake interevent times, in which each earthquake occurrence depends upon only the occurrence time of the last one and not from the full seismic history.

Keywords: earthquake interevent times; Greek seismicity; temporal correlations; statistical distributions

Citation: Kourouklas, C.; Tsaklidis, G.; Papadimitriou, E.; Karakostas, V. Analyzing the Correlations and the Statistical Distribution of Moderate to Large Earthquakes Interevent Times in Greece. *Appl. Sci.* **2022**, *12*, 7041. https://doi.org/10.3390/app12147041

Academic Editor: Igal M. Shohet

Received: 7 June 2022
Accepted: 11 July 2022
Published: 12 July 2022

Publisher's Note: MDPI stays neutral with regard to jurisdictional claims in published maps and institutional affiliations.

1. Introduction

The study of temporal properties of earthquakes contributes to analyzing the seismicity of a specific region in order to develop earthquake occurrence models. Earthquake occurrence is neither periodic nor completely random but often is clustered in both short- and long-term time scales [1,2] represented by a complex summation of triggered (e.g., aftershocks of a strong earthquake or swarm-like excitations) and independently distributed (spontaneous) events.

As a complex process, seismogenesis is characterized by scaling behavior [3], which is described by well-known empirical laws concerning the earthquake's magnitude distribution (the Gutenberg–Richter Law; [4]) and the aftershock's decay rate (the Omori Law; [5]). Focusing on the earthquake interevent time (ΔT), Bak et al. [6] stated that seismicity follows a universal scaling law, independently of the magnitude cut-off and the length of its spatial distribution. Corral [7,8] further studied this result and suggested that ΔT is optimally described by the generalized Gamma distribution.

Several studies (e.g., [9,10], among others) questioned these findings and concluded that ΔT distribution is deviated from universality. For example, Touati et al. [11] examined both real and synthetic ΔT and showed that a short ΔT between consecutive earthquakes deviated from the unified scaling behavior as a result of the interaction between the triggered and spontaneous earthquakes. They concluded that the distribution of ΔT is bimodal in a way that the correlated and independent earthquakes are following the Gamma distribution and exponential decay, respectively. In order to overcome these departures, Godano [12] proposed a new expression of the ΔT distribution, based on the Gamma distribution and included an additional parameter in its equation.

The latter conclusions underline the impact of the correlations among ΔT, not only in terms of short-term clustering as the epidemic-type models (e.g., ETAS model; [13]) assume but in a more general way. This issue is becoming the main objective of several studies over the years. For instance, Livina et al. [14] have investigated ΔT data sets of six different earthquake catalogs worldwide with various magnitude thresholds and found that they are correlated, exhibiting behavior in which short ΔT intervals are followed by short intervals and long ΔT intervals are followed by large intervals. Applications of Detrended Fluctuations Analysis (DFA) and conditional probability methods in different study areas such as Northern and Southern California [15], Italy and Israel [16,17] concluded similar results. Especially for Greece, Gkarlaouni et al. [18] applied DFA techniques for studying the memory of ΔT for two distinctive fault zones of various magnitude thresholds, namely $M_{thr} \geq 1.7$, $M_{thr} \geq 2.3$, $M_{thr} \geq 4.0$ and $M_{thr} \geq 6.0$, and for the periods 2008–2014, 2008–2014, 1981–2014 and 1700–2014, respectively. They found that the ΔT of small to moderate earthquakes exhibits strong long-range correlations whereas the ΔT data sets of large earthquakes ($M \geq 6.0$) could be considered statistically independent. Parson and Geist [19] found that the ΔT of global earthquakes (M~8) that occurred between 1900 and 2011, which might be considered rare events, are characterized by a lack of memory and can be described by a Poisson process.

Given the aforementioned findings, the detection of any correlations between ΔT or the existence of memory among them is an important characteristic of their temporal behavior. The investigation of the possible influence of the different magnitude thresholds above which the samples of ΔT can be considered as statistically independent, or in other words the possibility of a crossover regime in magnitudes in which the respective ΔT change their temporal behavior from correlated to uncorrelated, is important information with many potential uses. For example, it is useful for the selection of adequate earthquake data sets for stochastic point processes modeling. For such applications, four characteristics must be qualified: (1) The numbers of events occurring in two different time intervals are mutually independent (thus the aforementioned threshold M_{thr} has to be detected); (2) The probability distribution of the number of events within a specific time interval depends only on the length of the interval; (3) Two or more events occur simultaneously; and (4) especially for renewal models, the occurrence of the next event depends only on the time elapsed from the last event and not on the full history [20].

The main objectives of the current study are two. Firstly, the identification of the possible correlations between ΔT in earthquake data sets with different magnitude thresholds (M_{thr}). This concerns moderate to large earthquake ($4.1 \leq M_w < 5.2$) time intervals in 10 distinctive sub-areas of the Greek territory, and the detection of possible M_{thr} which designates the temporal crossover between correlated and statistically independent earthquakes. Both procedures are implemented via the application of time series analysis tools, which are widely used in seismicity studies (e.g., [21]). Specifically, in the current study, the Autocorrelation and Partial Autocorrelation Functions along with the Ljung–Box Q-test are applied. For further investigation of the memory of the ΔT data sets, Detrended Fluctuations Analysis is also applied. The second goal is seeking the statistical distribution that optimally fits the ΔT data sets for each sub-area for earthquakes with $M \geq M_{thr}$ which emerge from the application of the previous methodology. For this purpose, we tested four statistical distributions (Weibull, Lognormal, Gamma and Exponential) that are the most commonly used in investigations of earthquake ΔT data sets for the identification of the best-performing distribution.

2. Tectonic Setting

The Aegean region exhibits seismotectonic complexity, with the dominance of the subduction of the Eastern Mediterranean oceanic lithosphere beneath the Eurasian continental lithospheric plate in the southern Aegean Sea. This process forms the Hellenic Arc and its extensional back arc Aegean area [22] due to the slab roll back [23]. The Hellenic Arc is bounded to its northwestern edge by the right-lateral Kefalonia Transform Fault Zone (KTFZ) and to its southeastern edge by the left-lateral Rhoades Transform Fault (RTF). To the north of KTFZ, the continental collision of the Adriatic microplate and the Eurasia plate takes place, resulting in a compressional seismic zone running parallel both onshore and offshore to the western coastal areas of Greece and Albania. The westward motion of Anatolia as a rigid block results in the formation of the right lateral North Anatolian Fault Zone (NAFZ) extending through the Marmara Sea into the northern Aegean Sea along the North Aegean Trough (NAT), which is the boundary between the Eurasian plate and the Aegean microplate [24] (Figure 1).

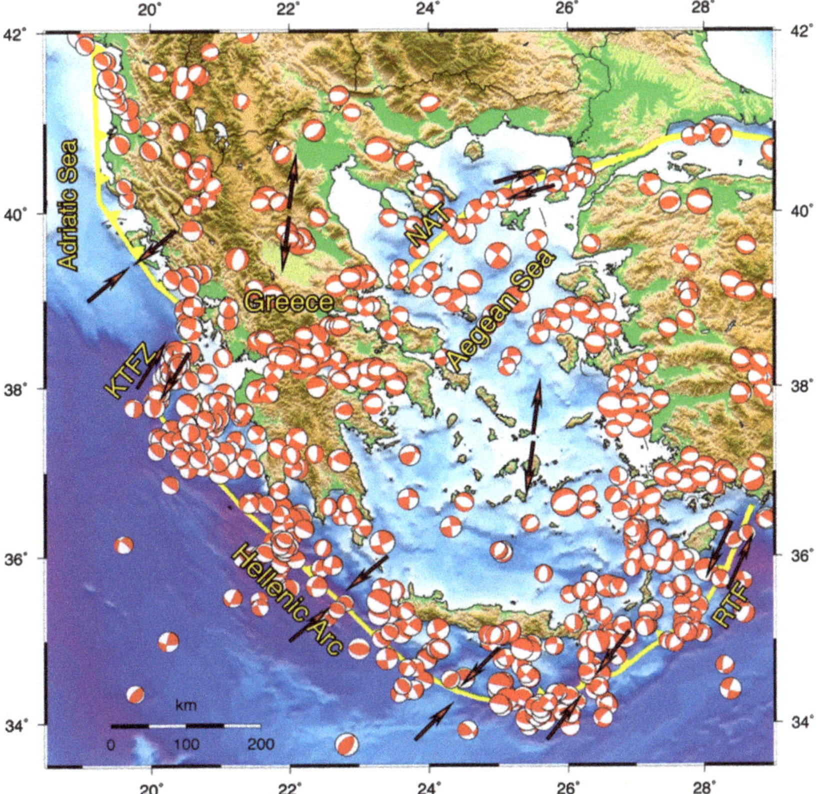

Figure 1. The major active boundaries (solid yellow lines) and their relative motions (red arrows) in the broader Aegean Sea area, along with the available focal mechanisms of earthquakes that occurred since 1976 as derived from the Global CMT database (Available online: www.globalcmt.org (last accessed on 30 April 2022)). Fault plane solutions are shown as equal area lower hemisphere projection (compressional quadrants are depicted in red).

The available focal mechanisms of the Global Centroid Moment Tensor (Global CMT; Available online: www.globalcmt.org (accessed on 30 April 2022)) since 1976 highlighted that the vast amount of seismicity is associated with the above-mentioned and fast deformed active boundaries (Figure 1). Specifically, it is derived that the majority of the thrust fault plane solutions (fps) are concentrated in the proximity of continental collision of the Adriatic microplate and Eurasia in northwest Greece and along the Hellenic Subduction Zone active boundaries. The right lateral KTFZ and NATFZ are related to strike-slip fault plane solutions, and to a lesser extent in the southeastern Aegean area, where the left-lateral Rhoades Transform Fault is activated. Fast extension in N–S direction dominates in the back arc area, along with an E–W extension in the transition zone running parallel to both continental collision and oceanic subduction.

The spatial extension of the active boundaries, along with the available fault plane solutions allow the definition of seismic zones with distinctive seismotectonic properties in the study area, and then the detection of the appropriate M_{thr} above which the temporal crossover between correlated and statistically independent earthquakes is seeking. The 10 sub-areas that are defined, based on the aforementioned criteria, are shown in the map of Figure 2 (red polygons), along with the $M_w \geq 4.1$ crustal ($h \leq 40$ km) seismicity covering the time period 1975–2021. The compressional stress field of western Greece and Albania constitutes the sub-area 1, in which seismicity is concentrated along the main low angle thrust faulting structures with an NW–SE strike parallel to the coast [24] and the corresponding P-axes to be generally oriented perpendicular to the collision front (Figure 1). The dextral strike-slip KTFZ is the sub-area 2. KTFZ is a well-defined fault zone characterized by right-lateral faulting with a considerable thrust component [25]. It consists of the most active fault zone in Greece with the frequent occurrence of large ($M_w \geq 6.0$) earthquakes.

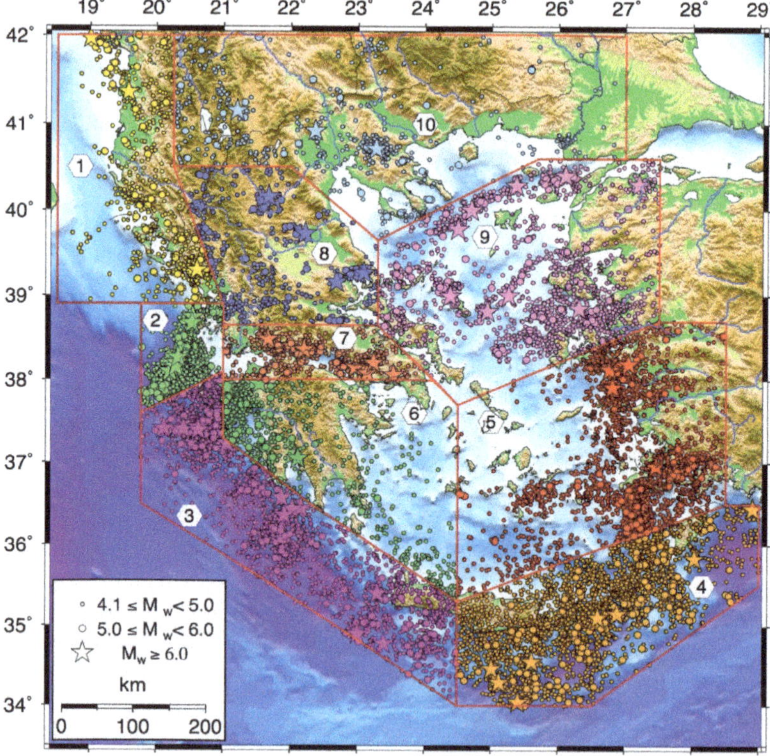

Figure 2. Epicentral distribution of Greek seismicity with $M_w \geq 4.1$ since 1975. Each different color illustrates the earthquakes of the 10 distinctive sub-areas that the broader Aegean region is divided (the borders of each sub-area are denoted with red polygons).

We divided the Hellenic Subduction Zone into two distinctive sub-areas, namely sub-areas 3 and 4, comprising the western and eastern parts of the Hellenic Subduction Zone, respectively. The Western Hellenic Arc is characterized by pure thrust fault plane solutions (Figure 1), in which the maximum compression, P-axis is oriented NNE–SSW along the Hellenic Arc, keeping the same orientation without being normal to the orientation of the subduction front. In the eastern part of the Hellenic Subduction Zone (sub-area 4) the available focal mechanisms evidence a more oblique motion.

The central part of the Aegean Sea, belonging to the extensional back arc area, is also separated into the eastern and western (including Peloponnese) parts, forming sub-areas 5 and 6, respectively. The Corinth Gulf, which is one of the most rapidly extending rifts worldwide, constitutes sub-area 7, while the central Greek mainland, is sub-area 8. Although all the four sub-areas belong to the extensional N–S back arc stress field, they are separated due to the fact that each of them constitutes distinctive and interconnected fault zones with different deformation rates [26].

The North Aegean Sea area is sub-area 9, comprising the right-lateral strike-slip North Aegean Trough Fault Zone (NATFZ), constituting the continuation of the North Anatolian Fault into the Aegean Sea [27], along with its major sub-parallel strands to the south, which are terminated to the west by the normal faults on the Greek mainland. It is characterized by dextral focal mechanisms with extensional components as well [28].

Northern Greece and the broader Balkans area constitute the 10th sub-area. It is a comparatively low seismicity area with the axis of maximum extension, T, to rotate from the N–S to an NE–SW orientation as one moves from its easternmost part to the west.

3. Data and Methodology

For the statistical correlations and the distribution fitting investigations on the 10 sub-region seismicity levels, we selected an earthquake catalog comprising the $M_w \geq 4.1$ events that occurred in the territory of Greece during 1975–2021, which we have taken from the regional catalog compiled by the Geophysics Department of the Aristotle University of Thessaloniki (Available online: http://geophysics.geo.auth.gr/ss (accessed on 30 April 2022)). The earthquake magnitudes are expressed in moment magnitude scale (M_w), as obtained directly from waveform modeling or equivalent M_w based on scaling relations suggested by Papazachos et al. [29]. We divided the catalog into 10 sub-catalogs corresponding to the 10 distinctive sub-areas. Figure 2 shows the epicentral distribution of the earthquakes comprised in each sub-catalog with different colors for each sub-area for the reader's ease. For each sub-catalog the samples of the earthquake interevent times, ΔT, data are then defined for earthquakes with magnitude thresholds of 0.1 bin increment, starting from $M_w{\geq}4.1$ (ΔT_1, ΔT_2, ... for the magnitude thresholds of $M_w \geq 4.1$, $M_w \geq 4.2$, ... and so on for each sub-area). These data sets are the inputs for applying the methodology, which is described below.

We seek for the threshold magnitude, M_{thr}, of a certain data set, above which the interevent times are proved to be independent. For this purpose, we investigated the correlation of ΔT in each subset, with the subsets being created for certain magnitude bins. We examined the correlations by calculating the Autocorrelation (ACF) and Partial Autocorrelation ($PACF$) functions, a widely used approach for the same purpose (e.g., [30] for global earthquake data; [31] in induced seismicity of Poland).

The Autocorrelation Function (ACF),

$$\rho_\kappa = \frac{\sum_{i=1}^{n-k}(x_i - \overline{x})(x_{i+k} - \overline{x})}{\sum_{i=1}^{n}(x_i - \overline{x})^2} \tag{1}$$

where n is the number of the observations, k is the number of the lags and $\overline{x}$ is the mean value of the observations, and is used for the investigation of correlations between past and future values of a given time series by assuming a confidence level (the 95% confidence level in the current application). If a certain ρ_κ value at lag k exceeds the $\pm$95% confidence interval then the given time series could be considered as correlated for the specific lag, otherwise, it could be considered statistically independent. Then, the possible correlations detected by the *ACF* can be confirmed in the same way based on the values of the Partial Autocorrelation Function that are given by

$$r_{k,k} = \frac{\rho_k - \sum_{i=1}^{k-1} r_{k-1,j}\rho_{k-j}}{1 - \sum_{i=1}^{k-1} r_{k-1,j}\rho_j} \tag{2}$$

where $r_{k,j} = r_{k-1,j} - r_{k,k}r_{k-1,j-i}$ for $j = 1, 2, \dots, k$.

Additionally, the Ljung–Box Q-test [32], which is a modified version of the Box–Pierce Portmanteau test, is applied as an alternative method to investigate correlations among ΔT. This test is implemented under the null hypothesis that the time series exhibits no autocorrelation for a fixed number of lags, L, against the alternative that some (statistically significant) autocorrelation exists. The test statistic is given by

$$Q = n(n+2) \sum_{i=1}^{L} \left(\frac{\rho_k^2}{n-k} \right). \tag{3}$$

The asymptotic distribution of Q is Chi-square (x^2) with L degrees of freedom. If the statistic of the test for the given lag L is less than the critical value of the Chi-square ($Q_{stat} < Q_{critical}$), the null hypothesis cannot be rejected.

For further confirmation of the possible crossover M_{thr} and for comparing our results with an independent method, we applied the Detrended Fluctuations Analysis [33] in all the 10 sub-regions. DFA is implemented under the hypothesis that if a given time series is correlated then the fluctuation function, $F(n)$, follows an increasing power-law trend according to the factor n^{α} ($F(n) \sim n^{\alpha}$), where n is the size of the window and α is the scaling exponent. If the value of the exponent α is larger than 0.5 ($\alpha > 0.5$) then the data sample is positively correlated, whereas if α is smaller than 0.5 ($\alpha < 0.5$) the data are negatively correlated (anti-correlated). For the case when the exponent α is equal to 0.5 ($\alpha = 0.5$), the sample could be considered statistically independent. The exponent α is calculated as the slope of the linear regression of the logarithm of the fluctuation function, $F(n)$, against the logarithm of the window size n.

The next step after the detection of the M_{thr}, above which the ΔT becomes statistically uncorrelated, is the determination of the best performed statistical distribution that each ΔT sample follows. Over the years, many statistical distributions such as the Weibull [34], the Lognormal [35,36], the Stretched Exponential [37] and the Gamma distribution [7,38] have been proposed. In this study, we applied the four most popular statistical distributions in the relevant studies (e.g., [39,40]), namely the Weibull, the Gamma, the Lognormal and the Exponential, to each ΔT data set for the 10 sub-areas. The probability density functions (*pdf*) of the four candidate distributions are given in Table 1. The parameter estimation for each distribution is achieved via the application of the Maximum Likelihood Estimation (MLE) method using the respective log-likelihood formulae [41].

In order to compare the distribution's performance and select the one that best fits the observations, we applied the Anderson–Darling non-parametric goodness of fit test (A–D test) [42]. The A–D test is implemented by calculating the distance, A^2, between the empirical cumulative distribution function (*ecdf*) and the cumulative distribution function (*cdf*) of the distribution applied to the data. The A–D test statistic is given by

$$A^2 = -n - \sum_{i=1}^{n} \frac{2i-1}{n} [lnF(x_i) + ln(1 - F(x_{n+1-1}))], \tag{4}$$

where $\{x_1, x_2, \dots, x_n\}$ are the ordered sample data points, n is the number of observations and F is the cdf of the distribution under study. The test compares the quantity A^2 with a critical value, c, under the null hypothesis that the data are distributed according to F.

If A^2 is less than or equal to the critical value, then the null hypothesis cannot be rejected. The decision of rejecting or not the null hypothesis in the present study is based on the obtained p-values, compared with the significance level, which is equal to 0.05 ($\alpha = 0.05$). If the p-value is greater than α (p-value $> \alpha$) then the null hypothesis can not be rejected. On the contrary, if the p-value is lower than α (p-value $< \alpha$) the null hypothesis can be rejected.

Table 1. Definitions of the probability density functions (pdf) of the Gamma, Lognormal, Weibull and Exponential distributions applied in the statistically independent interevent times data, ΔT, of the 10 sub-areas of broader Aegean region.

Distribution	Probability Density Function	Parameters
Gamma	$f(x\|\kappa,\theta) = \frac{1}{\theta^\kappa \Gamma(\kappa)} x^{\kappa-1} \exp\left\{\frac{-x}{\theta}\right\}$	κ (shape) θ (scale)
Lognormal	$f(x\|\mu,\sigma) = \frac{1}{x\sigma\sqrt{2\pi}} \exp\left\{-\frac{(lnx-\mu)^2}{2\sigma^2}\right\}$	μ (natural logarithm of mean value) σ (natural logarithm of standard deviation)
Weibull	$f(x\|\alpha,b) = \frac{b}{\alpha}\left(\frac{x}{b}\right)^{b-1} \exp\left\{-\left(\frac{x}{\alpha}\right)^b\right\}$	α (scale) b (shape)
Exponential	$f(x\|\mu) = \frac{1}{\mu} \exp\left\{-\frac{x}{\mu}\right\}$	μ (mean value)

To further compare the four distributions, we calculated the Akaike [43] and Bayesian [44] Information Criteria (AIC and BIC, respectively). From these calculations, the distribution that displays the best performance is the one with the minimum value of the criterion in both cases. The values of AIC and BIC are given by

$$AIC = -2lnL + 2k \tag{5}$$

and

$$BIC = -2lnL + kln(n) \tag{6}$$

where lnL stands for the value of the log-likelihood function obtained from the MLE approach, k is the number of parameters of the distribution and n is the number of the observations.

4. Application

4.1. Identifying the Earthquake Interevent Time Correlations

The detection of the correlations is investigated through time series analysis. Specifically, the *ACF* and *PACF* values, along with the statistics of Ljung–Box Q-test are sequentially applied for each ΔT_i data sample (ΔT_1, ΔT_2, ... , ΔT_n for $M_w \geq 4.1$, $M_w \geq 4.2$, ... , respectively) for a given sub-area until the temporal crossover between correlated and statistically independent events is determined. It is worth mentioning that the 95% confidence intervals of *ACF* and *PACF* are calculated separately for each data sample ΔT_i, which obtains a different length.

Figures 3–5 illustrate the results of the correlation analysis for the ΔT data set for the central Ionian Islands area (sub-area 2). The values of *ACF* (Figure 3) exhibit strong positive autocorrelation until the magnitude threshold $M_{thr} \geq 4.6$ (Figure 3a–f). From this M_{thr} onwards, the autocorrelation becomes weaker (Figure 3g,h) until $M_{thr} \geq 4.9$ (Figure 3i) when no significant autocorrelations of any lag are recorded, and the relevant samples of ΔT can be considered statistically uncorrelated. The values of *PACF* (Figure 4) confirm this crossover between correlated and statistically independent events at the magnitude threshold of $M_{thr} \geq 4.9$.

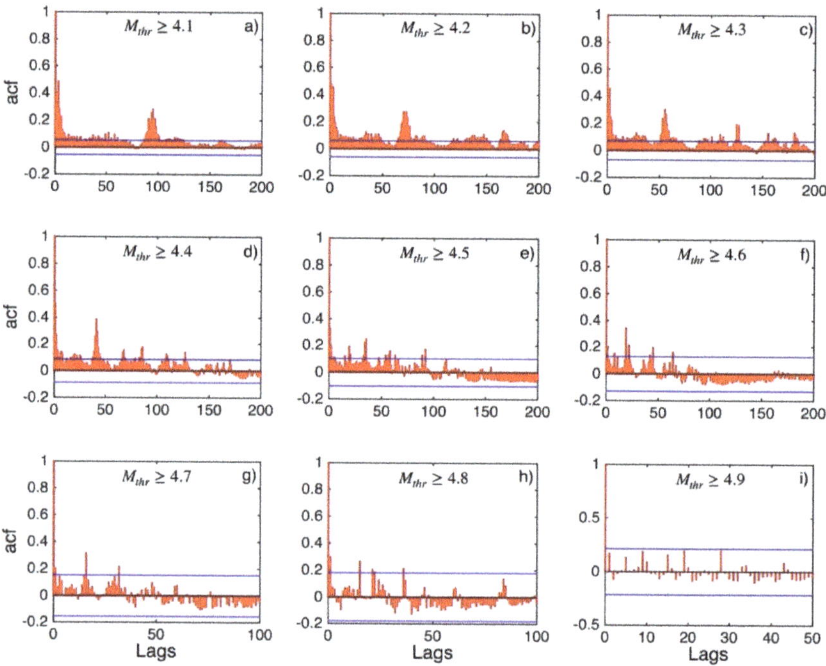

Figure 3. *ACF* values (red stems) and their 95% confidence intervals (blue lines) for central Ionian Islands (sub-area 2) for magnitude thresholds (M_{thr}) varying from 4.1 to 4.9 (subplots **a–i**).

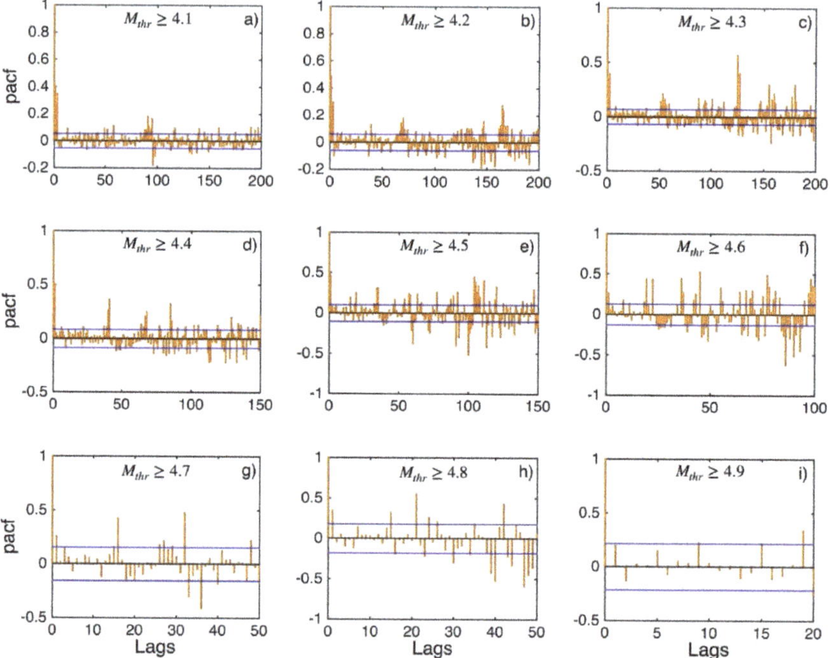

Figure 4. *PACF* values (orange stems) and their 95% confidence intervals (blue lines) for central Ionian Islands (sub-area 2) for magnitude thresholds (M_{thr}) varying from 4.1 to 4.9 (subplots **a–i**).

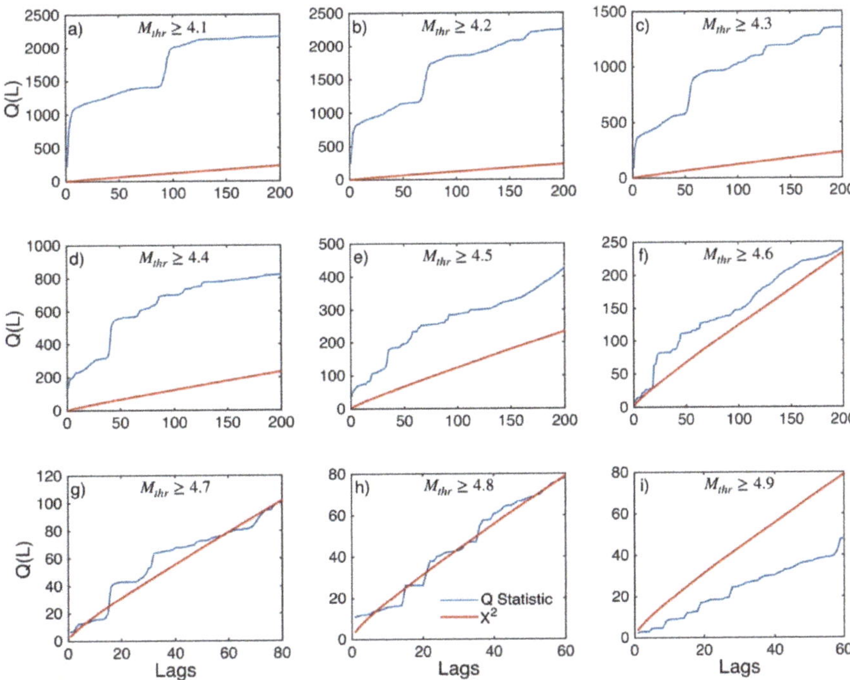

Figure 5. Statistics of Ljung–Box Q-test (blue lines) and the critical values for the X^2 distribution (red lines) for central Ionian Islands (sub-area 2) for magnitude thresholds (M_{thr}) varying from 4.1 to 4.9 (subplots **a–i**).

For securing that the finding of the memory weakening as the magnitude threshold increases is not biased because of the progressive decrease in the data samples length, the computation of the number of lags exceeding the respective level of significance over the total number of lags for both the *ACF* and *PACF* values is implemented. The ΔT data samples of this sub-region exhibit an increase in the ratio for the ACF values from 0.33 to 0.60 for the $M_{thr} \geq 4.1$ and $M_{thr} \geq 4.2$, respectively. From an $M_{thr} \geq 4.3$ and above we observe a systematic decrease in this ratio for *ACF* values. We obtained the same results for the ratio related to the respective values of *PACF*. Taking into account these results, it could be stated that the smaller length of the respective data samples as the M_{thr} values increase does not bias the results of the *ACF* and *PACF* calculations, implying weakening of the memory with respect to the increase in magnitude threshold.

The results of the Ljung–Box Q-test (Figure 5) unveil the same temporal behavior of the ΔT data samples. Specifically, it is observed that the statistic of the test (Q_{stat}; blue lines in Figure 5) is much larger than the critical value ($Q_{critical}$; red lines in Figure 5) of the Chi-square distribution, and consequently, the null hypothesis that the ΔT data samples are independent can be rejected up to $M_{thr} \geq 4.8$ (Figure 5a–h). The ΔT data samples in which the Q statistic is clearly smaller than the critical value is again one for the $M_{thr} \geq 4.9$ earthquakes (Figure 5i), confirming the results of the *ACF* and *PACF* calculations. In this respect, the magnitude threshold above which the ΔT data of the central Ionian Islands appear statistically independent is for the $M_{thr} \geq 4.9$ earthquakes.

Table 2 presents the results of the aforementioned workflow for the detection of the temporal crossover among correlated and statistically independent earthquakes for the 10 sub-areas. In all the 10 cases, the results of *ACF–PACF* analysis are in excellent agreement with those of the Ljung–Box Q-test (the detailed results of both *ACF–PACF* analysis and the Ljung–Box Q-test are given graphically in Appendix A). The temporal crossover of the earthquakes interevent times data ranges from $M_{thr} \geq 4.7$ up to $M_{thr} \geq 5.1$. The lower magnitude threshold ($M_{thr} \geq 4.7$) is observed in the southwest Aegean Sea and Peloponnese (sub-area 6), whereas the largest ($M_{thr} \geq 5.1$) in western Greece and Albania (sub-area 1). Two out of the ten sub-areas (western Hellenic Arc and northern Greece and Balkans) are exhibiting temporal crossover for $M_{thr} \geq 4.8$ threshold. Similarly, the ΔT data samples of sub-areas 2 and 7, namely the central Ionian Islands and Corinth Gulf, respectively, are statistically independent at $M_{thr} \geq 4.9$ thresholds. The eastern Hellenic Arc, southeast Aegean Sea, central Greece and North Aegean Sea (sub-areas 4, 5, 8 and 9, respectively) are exhibiting a temporal crossover for earthquakes with $M_{thr} \geq 5.0$.

Table 2. Summary of the identification of the magnitude threshold (M_{thr}) above which interevent time data, ΔT, are statistically independent for the 10 sub-areas of the broader Aegean region.

Sub-Area	M_{thr} Identified Using the ACF and PACF Values	M_{thr} Identified Using the Ljung–Box Q-Test	Sample Size (Interevent Times)
1-Western Greece and Albania	5.1	5.1	50
2-Central Ionian Islands	4.9	4.9	88
3-Western Hellenic Arc	4.8	4.8	260
4-Eastern Hellenic Arc	5.0	5.0	130
5-Southeast Aegean Sea	5.0	5.0	80
6-Southwest Aegean Sea and Peloponnese	4.7	4.7	100
7-Corinth Gulf	4.9	4.9	57
8-Central Greece	5.0	5.0	45
9-North Aegean Sea	5.0	5.0	111
10-Northern Greece and Balkans	4.8	4.8	69

To support these results, the Detrended Fluctuations Analysis (DFA) is performed for all sub-regions and M_{thr} values, in which the ACF–PACF analysis and Ljung–Box Q-test are applied. Figure 6 summarizes the results of the calculation of the exponent α for each sub-region versus the respective M_{thr} values. In all cases, a systematic decrease in the exponent α with an increase in the magnitude threshold can be observed. In all cases, α ranges from values that indicate strong positive correlations (around 0.8) to ones that reveal weaker correlations or statistically independent behavior (from 0.65 to 0.33). More specifically, exponent α is obtaining values near 0.5 for the highest M_{thr} values (except in the case of sub-region 9, which is equal to 0.67). These latter results are in very good agreement with the initially presented ones of the ACF–PACF analysis and Ljung–Box Q-test.

4.2. Distribution of the Statistically Independent Earthquake Interevent Times

After the detection of the magnitude thresholds above where the ΔT intervals are considered statistically independent, the statistical distribution of those data sets is investigated, with the application of the Weibull, Lognormal, Gamma and Exponential distributions. The parameters of each distribution are estimated via the MLE method according to their respective formulations, along with their 95% confidence intervals. Tables 3 and 4 present the estimation results, along with the respective 95% confidence intervals and the values of the negative log-likelihood functions for sub-areas 1–5 and 6–10, respectively. An interesting point arising from the MLE approach is that the shape parameters of both the Gamma and Weibull distributions, k and b, respectively, obtain values below one in all cases (in all 10 sub-areas). This is an indicator of a clustering-type behavior of earthquake occurrence [45].

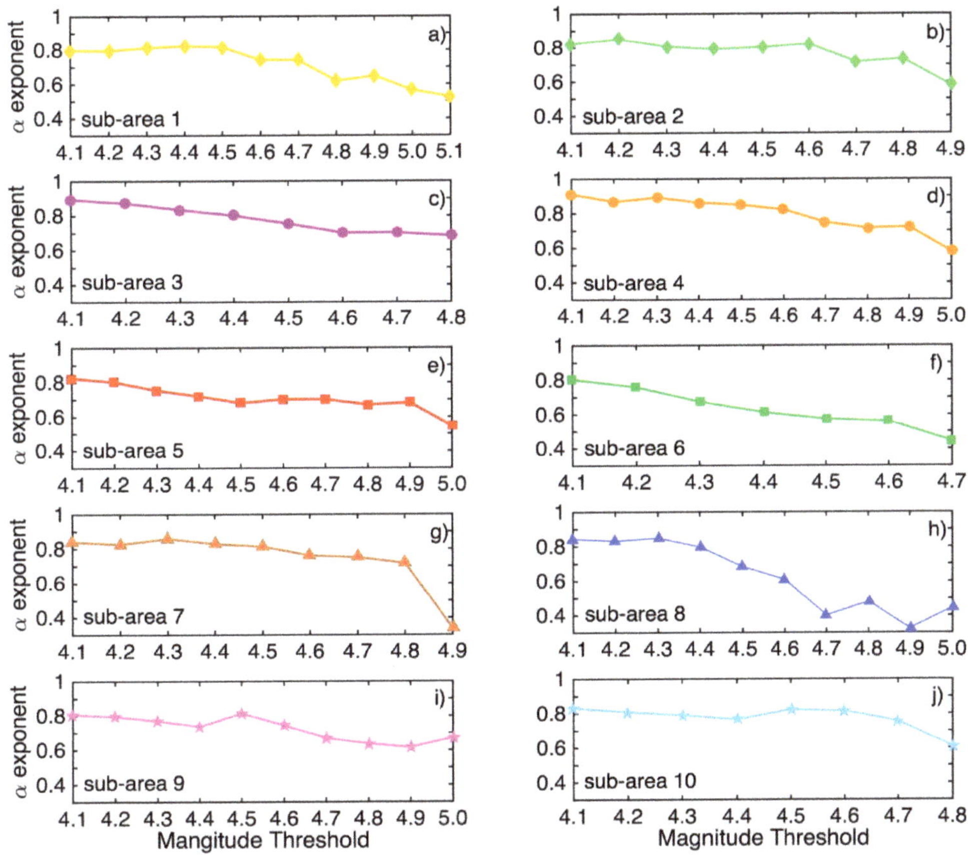

Figure 6. Exponent α of Detrended Fluctuations Analysis as a function of M_{thr} for the 10 sub-areas (subplots **a–j**).

Table 3. MLE parameters estimates, 95% confidence intervals, log-likelihood values calculation, Akaike and Bayesian Information Criteria calculated values (AIC and BIC, respectively), along with the estimated p-values of the Anderson–Darling Goodness of Fit test for the four applied statistical distributions on the statistically independent interevent times data, ΔT, of the sub-areas 1 to 5.

Sub-Area	Distribution	Parameters	-Log$L$	AIC	BIC	A–D Test p-Value
	Gamma	$k = 0.38$ [0.29, 0.53] $\theta = 864.34$ [497.05, 1503.01]	310.25	624.50	628.28	0.690
1	Lognormal	$\mu = 4.08$ [3.28, 4.87] $\sigma = 2.76$ [2.30, 3.45]	318.72	641.43	645.22	0.049
	Weibull	$\alpha = 192.99$ [111.07, 338.35] $b = 0.52$ [0.42, 0.65]	310.99	625.99	629.77	0.445
	Exponential	$\mu = 331.79$ [255.46, 448.48]	333.42	668.84	670.73	2.531×10^{-5}

Table 3. *Cont.*

Sub-Area	Distribution	Parameters	-Log*L*	AIC	BIC	A–D Test *p*-Value
2	Gamma	$k = 0.22$ [0.17, 0.28] $\theta = 1021.21$ [570.65, 1826.70]	313.14	630.2954	634.6747	0.092
	Lognormal	$\mu = 2.02$ [1.09, 3.14] $\sigma = 3.78$ [3.23, 4.57]	314.82	632.5410	636.9203	0.061
	Weibull	$\alpha = 47.69$ [21.23, 107.16] $b = 0.31$ [0.42, 0.65]	314.27	633.6488	638.0281	0.065
	Exponential	$\mu = 331.79$ [255.46, 448.48]	422.11	846.2366	848.4262	9.090×10^{-6}
3	Gamma	$k = 0.39$ [0.34, 0.45] $\theta = 164.12$ [129.27, 208.36]	1226.15	2456.31	2463.44	0.809
	Lognormal	$\mu = 2.48$ [2.15, 2.80] $\sigma = 2.65$ [2.44, 2.90]	1266.50	2537.01	2544.14	1.762×10^{-4}
	Weibull	$\alpha = 38.37$ [30.10, 49.91] $b = 0.52$ [0.47, 0.58]	1231.50	2467.01	2474.13	0.109
	Exponential	$\mu = 64.32$ [57.16, 72.90]	1342.59	2687.18	2690.74	2.308×10^{-5}
4	Gamma	$k = 0.48$ [0.39, 0.59] $\theta = 266.75$ [194.03, 366.71]	729.66	1463.31	1469.05	0.822
	Lognormal	$\mu = 3.53$ [3.11, 3.95] $\sigma = 2.43$ [2.16, 2.76]	758.40	1520.79	1526.52	0.004
	Weibull	$\alpha = 94.50$ [70.62, 126.46] $b = 0.62$ [0.54, 0.71]	733.45	1470.90	1476.64	0.335
	Exponential	$\mu = 128.60$ [109.07, 153.93]	761.38	1524.76	1527.62	4.617×10^{-6}
5	Gamma	$k = 0.29$ [0.23, 0.37] $\theta = 687.91$ [429.02, 1130.01]	433.65	871.29	876.06	0.605
	Lognormal	$\mu = 2.96$ [2.16, 3.75] $\sigma = 3.54$ [3.07, 4.20]	450.92	905.84	910.60	0.016
	Weibull	$\alpha = 90.73$ [51.92, 158.77] $b = 0.41$ [0.34, 0.49]	438.83	881.65	886.42	0.138
	Exponential	$\mu = 202.35$ [164.42, 255.19]	504.80	1011.60	1013.98	7.500×10^{-6}

Table 4. MLE parameters estimates, 95% confidence intervals, log-likelihood values calculation, Akaike and Bayesian Information Criteria calculated values (AIC and BIC, respectively), along with the estimated p-values of the Anderson–Darling Goodness of Fit test for the four applied statistical distributions on the statistically independent interevent times data, ΔT, of the sub-areas 6 to 10.

Sub-Area	Distribution	Parameters	-Log$L$	AIC	BIC	A–D Test p-Value
6	Gamma	$k = 0.54$ [0.43, 0.68] $\theta = 297.33$ [209.32, 422.38]	596.96	1197.91	1203.14	0.489
	Lognormal	$\mu = 3.91$ [3.43, 4.38] $\sigma = 2.40$ [2.11, 2.78]	626.13	1256.25	1261.48	0.001
	Weibull	$\alpha = 129.01$ [95.63, 174.04] $b = 0.68$ [0.34, 0.49]	600.18	1204.36	1209.59	0.261
	Exponential	$\mu = 160.18$ [133.02, 196.66]	613.71	1229.41	1232.03	0.004
7	Gamma	$k = 0.27$ [0.29, 0.36] $\theta = 1092.53$ [614.10, 1943.69]	320.67	645.33	649.42	0.628
	Lognormal	$\mu = 3.09$ [2.07, 4.12] $\sigma = 3.86$ [3.261, 4.73]	333.77	671.55	675.63	0.034
	Weibull	$\alpha = 118.51$ [58.18, 241.43] $b = 0.38$ [0.31, 0.47]	324.89	653.78	657.87	0.171
	Exponential	$\mu = 295.46$ [231.59, 390.10]	381.25	764.49	766.54	1.052×10^{-5}
8	Gamma	$k = 0.26$ [0.19, 0.36] $\theta = 1452.99$ [745.40, 2832.30]	252.93	509.86	513.43	0.315
	Lognormal	$\mu = 3.18$ [2.04, 4.31] $\sigma = 3.73$ [3.08, 4.72]	259.64	523.28	526.85	0.189
	Weibull	$\alpha = 130.99$ [55.36, 309.98] $b = 0.36$ [0.28, 0.46]	255.63	515.26	518.83	0.215
	Exponential	$\mu = 373.98$ [284.09, 514.69]	304.66	611.33	613.11	1.363×10^{-5}

Table 4. *Cont.*

Sub-Area	Distribution	Parameters	-LogL	AIC	BIC	A–D Test *p*-Value
9	Gamma	$k = 0.25$ [0.21, 0.31] $\theta = 589.20$ [386.32, 898.62]	533.32	1070.65	1076.07	0.090
	Lognormal	$\mu = 2.22$ [1.46, 2.98] $\sigma = 4.05$ [3.58, 4.66]	559.08	1122.17	1127.59	0.001
	Weibull	$\alpha = 56.14$ [32.47, 97.08] $b = 0.38$ [0.30, 0.42]	543.85	1091.70	1097.12	0.011
	Exponential	$\mu = 149.83$ [125.43, 182.13]	667.05	1336.10	1338.81	5.405×10^{-6}
10	Gamma	$k = 0.34$ [0.26, 0.45] $\theta = 638.99$ [394.00, 1035.81]	397.61	799.12	803.68	0.419
	Lognormal	$\mu = 3.42$ [2.64, 4.20] $\sigma = 3.24$ [2.78, 3.90]	414.74	833.47	837.94	0.012
	Weibull	$\alpha = 122.94$ [72.54, 208.36] $b = 0.47$ [0.38, 0.57]	402.44	808.88	813.35	0.127
	Exponential	$\mu = 219.10$ [157.37, 281.60]	440.87	883.75	885.99	8.695×10^{-6}

The Anderson–Darling (A–D) test is then applied to each sample in order to compare the distributions derived via the MLE parameter estimates and empirical cdf (Figures 7 and 8 for sub-areas 1–5 and 6–10, respectively). Results of the A–D goodness of fit test (Tables 3 and 4) show that in all cases the Exponential distribution can be rejected since the respective *p*-values are getting lower than the 0.05 level of significance (*p*-value < 0.05). Similarly, the Lognormal distribution is also rejected in 8 out of 10 cases. The rejected distributions report *p*-values larger than 0.05 for the central Ionian Islands and central Greece areas (sub-areas 2 and 8; Table 3). The Gamma and Weibull distributions perform better than the others in all the 10 ΔT data sets. Gamma distribution exhibits slightly better performance than the Weibull since the respective *p*-values are always larger than the ones of Weibull. These results are also confirmed by the obtained values of both Information Criteria (AIC and BIC; Tables 3 and 4). In all 10 cases, the Gamma distribution is the one that reports the minimum values for both criteria.

Summarizing the results of the comparison of the distributions, although both the Gamma and Weibull distributions can be accepted according to the A–D test in all cases (also the Lognormal one in the data sampled of sub-areas 2 and 8), the *p*-values of the test are smaller for the Gamma distribution. The values of AIC and BIC further confirm that Gamma distribution best fits the data since the relevant values are the lowest among the four. By combining the results of the A–D test with those of the information criteria, we found that the Gamma distribution fits better than the other three distributions for all data sets. This result is in agreement with Corral's analysis [7,8], where the earthquake interevent times follow a universal scaling law independent of the region and the M_c, and can be modeled by the Gamma distribution.

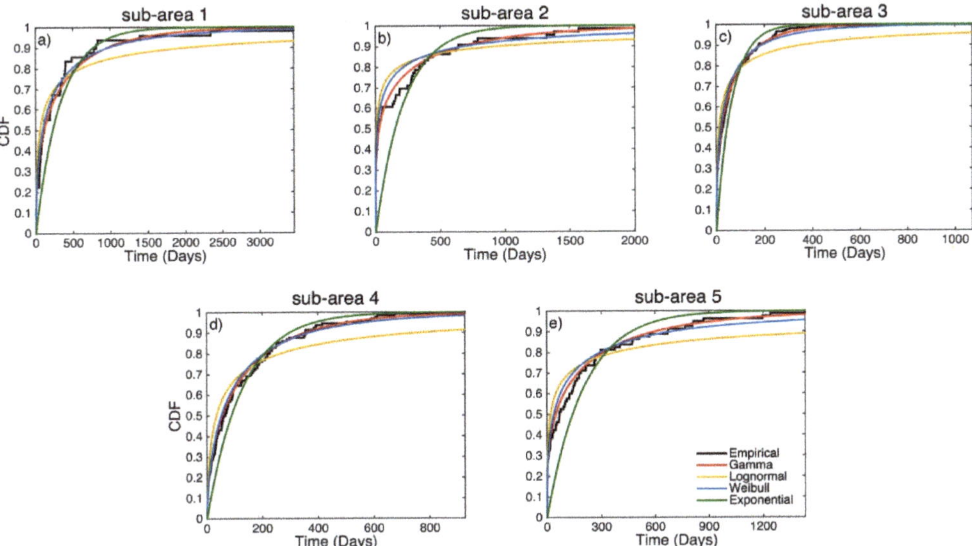

Figure 7. Comparison of empirical cdf (black lines) and the cdfs of Gamma (red lines), Lognormal (yellow lines), Weibull (blue lines) and Exponential (green lines) applied distributions for sub-areas 1 to 5 (subplots **a–e**).

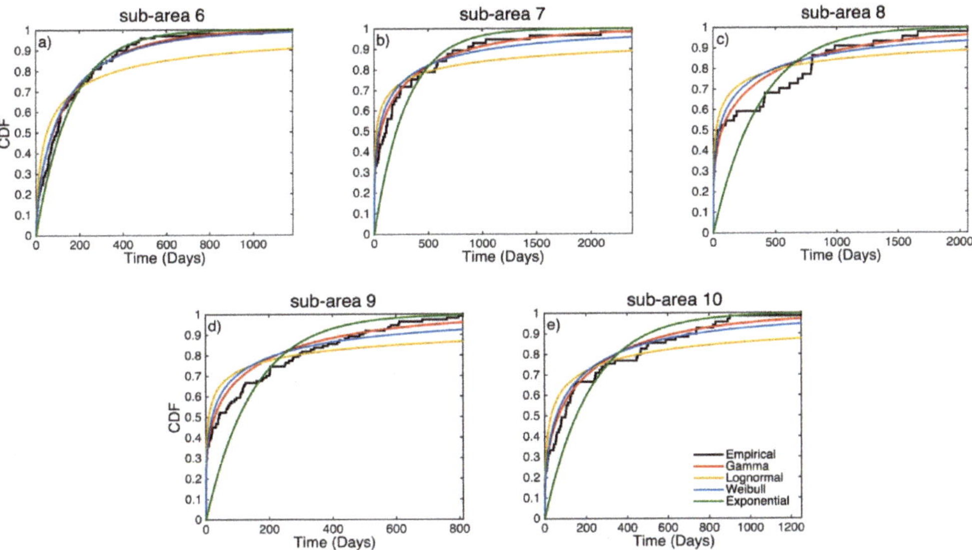

Figure 8. Comparison of empirical cdf (black lines) and the cdfs of Gamma (red lines), Lognormal (yellow lines), Weibull (blue lines) and Exponential (green lines) applied distributions for sub-areas 6 to 10 (subplots **a–e**).

As already stated, the estimated value of shape parameter, k, of Gamma distribution is found to be smaller than one in all the 10 ΔT data samples, independent of the temporal crossover magnitude threshold. Specifically, k ranges from 0.22 to 0.54 in the 10 sub-areas. This is an interesting finding from a seismological point of view because k plays an important role in earthquake occurrence applications; it could be considered the regulation parameter of the earthquake occurrence process [45]. Specifically, if $k < 1$, the temporal behavior of seismicity could be considered clustered, whereas if $k > 1$, the process tends to be quasi-periodic. In the case of $k = 1$, the process is neither periodic nor clustered, representing the memory-less Poisson process. In this respect, the estimated values of k indicate that although the ΔT data are statistically independent, there is still a weak inherent memory. This implies that earthquake interevent times above the given magnitude threshold of the temporal crossover are members of a renewal process, instead of the memory-less Poisson one.

5. Conclusions

The detection of the threshold magnitude, M_{thr}, above which the earthquake interevent times, ΔT, might be considered statistically independent, is investigated through the *ACF* and *PACF* values along with the application of the Ljung–Box Q-test in 10 distinctive sub-areas of the Greek territory. The analysis revealed that in all cases the results of the Ljung–Box Q-test adequately agree with the results derived by *ACF* and *PACF*. Detrended Fluctuations Analysis results further confirm the weakening of the memory as the M_{thr} of earthquake interevent time data increases. The obtained temporal crossover of the ΔT data ranges from $M_{thr} \geq 4.7$ up to $M_{thr} \geq 5.1$ in the 10 sub-areas. These findings are statistically very important because one can determine above which magnitude threshold the property of mutual independence is fulfilled. This independency property concerns either the events occurring in two different time intervals or only the occurrence time of the last event but not the full seismicity history. The proposed workflow for the investigation of the crossover behavior of ΔT data is easy to apply and capable of providing reliable results aimed at fixing the methodological issue of statistical independence. In other words, it could be considered as a routine initial step before a certain stochastic model application in earthquake interevent time data.

Once the magnitude thresholds above which the ΔT data samples are statistically independent are identified, the Gamma, Weibull, Lognormal and Exponential distributions are applied and compared in the respective data samples in order to model their temporal behavior. The comparison, which was implemented via the combination of the A–D goodness of fit test and the values of AIC and BIC criteria, revealed that the Gamma distribution exhibits the best performance. This latter result agrees with the analysis performed by Corral [7,8], who claimed that the earthquake interevent times follow a universal scaling law which could be represented by the Gamma distribution. This fact implies that the ΔT can be described better with a general renewal process rather than the Markovian Poisson process.

The values of the shape parameter, k, of Gamma distribution, which characterizes the seismic process, in all cases are estimated as $k < 1$ implying a clustering behavior of seismicity. In such cases, the earthquake occurrence probability soon after a strong earthquake is high, and most events occur at times less than the mean interevent time.

It is notable that sub-areas with low seismic activity are associated with larger values of k. In more detail, the largest estimated value of k is equal to 0.54 in the southwestern Aegean Sea and Peloponnese (sub-area 6), an area with considerably lower seismicity than the Corinth Gulf (sub-area 7) and North Aegean Sea (sub-area 9), in which the estimated k values are equal to 0.27 and 0.25, respectively. The smallest value of k ($k = 0.22$) is estimated in the central Ionian Islands (sub-area 2), which exhibits the highest moment rate in the Mediterranean region.

This study's results provide information on the temporal behavior of the ΔT of moderate to large earthquakes over a sufficiently long period (1975–2021). These findings might be potentially used for both additional statistical applications (e.g., Stress Release applications), in which the need for documenting the independency magnitude cut-off is necessary, and as initial earthquake occurrence forecast models in intermediate time scales, which can be considered as inputs for improving regional seismic hazard assessment investigations (e.g., [46]).

Author Contributions: Conceptualization, C.K., G.T., E.P. and V.K.; data curation, C.K., E.P. and V.K.; formal analysis, C.K., G.T., E.P. and V.K.; investigation, C.K.; methodology, C.K. and G.T.; Validation, C.K.; writing—original draft, C.K.; writing—review and editing, C.K., G.T., E.P. and V.K. All authors have read and agreed to the published version of the manuscript.

Funding: This research received no external funding.

Institutional Review Board Statement: Not applicable.

Informed Consent Statement: Not applicable.

Data Availability Statement: The regional earthquake catalog of the Geophysics Department of the Aristotle University of Thessaloniki is accessible from http://geophysics.geo.auth.gr/ss (last accessed on 30 April 2022). Fault plane solution data used in Figure 1 came from https://www.globalcmt.org/ (last accessed on 30 April 2022).

Acknowledgments: We greatly appreciate the editorial assistance of A. Zavyalov as well as the constructive comments of three anonymous reviewers who contributed to the improvement of the manuscript. The GMT software [47] is used for creating the maps. Some figures and calculations are produced using the MATLAB software (Available online: http://www.mathworks.com/products/matlab (last accessed on 30 April 2022)). Geophysics Department Contribution 962.

Conflicts of Interest: The authors declare no conflict of interest.

Appendix A

In this section the detailed results of both *ACF–PACF* analysis and the Ljung–Box Q-test are presented graphically for the nine sub-areas, except the ones in the Central Ionian Islands (sub-area 2), which are shown in the main body of the study.

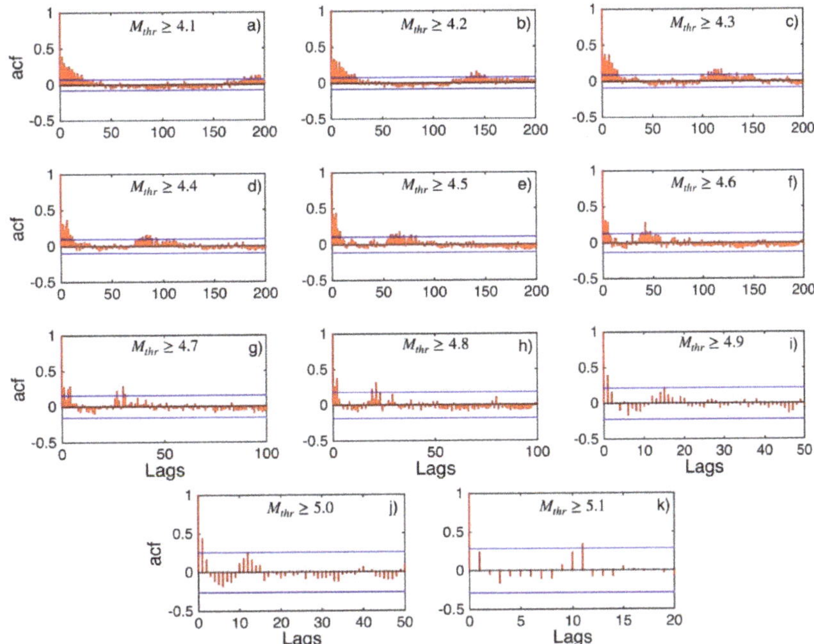

Figure A1. *ACF* values (red stems) and their 95% confidence intervals (blue lines) for Western Greece and Albania (sub-area 1) for magnitude thresholds (M_{thr}) varying from 4.1 to 5 (subplots **a–k**).

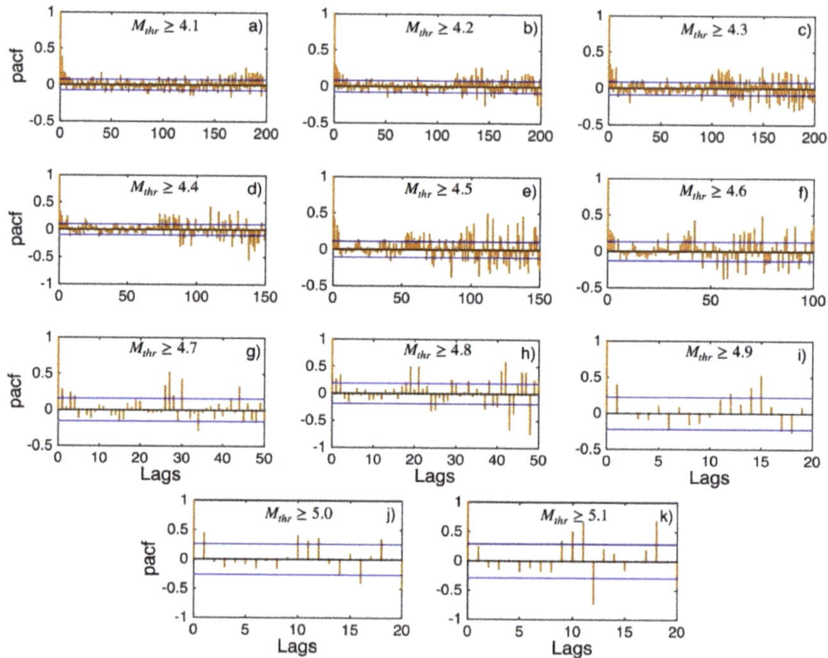

Figure A2. *PACF* values (orange stems) and their 95% confidence intervals (blue lines) for Western Greece and Albania (sub-area 1) for magnitude thresholds (M_{thr}) varying from 4.1 to 5.1 (subplots **a–k**).

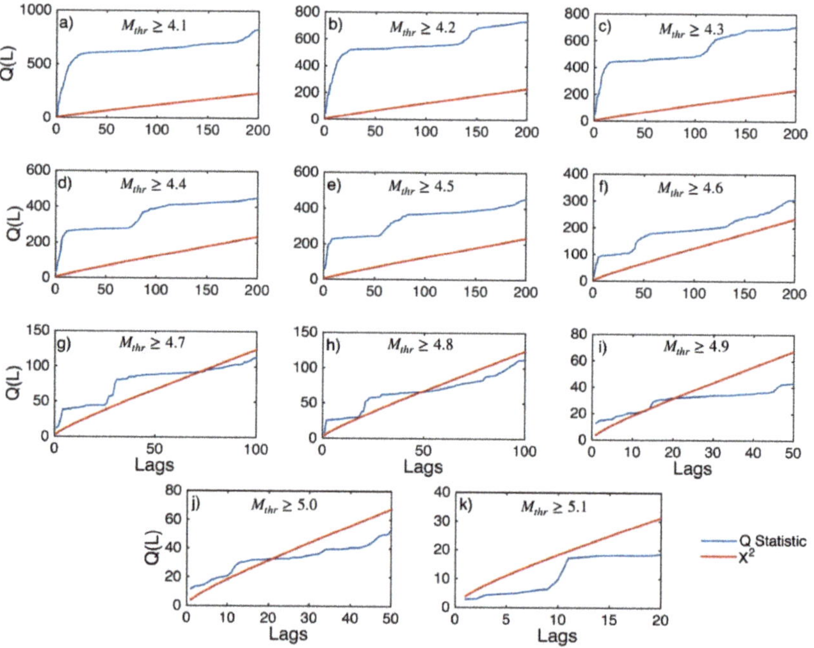

Figure A3. Statistics of Ljung–Box Q-test (blue lines) and the critical values for the X^2 distribution (red lines) for Western Greece and Albania (sub-area 1) for magnitude thresholds (M_{thr}) varying from 4.1 to 5.1 (subplots **a–k**).

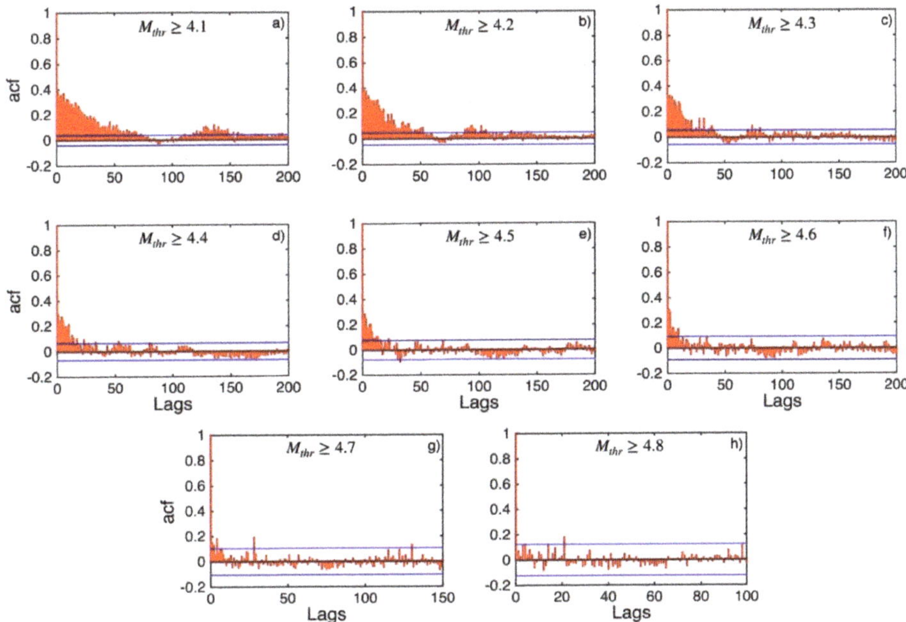

Figure A4. *ACF* values (red stems) and their 95% confidence intervals (blue lines) for Western Hellenic Arc (sub-area 3) for magnitude thresholds (M_{thr}) varying from 4.1 to 4.8 (subplots **a–h**).

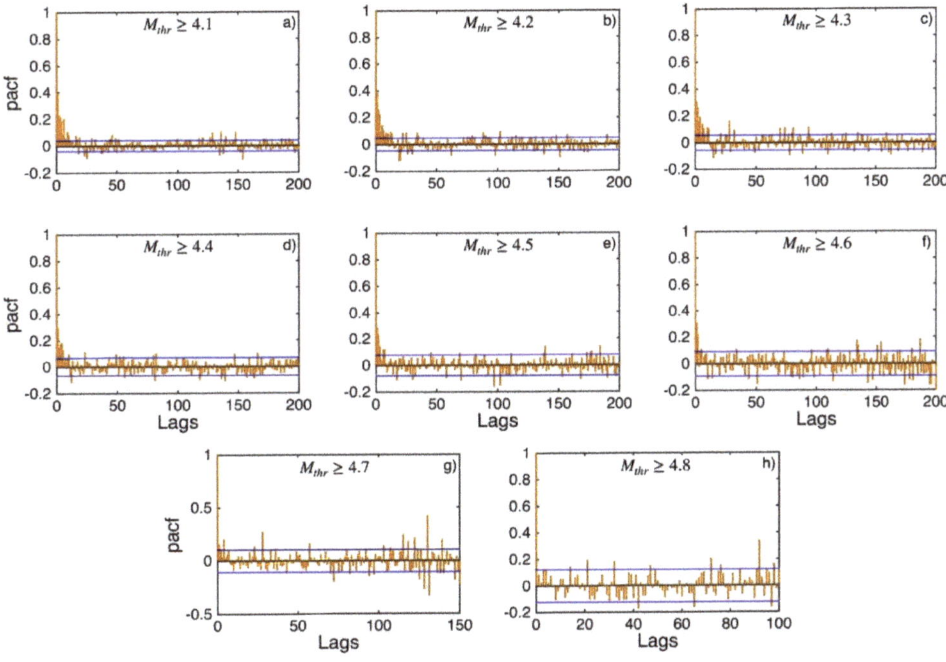

Figure A5. *PACF* values (orange stems) and their 95% confidence intervals (blue lines) for Western Hellenic Arc (sub-area 3) for magnitude thresholds (M_{thr}) varying from 4.1 to 4.8 (subplots **a–h**).

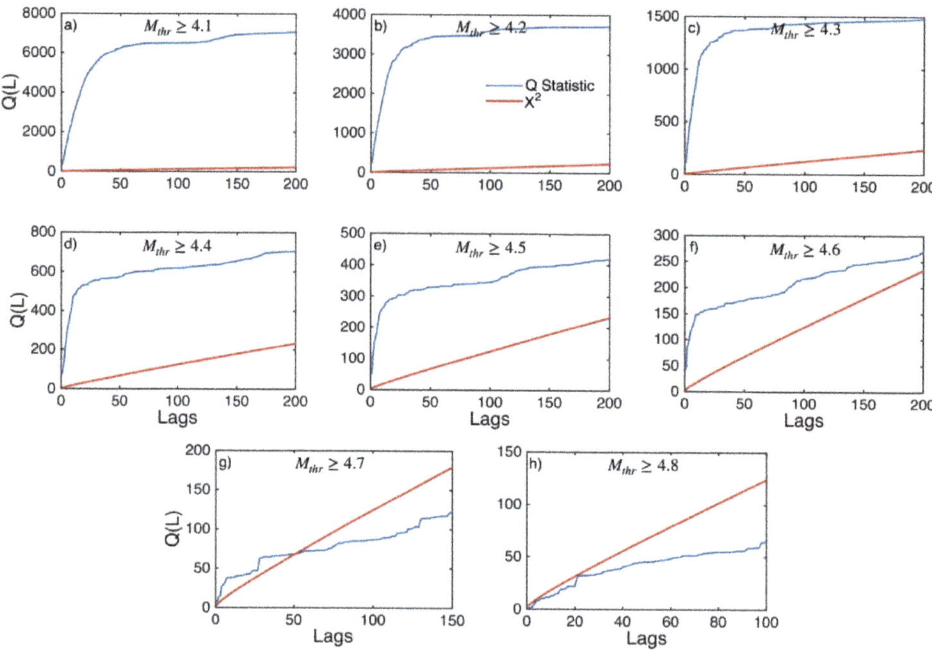

Figure A6. Statistics of Ljung–Box Q-test (blue lines) and the critical values for the X^2 distribution (red lines) for Western Hellenic Arc (sub-area 3) for magnitude thresholds (M_{thr}) varying from 4.1 to 4.8 (subplots **a–h**).

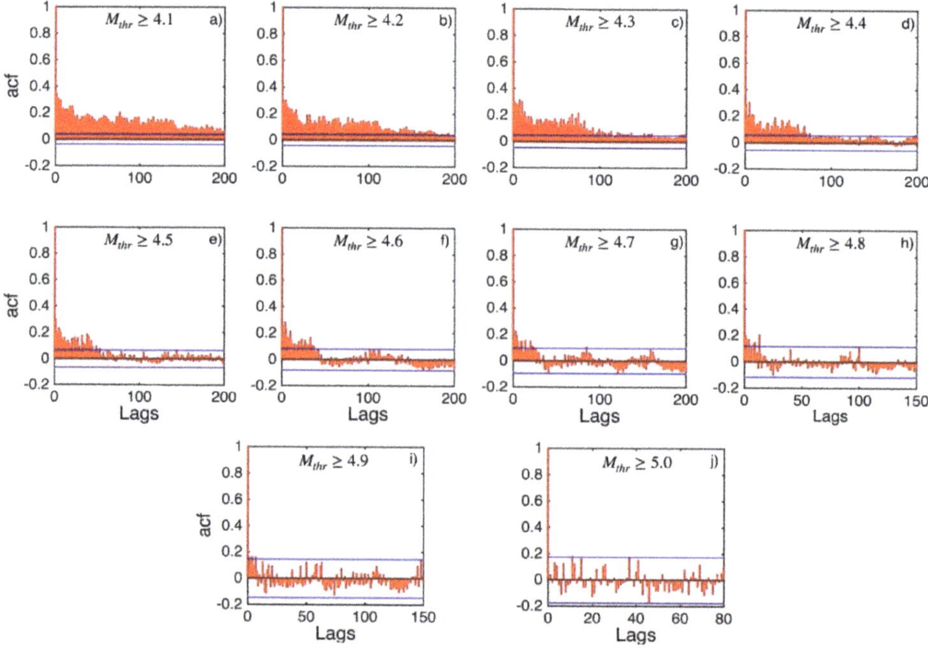

Figure A7. *ACF* values (red stems) and their 95% confidence intervals (blue lines) for Eastern Hellenic Arc (sub-area 4) for magnitude thresholds (M_{thr}) varying from 4.1 to 5.0 (sublots **a–j**).

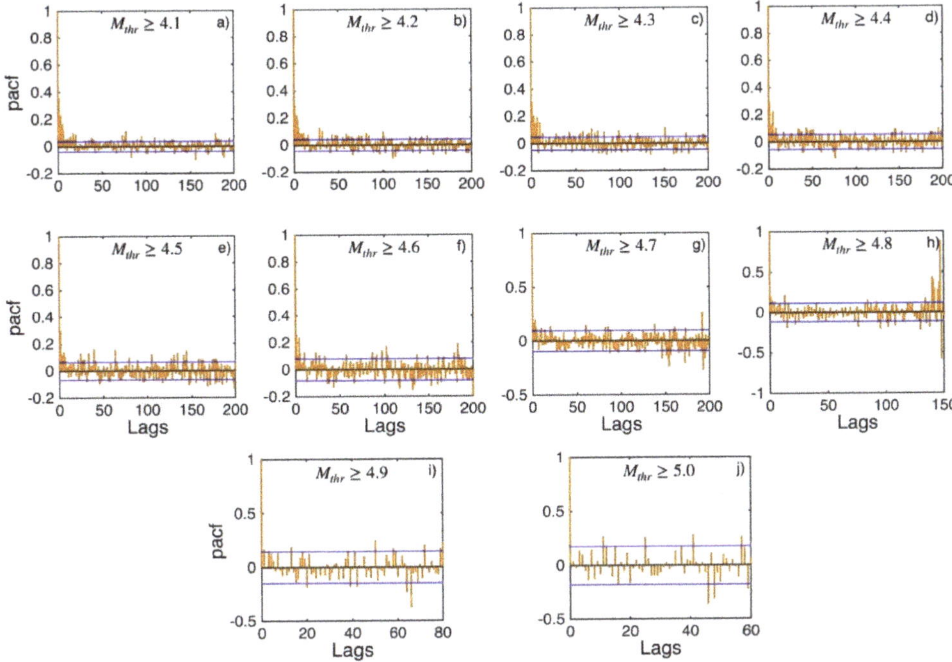

Figure A8. *PACF* values (orange stems) and their 95% confidence intervals (blue lines) for Eastern Hellenic Arc (sub-area 4) for magnitude thresholds (M_{thr}) varying from 4.1 to 5.0 (subplots **a–j**).

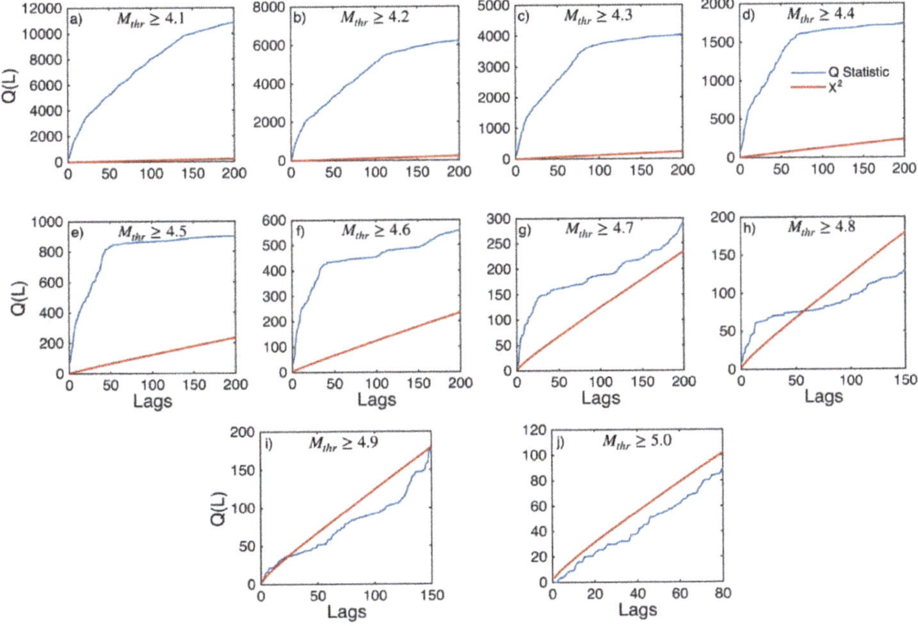

Figure A9. Statistics of Ljung–Box Q-test (blue lines) and the critical values for the X^2 distribution (red lines) for Eastern Hellenic Arc (sub-area 4) for magnitude thresholds (M_{thr}) varying from 4.1 to 5.0 (subplots **a–j**).

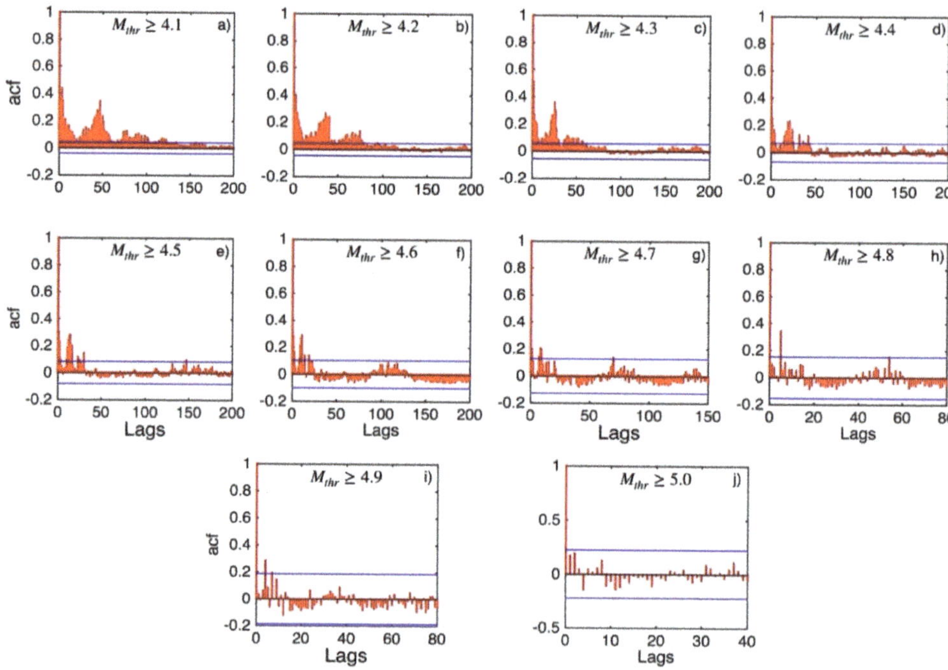

Figure A10. *ACF* values (red stems) and their 95% confidence intervals (blue lines) for Southeast Aegean Sea (sub-area 5) for magnitude thresholds (M_{thr}) varying from 4.1 to 5.0 (subplots **a–j**).

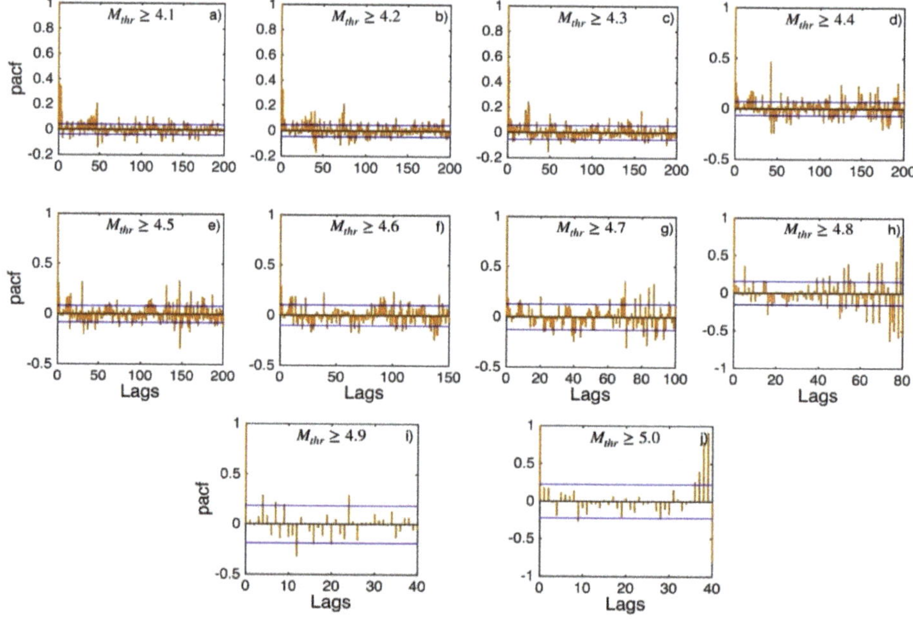

Figure A11. *PACF* values (orange stems) and their 95% confidence intervals (blue lines) for Southeast Aegean Sea (sub-area 5) for magnitude thresholds (M_{thr}) varying from 4.1 to 5.0 (subplots **a–j**).

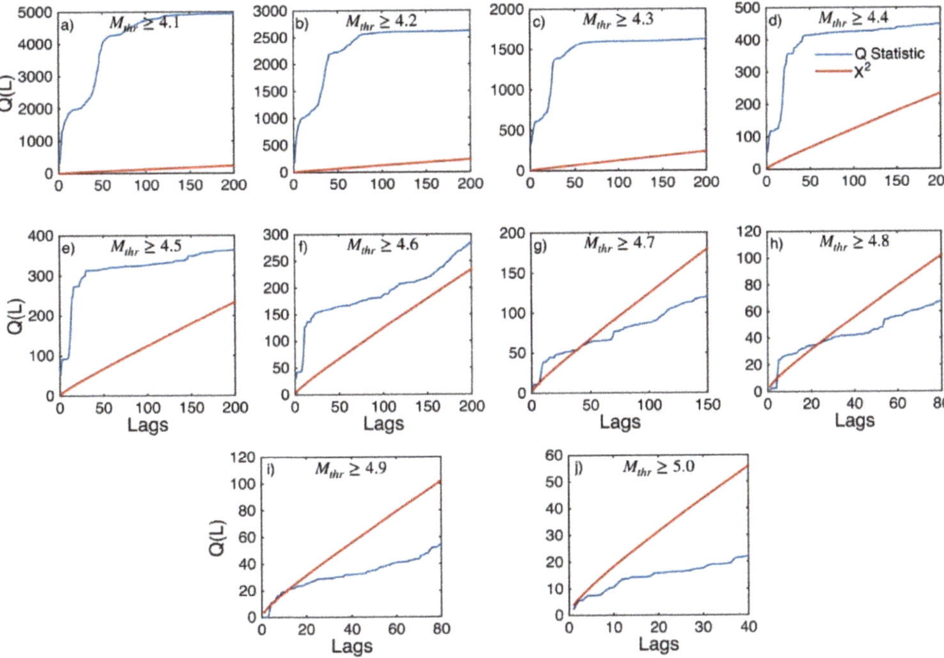

Figure A12. Statistics of Ljung–Box Q-test (blue lines) and the critical values for the X^2 distribution (red lines) for Southeast Aegean Sea (sub-area 5) for magnitude thresholds (M_{thr}) varying from 4.1 to 5.0 (subplots **a–j**).

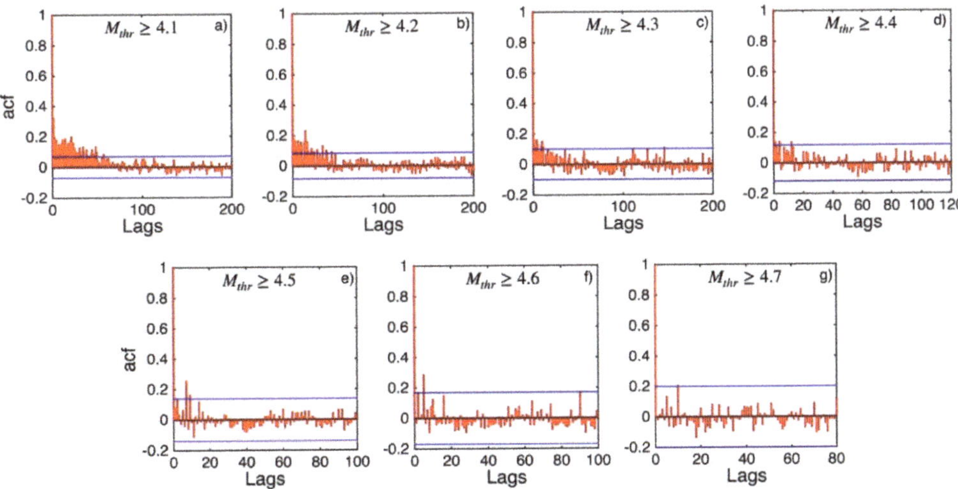

Figure A13. *ACF* values (red stems) and their 95% confidence intervals (blue lines) for Southwest Aegean Sea and Peloponnese (sub-area 6) for magnitude thresholds (M_{thr}) varying from 4.1 to 4.7 (subplots **a–g**).

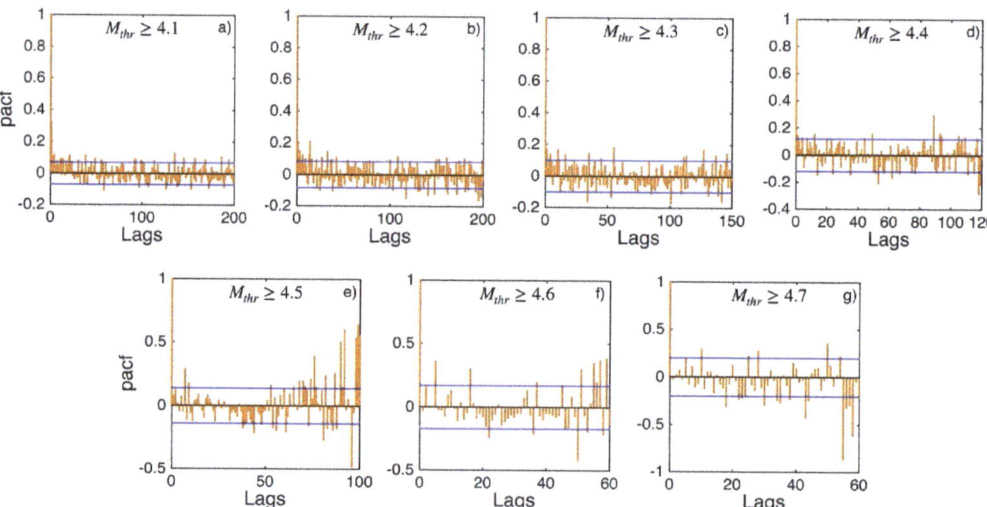

Figure A14. *PACF* values (orange stems) and their 95% confidence intervals (blue lines) for Southwest Aegean Sea and Peloponnese (sub-area 6) for magnitude thresholds (M_{thr}) varying from 4.1 to 4.7 (subplots **a–g**).

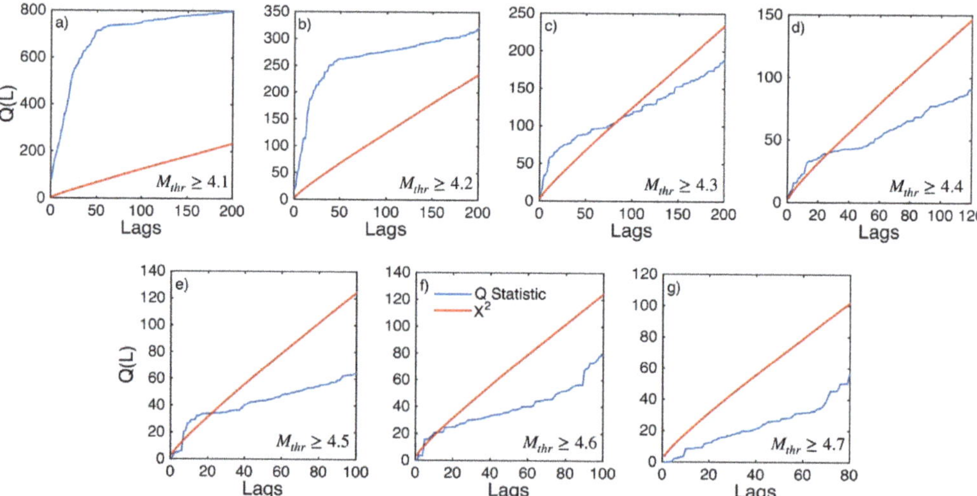

Figure A15. Statistics of Ljung–Box Q-test (blue lines) and the critical values for the X^2 distribution (red lines) for Southwest Aegean Sea and Peloponnese (sub-area 6) for magnitude thresholds (M_{thr}) varying from 4.1 to 4.7 (subplots **a–g**).

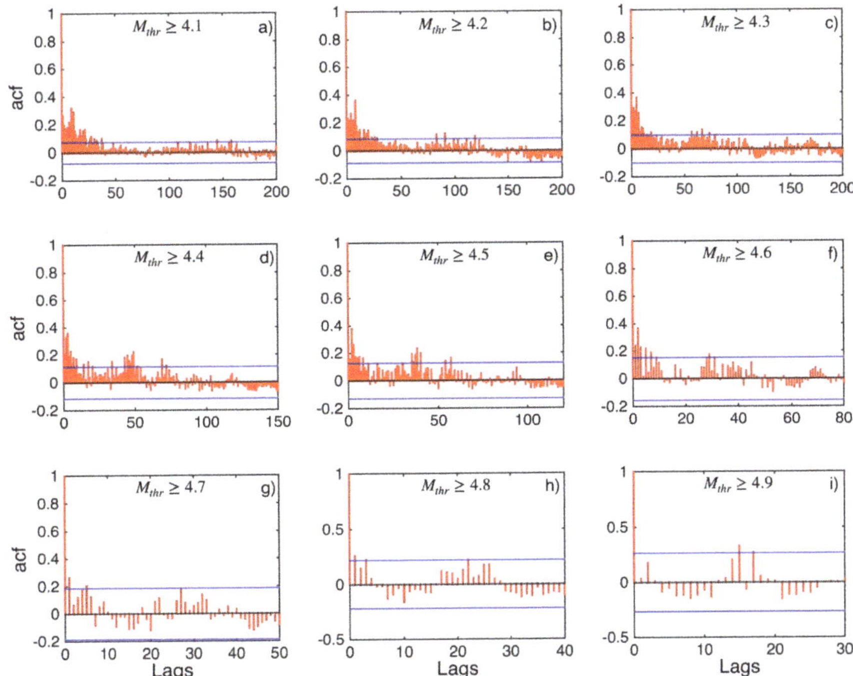

Figure A16. *ACF* values (red stems) and their 95% confidence intervals (blue lines) for Corinth Gulf (sub-area 7) for magnitude thresholds (M_{thr}) varying from 4.1 to 4.9 (subplots **a**–**i**).

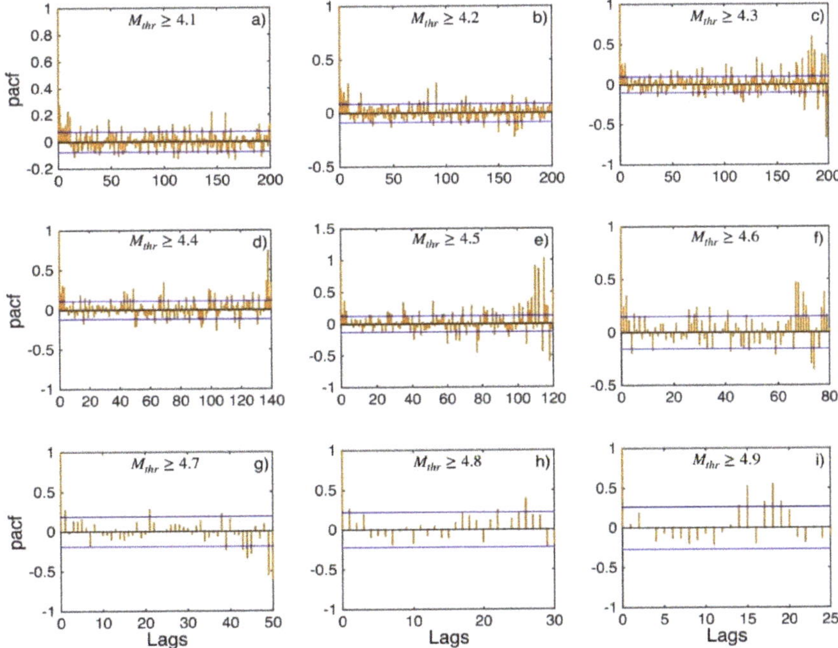

Figure A17. *PACF* values (orange stems) and their 95% confidence intervals (blue lines) for Corinth Gulf (sub-area 7) for magnitude thresholds (M_{thr}) varying from 4.1 to 4.9 (subplots **a**–**i**).

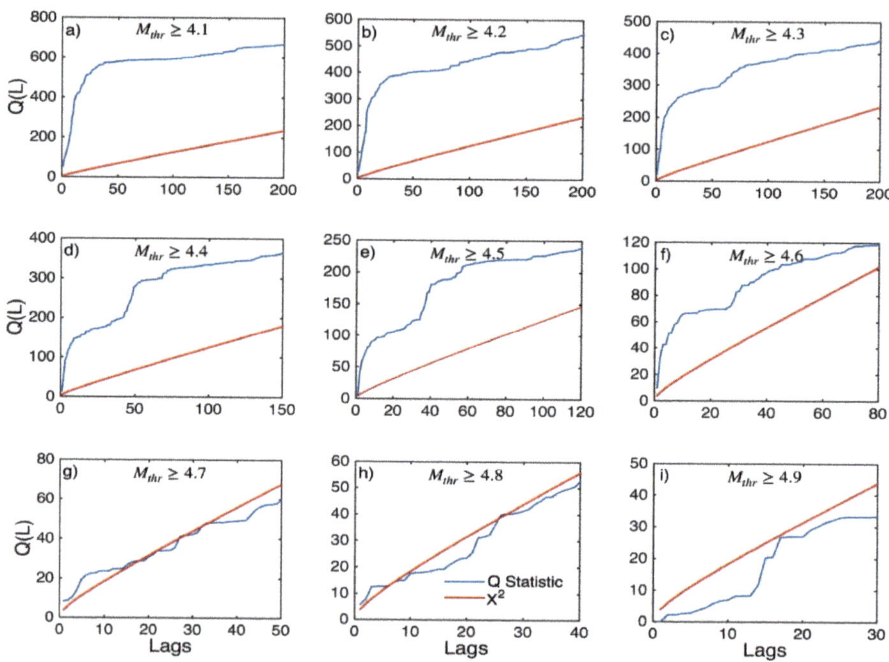

Figure A18. Statistics of Ljung–Box Q-test (blue lines) and the critical values for the X^2 distribution (red lines) for Corinth Gulf (sub-area 7) for magnitude thresholds (M_{thr}) varying from 4.1 to 4.9 (subplots **a–i**).

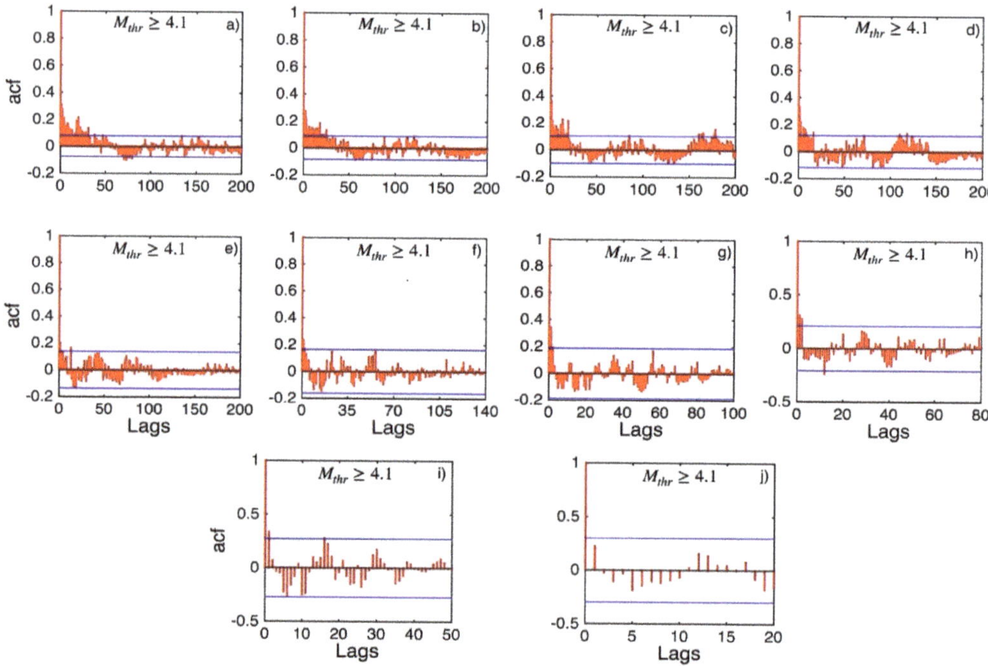

Figure A19. *ACF* values (red stems) and their 95% confidence intervals (blue lines) for Central Greece (sub-area 8) for magnitude thresholds (M_{thr}) varying from 4.1 to 5.0 (subplots **a–j**).

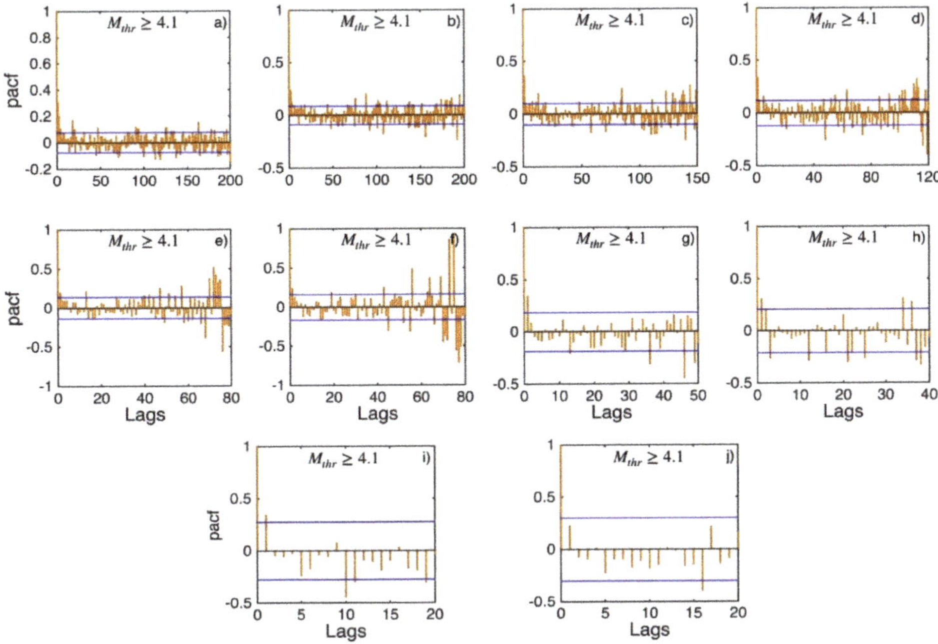

Figure A20. *PACF* values (orange stems) and their 95% confidence intervals (blue lines) for Central Greece (sub-area 8) for magnitude thresholds (M_{thr}) varying from 4.1 to 5.0 (subplots **a–j**).

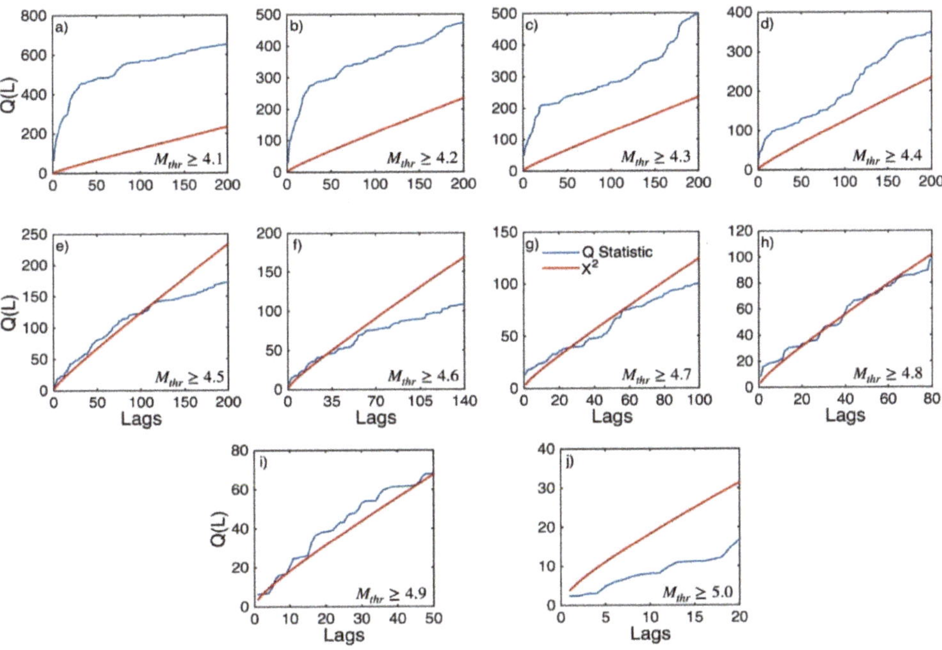

Figure A21. Statistics of Ljung–Box Q-test (blue lines) and the critical values for the X^2 distribution (red lines) for Central Greece (sub-area 8) for magnitude thresholds (M_{thr}) varying from 4.1 to 5.0 (subplots **a–j**).

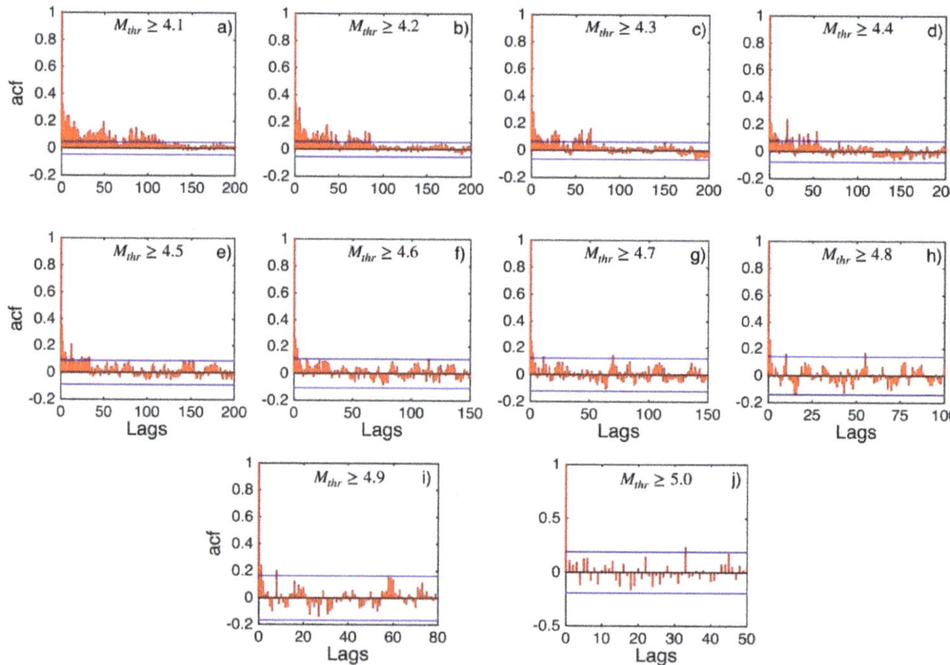

Figure A22. *ACF* values (red stems) and their 95% confidence intervals (blue lines) for North Aegean Sea (sub-area 9) for magnitude thresholds (M_{thr}) varying from 4.1 to 5.0 (subplots **a–j**).

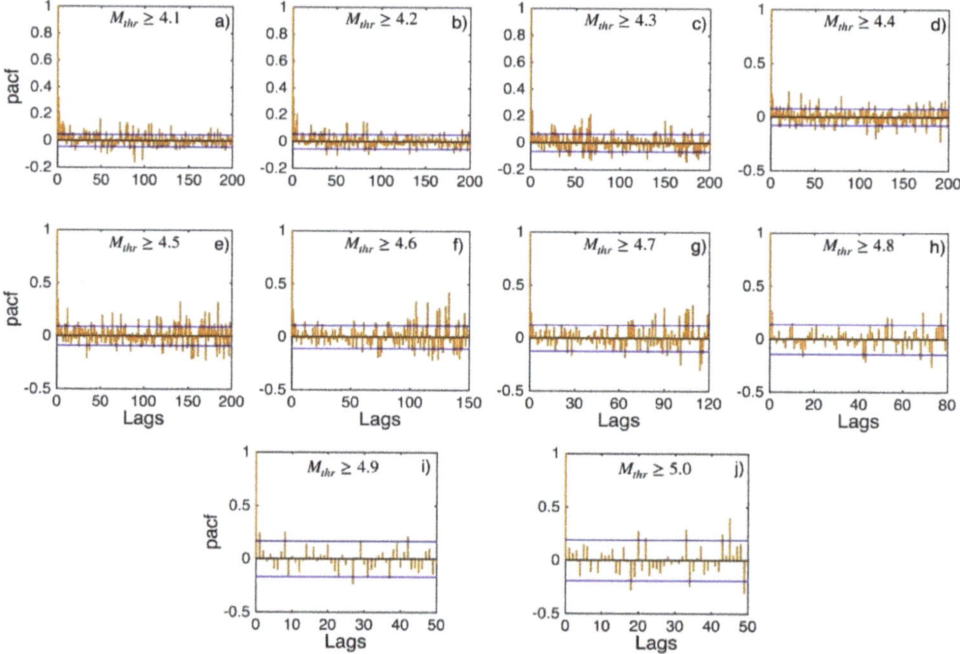

Figure A23. *PACF* values (orange stems) and their 95% confidence intervals (blue lines) for North Aegean Sea (sub-area 9) for magnitude thresholds (M_{thr}) varying from 4.1 to 5.0 (subplots **a–j**).

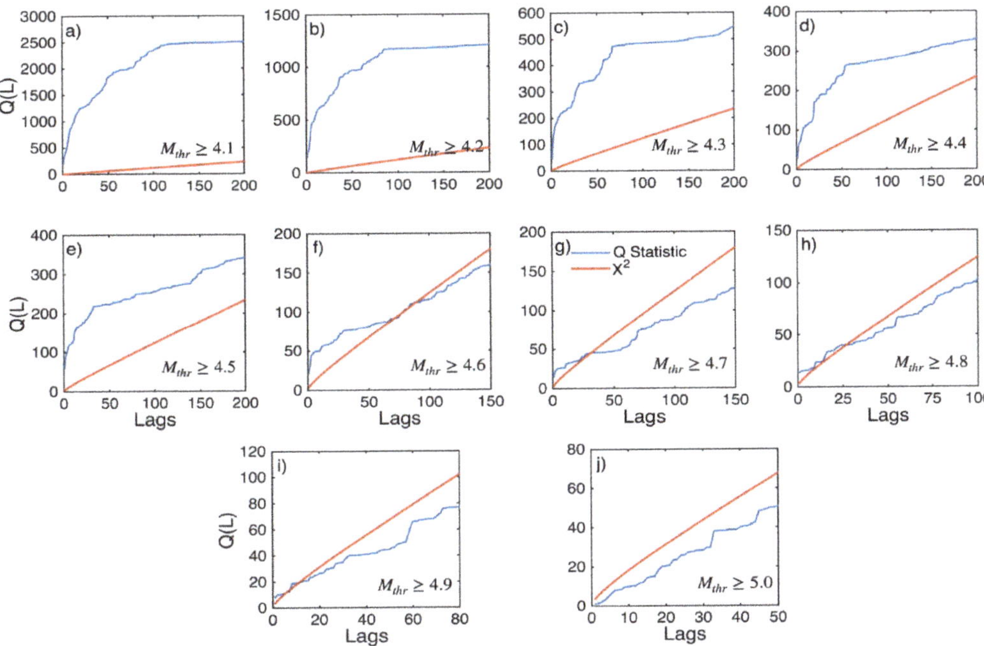

Figure A24. Statistics of Ljung–Box Q-test (blue lines) and the critical values for the X^2 distribution (red lines) for North Aegean Sea (sub-area 9) for magnitude thresholds (M_{thr}) varying from 4.1 to 5.0 (subplots **a**–**j**).

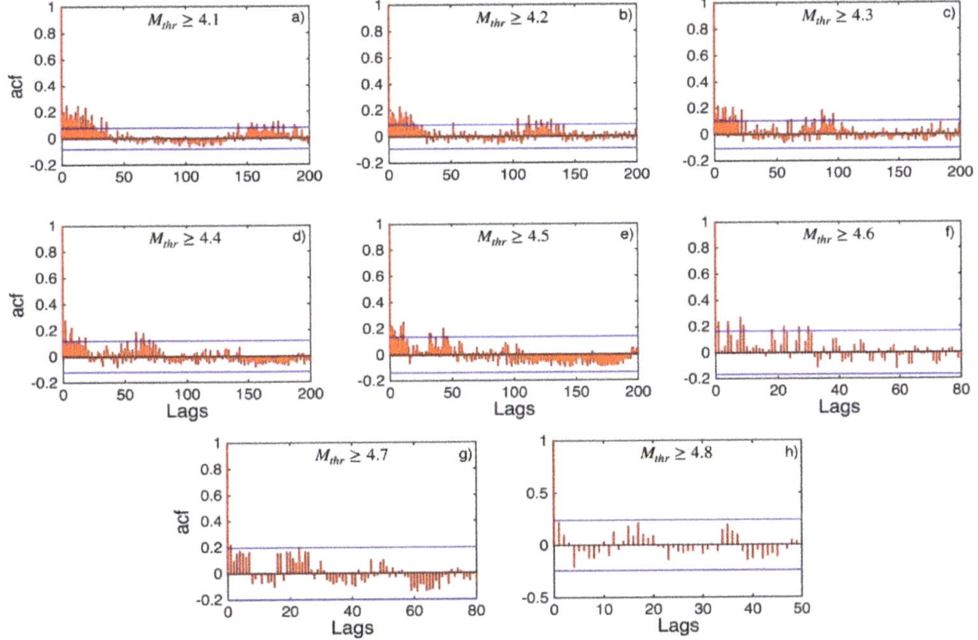

Figure A25. *ACF* values (red stems) and their 95% confidence intervals (blue lines) for Northern Greece and Balkans (sub-area 10) for magnitude thresholds (M_{thr}) varying from 4.1 to 4.8 (subplots **a**–**h**).

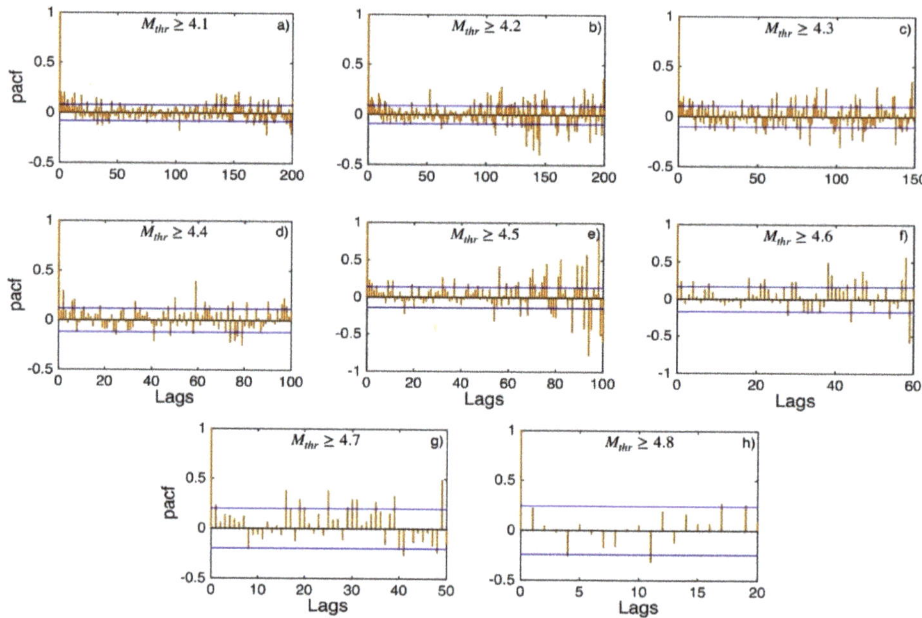

Figure A26. *PACF* values (orange stems) and their 95% confidence intervals (blue lines) for Northern Greece and Balkans (sub-area 10) for magnitude thresholds (M_{thr}) varying from 4.1 to 4.8 (subplots **a–h**).

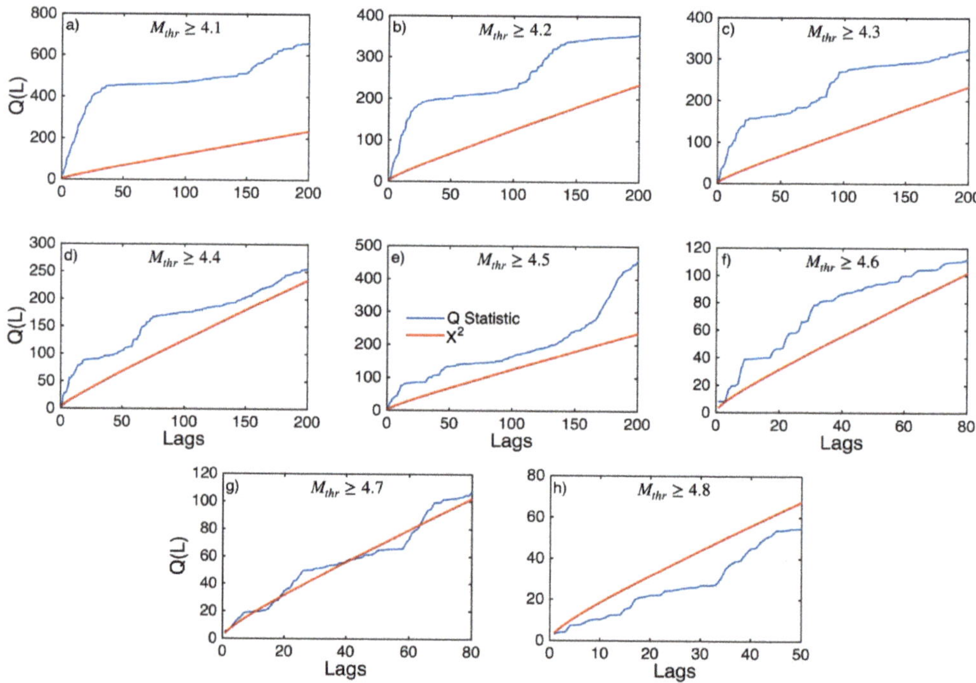

Figure A27. Statistics of Ljung–Box Q-test (blue lines) and the critical values for the X^2 distribution (red lines) for Northern Greece and Balkans (sub-area 10) for magnitude thresholds (M_{thr}) varying from 4.1 to 4.8 (subplots **a–h**).

References

1. Kagan, Y.Y.; Jackson, D.D. Long-term probabilistic forecasting of earthquakes. *J. Geophys. Res.* **1994**, *99*, 13685–13700. [CrossRef]
2. Dieterich, J. A constitutive law for rate of earthquake production and its application to earthquake clustering. *J. Geophys. Res.* **1994**, *99*, 2601–2618. [CrossRef]
3. de Arcangelis, L.; Godano, C.; Grasso, J.R.; Lippiello, E. Statistical physics approach to earthquake occurrence and forecasting. *Phys. Rep.* **2016**, *628*, 1–91. [CrossRef]
4. Gutenberg, B.; Richter, C. Frequency of earthquakes in California. *Bull. Seismol. Soc. Am.* **1944**, *34*, 185–188. [CrossRef]
5. Omori, F. On the aftershocks of earthquakes. *J. Coll. Sci. Imp. Univ. Tokyo* **1894**, *7*, 111–200.
6. Bak, P.; Christensen, K.; Danon, L.; Scanlon, T. Unified scaling law for earthquakes. *Phys. Rev. Lett.* **2002**, *88*, 178501. [CrossRef]
7. Corral, A. Local distributions and rate fluctuations in a unified scaling law for earthquakes. *Phys. Rev. E* **2003**, *68*, 035102-1–035102-4. [CrossRef]
8. Corral, A. Long-term clustering, scaling, and universality in the temporal occurrence of earthquakes. *Phys. Rev. Lett.* **2004**, *92*, 108501. [CrossRef]
9. Molchan, G. Interevent time distribution in seismicity: A theoretical approach. *Pure Appl. Geophys.* **2005**, *162*, 1135–1150. [CrossRef]
10. Saichev, A.; Sornette, D. Universal distribution of interearthquake times explained. *Phys. Rev. Lett.* **2006**, *97*, 078501. [CrossRef]
11. Touati, S.; Naylor, M.; Main, I.G. Origin and nonuniversality of the earthquake interevent time distribution. *Phys. Rev. Lett.* **2009**, *102*, 168501. [CrossRef] [PubMed]
12. Godano, C. A new expression for the earthquake interevent time distribution. *Geophys. J. Int.* **2015**, *202*, 219–223. [CrossRef]
13. Ogata, Y. Statistical models for earthquake occurrence and residual analysis for point process. *J. Am. Stat. Assoc.* **1988**, *83*, 9–27. [CrossRef]
14. Livina, V.N.; Havlin, S.; Bunde, A. Memory in the occurrence of earthquakes. *Phys. Res. Lett.* **2005**, *95*, 208501. [CrossRef]
15. Lennartz, S.; Livina, V.N.; Bunde, A.; Havlin, S. Long-term memory in earthquakes and the distribution of interoccurrence time. *EPL* **2008**, *81*, 69001. [CrossRef]
16. Fan, J.; Zhou, D.; Shekhtman, L.M.; Shapira, A.; Hofstetter, R.; Marzocchi, W.; Ashkenazy, Y.; Havlin, S. Possible origin of memory in earthquakes: Real catalogs and an epidemic-type aftershock sequence model. *Phys. Rev. E* **2019**, *99*, 042210. [CrossRef]
17. Zhang, Y.; Fan, J.; Marzocchi, W.; Shapira, A.; Hofstetter, R.; Havlin, S.; Ashkenazy, Y. Scaling laws in earthquake memory for interevent times and distances. *Phys. Rev. Res.* **2020**, *2*, 013264. [CrossRef]
18. Gkarlaouni, C.; Lasocki, S.; Papadimitriou, E.; Tsaklidis, G. Hurst analysis of seismicity in Corinth rift and Mygdonia graben (Greece). *Chaos Solit. Fractals* **2017**, *96*, 30–42. [CrossRef]
19. Parsons, T.; Geist, E.L. Were global $M \geq 8.3$ earthquake time intervals random between 1900 and 2011? *Bull. Seismol. Soc. Am.* **2012**, *102*, 1583–1592. [CrossRef]
20. Zhuang, J.; Harte, D.; Werner, M.J.; Hainzl, S.; Zhou, S. Basic models of seismicity: Temporal models. *Community Online Resour. Stat. Seism. Anal.* **2012**, *1*, 42. [CrossRef]
21. Rahimi, M.; Zamali, A.; Ghotbi, A.R. The study of seismicity of Alborz (Northern Iran) and Zagros (Southern Iran) regions by using time series analysis. *Acta Geophys.* **2022**, *70*, 27–37. [CrossRef]
22. Papazachos, B.C.; Comninakis, P.E. Geophysical and tectonic features of the Aegean arc. *J. Geophys. Res.* **1971**, *76*, 8517–8533. [CrossRef]
23. LePichon, X.; Angelier, J. The Hellenic Arc and Trench system: A key to the neotectonic evolution of the eastern Mediterranean area. *Tectonophysics* **1979**, *60*, 1–42.
24. Papazachos, B.C.; Papadimitriou, E.E.; Kiratzi, A.A.; Papazachos, C.B.; Louvari, E.K. Fault plane solutions in the Aegean Sea and the surrounding area and their tectonic implications. *Boll. Geofis. Teor. Appl.* **1998**, *39*, 199–218.
25. Scordilis, M.; Karakaisis, G.; Karakostas, V.; Panagiotopoulos, D.G.; Comninakis, P.E.; Papazachos, B.C. Evidence for transform faulting in the Ionian sea: The Cephalonia island earthquake sequence of 1983. *Pure Appl. Geophys.* **1985**, *123*, 388–397. [CrossRef]
26. Goldsworthy, M.; Jackson, J.; Haines, J. The continuity of active faults systems in Greece. *Geophys. J. Int.* **2002**, *148*, 596–618. [CrossRef]
27. Armijo, R.; Meyer, B.; Hubert, A.; Barka, A. Westward propaga- tion of the North Anatolian fault into the northern Aegean: Timing and kinematics. *Geology* **1999**, *27*, 267–270. [CrossRef]
28. Taymaz, T.; Jackson, J.; McKenzie, D. Active tectonics of the North and central Aegean Sea. *Geophys. J. Int.* **1991**, *106*, 433–490. [CrossRef]
29. Papazachos, B.C.; Kiratzi, A.A.; Karakostas, B.G. Towards a homogeneous moment magnitude determination for earthquakes in Greece and the surrounding area. *Bull. Seismol. Soc. Am.* **1997**, *87*, 474–483. [CrossRef]
30. Naylor, M.; Main, I.G.; Touati, S. Quantifying uncertainty in mean earthquake interevent times for a finite sample. *J. Geophys. Res.* **2009**, *114*, B01316. [CrossRef]
31. Weglarczyk, S.; Lasocki, S. Studies of short and long memory in mining-induced seismic processes. *Acta Geophys.* **2009**, *57*, 696–715. [CrossRef]
32. Ljung, G.; Box, G.E.P. On a Measure of Lack of Fit in Time Series Models. *Biometrika* **1978**, *66*, 67–72. [CrossRef]
33. Peng, C.K.; Buldyrev, S.V.; Havlin, S.; Simons, M.; Stanley, H.E.; Goldberger, A.L. Mosaic organization of DNA nucleotides. *Phys. Rev. E Stat. Phys. Plasmas Fluids Relat. Interdiscip. Topics* **1994**, *49*, 1685. [CrossRef]

34. Abaimov, S.G.; Turcotte, D.L.; Shcherbakov, R.; Rundle, J.B.; Yakovlev, G.; Goltz, C.; Newman, W.I. Earthquakes: Recurrence and interoccurrence Times. *Pure Appl. Geophys.* **2008**, *165*, 777–795. [CrossRef]
35. Nishenko, S.P.; Buland, R. A generic recurrence interval distribution for earthquake forecasting. *Bull. Seismol. Soc. Am.* **1987**, *77*, 1382–1399.
36. Werner, M.J.; Ide, K.; Sornette, D. Earthquake forecasting based on data assimilation: Sequential Monte Carlo methods for renewal point processes. *Nonlinear Process. Geophys.* **2011**, *18*, 49–70. [CrossRef]
37. Altmann, E.G.; Kantz, H. Recurrence time analysis, long-term correlations, and extreme events. *Phys. Rev. E* **2005**, *71*, 1–9. [CrossRef]
38. Udias, A.; Rice, J. Statistical analysis of microearthquake activity near San Andreas Geophysical Observatory, Hollister, California. *Bull. Seismol. Soc. Am.* **1975**, *65*, 809–827. [CrossRef]
39. Mesimeri, M.; Kourouklas, C.; Papadimitriou, E.; Karakostas, V.; Kementzetzidou, D. Analysis of microseismicity associated with the 2017 seismic swarm near the Aegean coast of NW Turkey. *Acta Geophys.* **2018**, *66*, 479–495. [CrossRef]
40. Coban, K.H.; Sayil, N. Conditional Probabilities of Hellenic Arc Earthquakes Based on Different Distribution Models. *Pure Appl. Geophys.* **2020**, *177*, 5133–5145. [CrossRef]
41. Johnson, N.L.; Kotz, S.; Balakrishnan, N. *Continuous Univariate Distributions*, 2nd ed.; Wiley: New York, NY, USA, 1994; Volume 1, pp. 1–784.
42. Anderson, T.W.; Darling, D.A. Asymptotic theory of certain 'goodness-of-fit' criteria based on stochastic processes. *Ann. Math. Stat.* **1952**, *23*, 193–212. [CrossRef]
43. Akaike, H. A new look at the statistical model identification. *IEEE Trans. Autom. Contr.* **1974**, *19*, 716–723. [CrossRef]
44. Schwarz, G. Estimating the dimension of a model. *Ann. Stat.* **1978**, *6*, 461–464. [CrossRef]
45. Convertito, V.; Faenza, L. Earthquake Recurrence. In *Encyclopedia of Earthquake Engineering*; Beer, M., Kougioumtzoglou, I.A., Patelli, E., Siu-Kui Au, I., Eds.; Springer: Berlin, Germany, 2014; pp. 1–22.
46. Mobarki, M.; Talbi, A. Spatio-temporal analysis of main seismic hazard parameters in the Ibero-Maghreb region using an M_w-homogenized catalog. *Acta Geophys.* **2022**, *70*, 979–1001. [CrossRef]
47. Wessel, P.; Smith, W.H.F.; Scharroo, R.; Luis, J.; Wobbe, F. Generic mapping tools: Improved version released. *EOS Trans. Am. Geophys. Union* **2013**, *94*, 409–410. [CrossRef]

Article

Evidencing Fluid Migration of the Crust during the Seismic Swarm by Using 1D Magnetotelluric Monitoring

Carlos A. Vargas [1], Alexander Caneva [1,*], Juan M. Solano [1], Adriana M. Gulisano [2,3,4] and Jaime Villalobos [5]

1 Department of Geosciences, Universidad Nacional de Colombia, Bogotá 111321, Colombia
2 Instituto Antártico Argentino, Dirección Nacional del Antártico, Buenos Aires B1650HMK, Argentina
3 Instituto de Astronomía y Física del Espacio, CONICET, Universidad de Buenos Aires, Buenos Aires C1053ABH, Argentina
4 Departamento de Física, Facultad Ciencias Exactas y Naturales, Universidad de Buenos Aires, Buenos Aires C1053ABH, Argentina
5 Department of Physics, Universidad Nacional de Colombia, Bogotá 111321, Colombia
* Correspondence: acanevar@unal.edu.co

Abstract: We applied multi–temporal 1D magnetotelluric (MT) surveys to identify space–time anomalies of apparent resistivity (ρ_a) in the upper lithosphere in the Antarctic Peninsula (the border between the Antarctic and the Shetland plates). We used time series over several weeks of the natural Earth's electric and magnetic fields registered at one MT station of the Universidad Nacional de Colombia (RSUNAL) located at Seymour–Marambio Island, Antarctica. We associated resistivity anomalies with contrasting earthquake activity. Anomalies of ρ_a were detected almost simultaneously with the beginning of a seismic crisis in the Bransfield Strait, south of King George Island (approximately 85.000 events were reported close to the Orca submarine volcano, with focal depths < 20 km and M_{WW} < 6.9). We explained the origin of these anomalies in response to fluid migration near the place of the fractures linked with the seismic swarm, which could promote disturbances of the pore pressure field that reached some hundreds of km away.

Keywords: apparent resistivity; earthquakes; magnetotellurics; electromagnetic anomalies; Antarctic Peninsula; Seymour–Marambio Island; Orca submarine volcano

Citation: Vargas, C.A.; Caneva, A.; Solano, J.M.; Gulisano, A.M.; Villalobos, J. Evidencing Fluid Migration of the Crust during the Seismic Swarm by Using 1D Magnetotelluric Monitoring. *Appl. Sci.* **2023**, *13*, 2683. https://doi.org/10.3390/app13042683

Academic Editors: Alexey Zavyalov and Eleftheria E. Papadimitriou

Received: 20 December 2022
Revised: 15 January 2023
Accepted: 20 January 2023
Published: 19 February 2023

1. Introduction

Physical property anomalies of the lithosphere, for instance, electrical resistivity, were hypothetically associated with earthquake activity in recent decades. In addition, it has been suggested that there are relationships between the magnitude of the earthquakes and the amplitude duration of apparent resistivity (ρ_a) anomalies [1,2]. In other cases, several electromagnetic (EM) anomalies were also detected in the ionosphere and subsurface and linked to large–magnitude seismic events [3,4].

Intense seismic activity has been related to changes in pore pressure [5]. These changes possibly are due to the fluid flux as a consequence of variations in the stress field. Hence, the ρ_a changes are expected to reflect temporal pore pressure changes.

Deep mapping of ρ_a is reached by using MT deployments. This method uses natural electrical and magnetic signals recorded at the Earth's surface, allowing us to estimate vertical electrical resistivity profiles in the subsoil by using relationships between orthogonal components of the electric and magnetic fields [6]. MT is not frequently used for monitoring due to the time consumption of data acquisition. However, it appears to be a reliable method for reaching depth resolutions similar to those associated with the earthquake sources in the brittle lithosphere.

However, the Bransfield Strait–Antarctic Peninsula is a region of active tectonics in response to a subduction process of the Shetland Plate under the Antarctic Plate [7] (Figure 1). The Bransfield Strait defines the geometry of the Bransfield Basin, a back–arc rift basin limited by the Shackleton Fracture Zone and the Hero Fracture Zone located north and south of the Shetland Plate, respectively. Swarms of seismic events near submarine volcanoes suggest current magmatic activity. In addition, normal faulting inferred from focal mechanisms in the strait suggests a transtensional process confirmed by different active rifts.

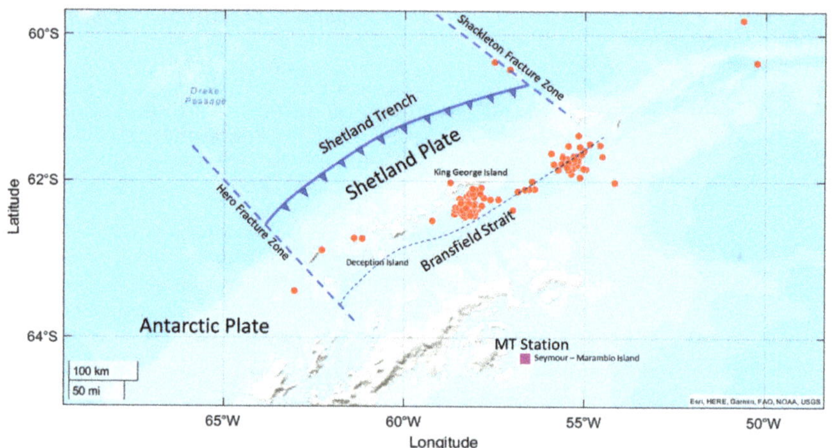

Figure 1. The main tectonic features are in the influence area of the Shetland and Antarctic plates. Red points correspond to earthquakes reported by the USGS with $m_b > 4.0$, which have epicentral distances between 200 and 300 km from the MT station installed in the Seymour–Marambio Island (purple square). The axis of the Bransfield Basin is suggested by a thin–dashed line, following the trend of the Bransfield Strait.

Taking advantage of the EM signals generated naturally in this region with low anthropogenic interference, in this work, we estimated space–time variations of the ρ_a in the crust to infer possible fluid migration and the changes in the pore pressure linked to the tectonic source that generated a seismic swarm in the Bransfield Strait, south of King George Island–25 de Mayo Island (Figure 1). According to the German Research Centre for Geosciences Potsdam [8], approximately 85.000 events were detected between August and November 2020 under the Orca submarine volcano, with focal depths < 20 km and a maximum of $m_b = 6.9$.

2. Materials and Methods

2.1. Magnetotelluric Station in the Seymour–Marambio Island

The ρ_a variations in time and space (depth) in the upper lithosphere were estimated using a permanent MT station located north of Seymour–Marambio Island, Antarctica, near the Multidisciplinary Antarctic Laboratory of the Marambio Island (LAMBI) (Figure 2a,b). It contains sensors to measure the magnetic and electric fields (Figure 2c), such as a Bartington Mag648L triaxial magnetometer with low noise, a range of ±60 μT (resolution of 0.012 nT/count), and four copper ground electrodes of 70 cm each, all located approximately at the same level on a plateau 200 m high. The magnetometer was buried inside the permafrost in a hole dug 75 cm deep to avoid drastic variations in surface temperature. In comparison, the electrodes were percussion buried 80 cm deep with their top connection protected by silicone to prevent corrosion. Each electrode was connected to a 1.2 cm diameter copper cable protected by a polymer resistant to extreme temperatures. The arrangement of the four electrodes made up the two almost orthogonal dipoles NNE (124 m) and EEN (80 m), which maintained the same direction as the magnetometer components. In addition, light gases such as CH_4 and CO_2 were measured to evaluate possible new evidence of their massive escape during the seismic swarm.

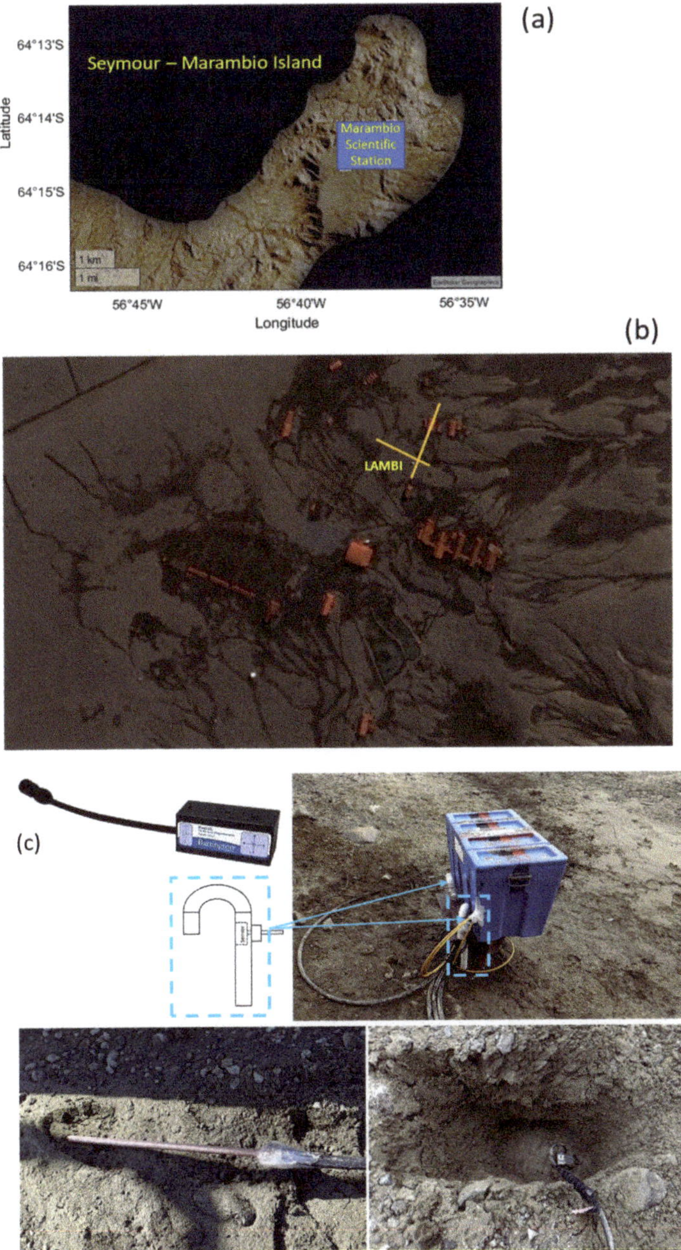

Figure 2. (**a**) MT deployment at the north Seymour–Marambio Island. (**b**) NNE (124 m) and EEN (80 m) are the two orthogonal dipoles' locations. (**c**) Picture of the triaxial magnetometer Bartington Mag648L and the snorkel tube where CH_4 and CO_2 are sensed (upper left). The blue box contains the coupling and digital recording system (upper right). The four dipole electrodes were percussion buried 80 cm deep (lower images). The magnetometer and electrodes were buried in the permafrost.

2.2. Magnetotelluric Sounding (MT)

The MT method studied the penetration and propagation of electromagnetic waves inside the Earth, associated with the action of electrical storms and/or the incidence of the solar wind on the Earth [6]. The method was based on measurements taken on the Earth's surface of the natural electric field (employing two perpendicular electric dipoles, Ex and Ey) and the magnetic field (in our case, using a triaxial flux magnetometer with components Bx, By, and Bz).

The Earth's surface partially reflected the fluctuating electromagnetic fields that originated in the ionosphere, and the ionosphere again reflected the returning fields due to their conductive characteristics. It repeatedly happened so that the fields eventually had a strong vertical component and could be considered vertically propagating plane waves, characterized by covering a broad spectrum of frequencies. These fields penetrated the ground and induced telluric (electric) currents, generating secondary magnetic fields. The telluric currents, detected by two pairs of electrodes, each pair of which composed a dipole, were perpendicularly oriented. The three components of the magnetic fields were measured: the vertical component and two horizontal components, parallel to each one of the electrical components [9,10].

This method provided information about the resistivity (conductivity) values for much greater depths than artificial source induction methods. Using long–period signals ranging from tens to thousands of seconds, the MT method reached investigation depths that may sample the entire lithosphere [9]. In this work, the permanent station guaranteed continuous datasets for years and, consequently, large depths of investigation.

Details for estimating the apparent resistivity structure of the subsoils for an instant and using the MT method can be found in [10] and [6]. Because an electromagnetic wave in the subsoil decays its amplitude due to the resistivity of the medium (assuming a homogeneous half–space), the depth at which the amplitude reaches the factor $1/e \approx 0.37$ (skin depth, δ) can be estimated using the expression $\delta(\omega) = \sqrt{2/\omega\mu\sigma}$ [6,10], which can be simplified to:

$$\delta \approx 503\sqrt{\rho T},\tag{1}$$

where ω is the angular frequency under the assumption of $e^{i\omega t}$ time dependence, μ is the magnetic permeability, σ is the electrical conductivity, ρ is the resistivity, and T is the period evaluated. Thus, the procedure implemented with the data acquired for this work is briefly described as follows:

(1) Time windows of 5000 s were chosen. An overlap of 500 s was used to estimate the resistivity's temporal evolution. Each time window was tapered with a Hanning window, and its frequency spectrum was calculated using the FFT.

(2) The orthogonal components of the natural electric (**E**) and magnetic (**B**) fields were related to the impedance tensor **Z** of the subsoil in the following way [10]:

$$\begin{pmatrix} E_x \\ E_y \end{pmatrix} = \begin{pmatrix} Z_{xx} & Z_{xy} \\ Z_{yx} & Z_{yy} \end{pmatrix} \begin{pmatrix} \frac{B_x}{\mu_0} \\ \frac{B_y}{\mu_0} \end{pmatrix},\tag{2}$$

where $\mathbf{E} = \mathbf{Z}\frac{\mathbf{B}}{\mu_0}$ and μ_0 is the magnetic permeability of free space $\left[\mathrm{V\,s\,A^{-1}\,m^{-1} = H\,m^{-1}} \right]$. From the complex coefficients of the frequency spectrum, the Z_{xy} component under the station was estimated.

(3) The Schmucker–Weidelt (C) transfer function was estimated as follows:

$$C(\omega) = \frac{E_x}{i\omega B_y} = \frac{Z_{xy}}{i\omega}.\tag{3}$$

(4) The ρ_a was calculated using the following expression:

$$\rho_a(\omega) = |C(\omega)|^2\mu_0\omega.\tag{4}$$

(5)   The ρ_a matrix was graphically represented for a given moment and different periods (depths according to Equation (1)), thus guaranteeing a space–time representation of the resistivity field that allowed detecting anomalies concerning a line base of observations.

2.3. K_p Index

According to NOAA's Space Weather Prediction Center, this index quantifies the level of impact on the Earth's horizontal magnetic field measured on the surface due to geomagnetic activity, that is, the emission of charges and high solar radiation. It means a significant noise source for our time series, and its activity is classified by the NOAA as follows:

(1) $K_p < 4$ (weak solar activity);
(2) $K_p = 4$ (mean solar activity);
(3) $K_p > 4$ (high solar activity).

An index $K_p > 4$ (high solar activity) would imply a high disturbance in the Earth's magnetic field and a low reliability of the data taken in that time interval. The details of its estimation are presented in the literature on geomagnetism and solar–terrestrial physics [11]. Hence, in addition to the space–time ρ_a representation, we included the evolution of this index.

3. Results

Figure 3 synthesizes the mapping of the vertical structure (1D) of the ρ_a along the time, including the ρ_a anomalies ($\Delta\rho_a/\overline{\rho_a}$ %), the K_p index, the depth of the hypocentral solutions of earthquakes with $m_b > 4.0$, and measurements of the emissions of CH_4 and CO_2 in the Marambio Station. The first significant event of the seismic swarm in the Bransfield Strait, south of King George Island, occurred on 2020/8/29, 12:47:3.768 UTC, Lat 62.4437S, Lon 58.1777W, H = 10 km, and $M_{ww} = 4.9$ (moment W–phase [12]). In terms of real values of ρ_a or anomalies, we detected changes of ρ_a approximately four hours previous to the event. Thirty–one hours later, another event of $M_{ww} = 5.4$ occurred, approximately 25 km from the last event, preceded by a large ρ_a anomaly, which reached almost the surface (Figure 4). Other minor ρ_a anomalies were also detected in this time window with connection to the deeper 1D–ρ_a structure or related to the solar activity as suggested by the K_p indexes. The larger ρ_a anomalies (A and B) did not match with large K_p indexes, meaning that the anomalies were not influenced by solar activity. High values of ρ_a near the surface (upper panel in Figure 3) could be interpreted as the not–well–consolidated permafrost and ice, and the permanent low values of the ρ_a anomalies around 10 km depth could be related to a lithological transition in the local structure of the upper lithosphere near the measurement instrument, which may connect other deeper and shallower fractures that facilitate the fluid migration.

Figure 4 shows details after starting the large anomaly (A) appreciated in Figure 3 on 2020/8/30, where it is possible to observe a shallowing of the ρ_a anomalies that match the beginning of more significant emissions of CH_4 and CO_2 in the Marambio Station. It suggests that pore pressure perturbations generated by the tectonic stress field disturbances in the region, meaning those that triggered the earthquakes, e.g., events of 2020/8/29, 12:47:3.768 UTC, and 2020/8/30, 19:42:27.389 UTC, with magnitudes $M_{ww} = 4.9$ and $M_{ww} = 5.4$, respectively, could be responsible for the migration of fluids and gases that reached the surface, which may be related to changes in the ρ_a (from deeper sources up to the surface, as suggested by the yellow arrows). Gasses detected by the Marambio Station are typically stored in the permafrost near the measurement instrument [13]. Thus, pore pressure perturbations from the seismic source in the Bransfield Basin could promote their trigger emission in Marambio Island. We also observed that there was a sudden increase in CO_2 gas. However, the variation of the CH_4 did not have the same trend. We speculate that contrasting concentrations of these gasses near the measurement instrument may explain this behavior.

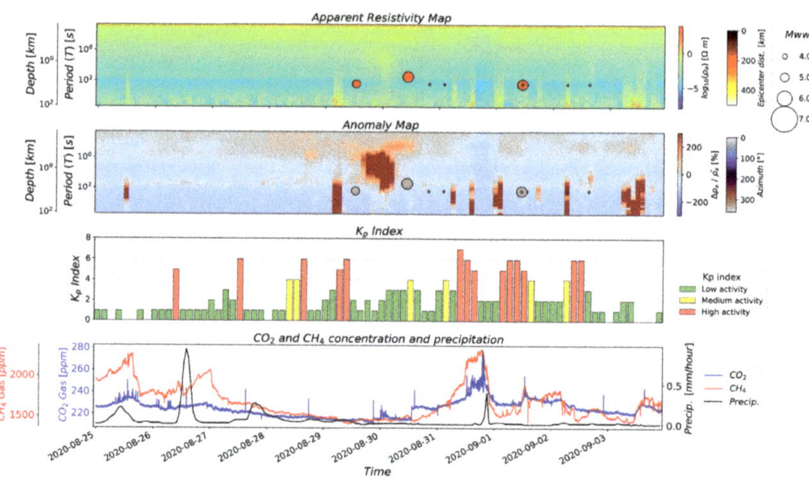

Figure 3. Evolution of the 1D–ρ_a structure. The upper panel includes the depth of the hypocentral solutions of earthquakes with $m_b > 4.0$ located in the Bransfield Strait, south of King George Island. The persistent low values of ρ_a in shallow depths may correspond to not–well–consolidated permafrost and/or surficial ice cap (upper panel). The second panel shows anomalies of ρ_a. Earthquakes reported by the USGS are in the same seismic swarm (epicentral distances and azimuth angles are almost similar). The third panel shows the K_p index, suggesting that larger ρ_a anomalies (A and N) may not be linked to solar activity. The lower panel shows temporal variations of CO_2 and CH_4 gas emissions. The extremely low precipitation at the Marambio Station does not seem to affect gas concentrations. Precipitation data are taken from https://power.larc.nasa.gov/data--access--viewer/ (accessed on 15 December 2022).

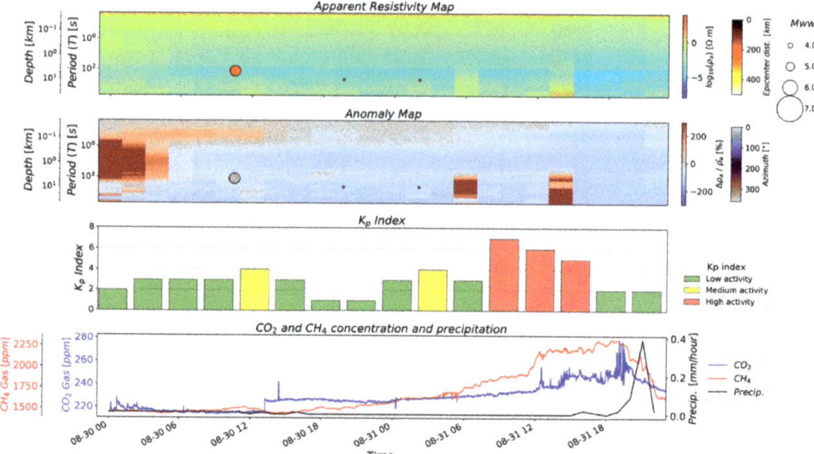

Figure 4. Time window after starting large anomaly (A) and presented in Figure 3, which shows a shallowing of the ρ_a anomalies (suggested by yellow arrows in the second panel) that coincide with the beginning of more significant emissions of CH_4 and CO_2 in the Marambio Station. These gasses are typically stored in the permafrost near the measurement instrument [13], but regional pore pressure perturbances promote their trigger emission. Deeper and strong anomalies in the 1D–ρ_a structure seem to be influenced by solar activity, as shown the K_p index distribution. Precipitation data are taken from https://power.larc.nasa.gov/data--access--viewer/ (accessed on 15 December 2022).

4. Discussion

The current geological setting and the tectonic evolution of the Antarctic Peninsula have been addressed during the last decades with important advances. However, heated debates have arisen about the lithospheric structure and geodynamics that underlie this region, the tectono–magmatic activity surrounding it, and the implications of lithosphere–atmosphere interaction [7,13–15]. Figure 5 illustrates the distribution of earthquakes (M > 4.0, reported by the USGS) in the study zone. Focal mechanisms and other hypocentral solutions with depths <50 km are presented on a gravity anomaly map [16], whose resolution only defines the geometry of the Bransfield Basin. We also overlaid the faults interpreted with swath bathymetry data [17] to infer the faulting responsible for the seismic swarm. Because there is no bathymetric evidence that these earthquakes could reach the seabed, the focal mechanisms are the unique tool that informs about an extensional regime with a strike–slip component. This tectonic regime has previously controlled several volcanic emplacements in the Bransfield Strait (e.g., Deception Island and Orca submarine volcano) [14,15]. More recently, the intrusion of 0.26–0.56 km^3 of magma has been suggested for the seismic swarm analyzed in this work [8], which could be linked with dramatic fluid migration in the crust and the sea. This figure's heat flux distribution (lower map) [18] shows the epicentral location of the seismic swarm, just in the zone of thermal contrast and throughout the Bransfield basin. As in other regions of the world bordered by several plates, the lithosphere system has dramatic lateral changes in thickness and thermal response [19–21]. In this case, the thermal and thickness lithospheric structures could be related to intense fracturing, forming an efficient porosity system based on microfractures where fluids migrate significant distances or facilitate the strong pore pressure gradients.

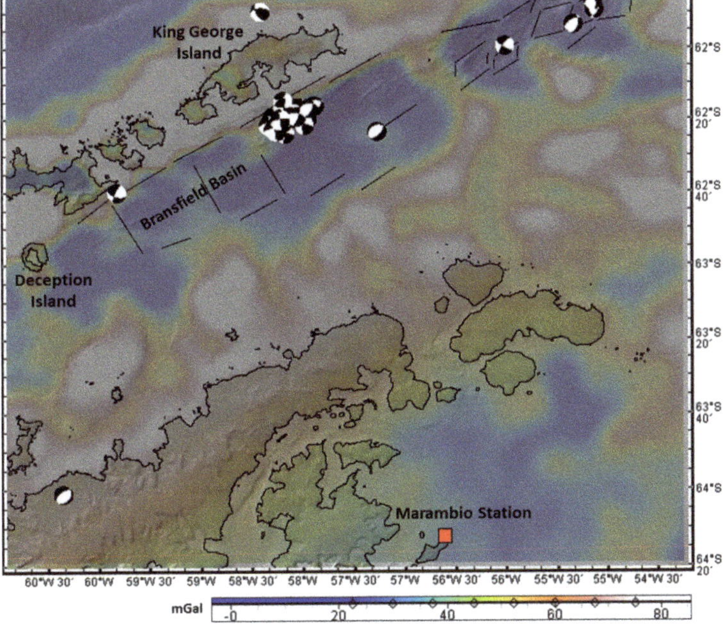

Figure 5. *Cont.*

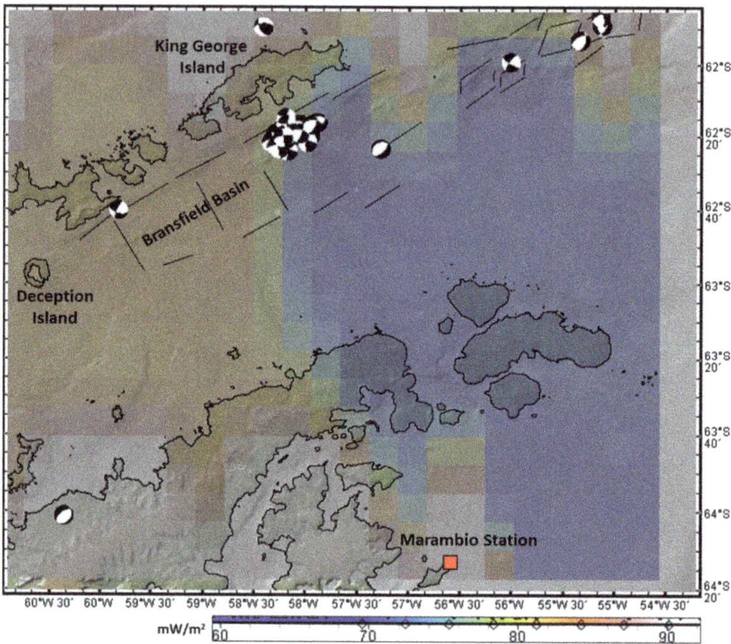

Figure 5. Antarctic Peninsula Map, including the archipelagos of the Shetland Plate (north of the Bransfield Basin). The beach balls represent focal mechanisms, and the green circles are some earthquakes with depths <50 km reported by the USGS. The free air anomalies [16] (upper map) define the geometry of the Bransfield Basin. The heat flow distribution [17] (lower map) in the area is <90 mW/m². Suggested faults inside the basin were interpreted with swath bathymetry data [18]. The Marambio station (red square) represents the location of the MT monitoring. Figure made with GeoMapApp (www.geomapapp.org, accessed on 15 December 2022).

However, it is suggested in the literature that anomalies in electromagnetic signals could be related to seismic events. Observational evidence is reported by analyzing radio, ionospheric, magnetic, or electric signals in diverse frequency bands, detected with a broad type of instruments at distances ranging from some few to thousands of km and linked to earthquakes with a wide range of magnitudes [22–52]. Diverse types of physical mechanisms have been proposed to explain them [53–61]. Some authors explain electromagnetic anomalies based on the electrokinetic effect, in which the solid rock becomes electrically charged. In contrast, the liquid phase in the rock's porosity acquires the opposite charge [53]. Even small changes in the thermal conditions inside the rock mass may promote convection of the liquid phase, generating an electric current that can induce magnetic fields. By observing the thermal contrasts on the surface in the study area (Figure 5), it is necessary to complement the previous explanation to account for the temporal behavior of the electromagnetic anomalies by tying together the stress field, which promotes variable gradients in fluid migration. As the pore pressure increases before intense seismic events, the rocks dehydrate and lose conductivity, which can be measured in significant depths using the MT method. These fractures may cause the breaking of ionic bonds, generating changes in the potential difference and, consequently, electromagnetic signals [53,61].

We hypothesize that these electromagnetic signals respond to changes in the resistivity field of the medium, which in turn is related to changes in the porosity and microfracture conditions and the volume of fluids contained, as suggested in Figure 6. Given changes in the deviatoric stress that generates earthquakes, changes in the porosity field, pore pressure gradients (up to distances greater than the size of the seismic source), and fluid mobility are expected. Thus, our reported changes in ρ_a and the latter observance of gas emissions on the surface (Figures 3 and 4) constitute possible evidence of the fluid migration in the upper lithosphere during the seismic swarm. Additional confirmation of the phenomena suggested in this work could come from studies that analyze the elastic and inelastic fields using seismic swarm data (e.g., [61–63]). We believe that permanent MT monitoring could be an interesting strategy for understanding the pore pressure conditions before fracture initiation in strong earthquakes or seismic swarm scenarios.

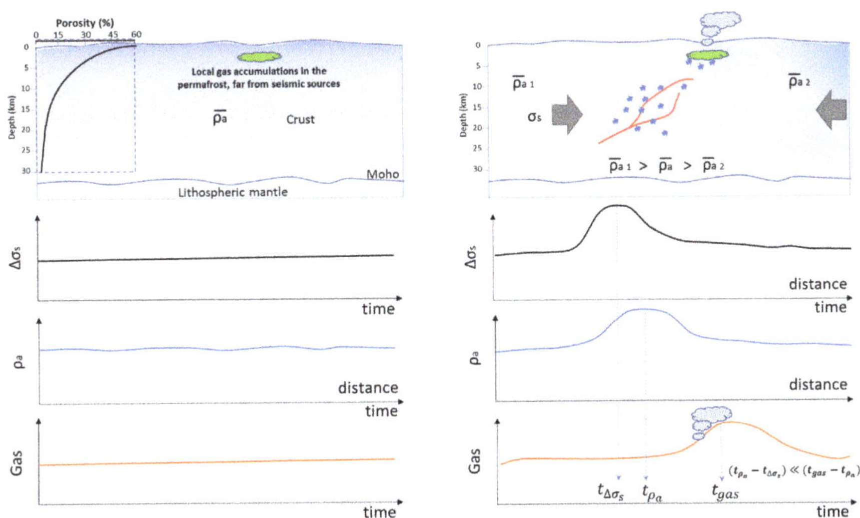

Figure 6. Schematic representation of the generation processes previous to (left) and after the seismic swarm. With no significant temporal variations in the tectonic stress ($\Delta\sigma_s$), no changes are expected in ρ_a nor in fluid migration that triggers pore pressure and gas emissions. In contrast, the relevant variations of the $\Delta\sigma_s$ in short times promote variations of porosity, ρ_a, pore pressure (suggested by the faded shift of the blue background to the right of the top right panel), fluid migration (blue arrows), and gas emissions from local gas accumulations in the permafrost (represented by a green cloud). Time onsets of these processes could meet the following rule: $\left(t_{\rho_a} - t_{\Delta\sigma_s}\right) \ll \left(t_{gas} - t_{\rho_a}\right)$.

In this work, we report an experiment near the Antarctic Peninsula, a region of very low anthropogenic electromagnetic noise and gas emission contamination, with low solar activity during the recorded dataset, and a relevant scenario generated by an isolated seismic swarm in an area of active tectonics. Even under these particular circumstances and when we are not inverting the 1D–ρ_a profile, uncertainties in the estimation of ρ_a could come from large periods analyzed, which means related to the deeper crustal structure. A future potential solution to this issue could be to stack signals from other near MT stations to consolidate space–time 1D–ρ_a maps that may reinforce the trusty anomalies. In addition, other gas measurements around the study region may clarify the role of the fluid migration paths involved in the dynamics of closing and opening of the porosity field.

Finally, we suggest several possible outlooks that the scientific community may address in the future for this type of research: (1) Deploy arrays of permanent MT stations in areas of high tectonic activity that allow the consolidation of datasets on the relationship between seismicity and crustal ρ_a anomalies. (2) In areas of active magmatism, in addition to deploying MT instruments at different distances, it could be necessary to install monitoring networks for fluid pressure and gas emission sensors to verify the hypothesis of fluid migration and pore pressure that trigger seismicity. (3) Design numerical experiments that allow inferring stress conditions, fluid volumes, and changes in the petrophysical properties of the crust necessary to reproduce ρ_a anomalies, such as those reported in this work.

5. Conclusions

One MT station located at Seymour–Marambio Island allowed estimating of the multi–temporal 1D–ρ_a structure in the upper lithosphere of the Antarctic Peninsula. The survey detected ρ_a anomalies that changed over time and were related to a surficial earthquake swarm in the Bransfield Strait, south of King George Island, under the Orca submarine volcano. We detected a shallowing of the ρ_a anomalies that matched with contrasting emission measurements of CH_4 and CO_2, suggesting that anomalies could be linked with fluid migration and propagation of pore pressure that triggered the release of gases. We hypothesize that before the occurrence of earthquakes, the stress field generates pore pressure gradients from sites close to the seismic source to distances greater than the size of the seismic source, promoting alterations in fluid migration that change the resistivity of the upper lithosphere.

Author Contributions: Conceptualization, C.A.V.; methodology, C.A.V. and A.C..; software, J.M.S.; validation, C.A.V., A.C. and J.M.S.; formal analysis, C.A.V., A.C., A.M.G., J.V. and J.M.S.; investigation, C.A.V., A.C., A.M.G., J.V. and J.M.S.; resources, C.A.V. and A.C.; data curation, C.A.V., A.C. and J.M.S.; writing—original draft preparation, C.A.V. and A.C.; writing—review and editing, C.A.V., A.C., A.M.G., J.V. and J.M.S.; visualization, C.A.V. and J.M.S.; supervision, C.A.V. and A.C.; project administration, C.A.V. and A.C.; funding acquisition, C.A.V. and A.C. All authors have read and agreed to the published version of the manuscript.

Funding: This research received no external funding.

Institutional Review Board Statement: Not applicable.

Informed Consent Statement: Not applicable.

Data Availability Statement: Once the paper is approved, the dataset will be openly available.

Acknowledgments: We thank the Universidad Nacional de Colombia for providing laboratories and informatic resources during data acquisition and for preparing this paper. We thank LAMBI (IAA/DNA) personnel. A.M.G. is a member of the Carrera del Investigador Científico CONICET. We thank the Special Issue Editors, Alexey Zavyalov and Eleftheria E. Papadimitriou, as well as two anonymous reviewers, for their valuable comments and suggestions.

Conflicts of Interest: The authors declare no conflict of interest.

References

1. Du, X.B.; Xue, S.Z.; Hao, Z.; Zhang, S.Z. On the relation of moderate–short term anomaly of earth resistivity to earthquake. *Acta Seismol. Sin.* **2000**, *13*, 393–403. [CrossRef]
2. Du, X.B.; Li, N.; Ye, Q.; Ma, Z.H.; Yan, R. A possible reason for the anisotropic changes in apparent resistivity near the focal region of strong earthquake. *Chin. J. Geophys.* **2007**, *50*, 1555–1565. (In Chinese) [CrossRef]
3. Fan, Y.Y.; Du, X.B.; Zlotnicki, J.; Tan, D.C.; An, Z.H.; Chen, J.Y.; Zheng, G.L.; Liu, J.; Xie, T. The electromagnetic phenomena before the Ms 8.0 Wenchuan earthquake. *Chin. J. Geophys.* **2010**, *53*, 997–1010. [CrossRef]
4. Liu, J.; Du, X.B.; Zlotnicki, J.; Fan, Y.Y.; An, Z.H.; Xie, T.; Zheng, G.L.; Tan, D.C.; Chen, J.Y. The changes of the ground and ionosphere electric/magnetic fields before several great earthquakes. *Chin. J. Geophys.* **2011**, *54*, 2885–2897. [CrossRef]
5. Sherif, M.A.; Ishibashi, I.; Tsuchiya, C. Pore–Pressure Prediction During Earthquake Loadings. *Soils Found.* **1978**, *18*, 19–30. [CrossRef] [PubMed]
6. Chave, A.; Jones, A. The Magnetotelluric Method. In *Theory and Practice*, 1st ed.; Cambridge University Press: New York, NY, USA, 2012; p. 603.
7. Robertson Maurice, S.D.; Wiens, D.A.; Shore, P.J.; Vera, E.; Dorman, L.M. Seismicity and tectonics of the South Shetland Islands and Bransfield Strait from a regional broadband seismograph deployment. *J. Geophys. Res.* **2003**, *108*, 2461. [CrossRef]
8. Cesca, S.; Sugan, M.; Rudzinski, Ł.; Vajedian, S.; Niemz, P.; Plank, S.; Petersen, G.; Deng, Z.; Rivalta, E.; Vuan, A.; et al. Massive earthquake swarm driven by magmatic intrusion at the Bransfield Strait, Antarctica. *Commun. Earth Environ.* **2022**, *3*, 89. [CrossRef]
9. Lowrie, W. *Fundamentals of Geophysics*, 2nd ed.; Cambridge University Press: New York, NY, USA, 2007; p. 391.
10. Simpson, F.; Bahr, K. *Practical Magnetotellurics*; Cambridge University Press: New York, NY, USA; Cambridge, UK, 2005; 254p. [CrossRef]
11. Menvielle, M.; Berthelier, A. The K–derived planetary indices: Description and availability. *Rev. Geophys.* **1991**, *29*, 415–432. [CrossRef]

12. Hayes, G.P.; Rivera, L.; Kanamori, H. Source inversion of the W–Phase: Real–time implementation and extension to low magnitudes. *Seismol. Res. Lett.* **2009**, *80*, 817–822. [CrossRef]

13. Vargas, C.A.; Solano, J.M.; Gulisano, A.M.; Santillana, S.; Casallas, E.A. Seismic properties of the permafrost layer using the HVSR method in Seymour–Marambio Island, Antarctica. *Earth Sci. Res. J.* **2022**, *26*, 197–204. [CrossRef]

14. Haase, K.M.; Beier, C.; Fretzdorff, S.; Smellie, J.L.; Garbe–Schönberg, D. Magmatic evolution of the South Shetland Islands, Antarctica, and implications for continental crust formation. *Contrib. Mineral. Petrol.* **2012**, *163*, 1103–1119.

15. Almendros, J.; Carmona, E.; Jiménez, V.; Díaz–Moreno, A.; Lorenzo, F. Volcano–tectonic activity at Deception Island volcano following a seismic swarm in the Bransfield Rift (2014–2015). *Geophys. Res. Lett.* **2018**, *45*, 4788–4798.

16. Bonvalot, S.; Balmino, G.; Briais, A.; Kuhn, M.; Peyrefitte, A.; Vales, N.; Biancale, R.; Gabalda, G.; Reinquin, F.; Sarrailh, M. *World Gravity Map*; Commission for the Geological Map of the World: Paris, France, 2012; BGI–CGMW–CNES–IRD.

17. Fuchs, A.; Norden, B. *International Heat Flow Commission (2021): The Global Heat Flow Database: Release*; GFZ Data Services, 2021; Available online: https://dataservices.gfz-potsdam.de/panmetaworks/showshort.php?id=5797045d-993f-11eb-9603-497c926 95674 (accessed on 19 January 2023). [CrossRef]

18. Gràcia, E.; Canals, M.; Farràn, M.; Sorribas, J.; Pallàs, R. The Central and Eastern Bransfield basins (Antarctica) from high–resolution swath–bathymetry data. *Antarct. Sci.* **1997**, *9*, 168–180. [CrossRef]

19. Blanco, J.F.; Vargas, C.A.; Monsalve, G. Lithospheric thickness estimation beneath Northwestern South America from an S–wave Receiver Function analysis. *Geochem. Geophys. Geosyst.* **2017**, *18*, 1376–1387. [CrossRef]

20. Artemieva, I.M. Lithosphere structure in Europe from thermal isostasy. *Earth–Sci. Rev.* **2019**, *188*, 454–468. [CrossRef]

21. Chisenga, C.; Yan, J.; Yan, P. A crustal thickness model of Antarctica calculated in spherical approximation from satellite gravimetric data. *Geophys. J. Int.* **2019**, *218*, 388–400. [CrossRef]

22. Anderson, L.A.; Johnson, G.R. Application of the self–potential method to geothermal exploration in Long Valley, California. *J. Geophys. Res.* **1997**, *81*, 1527–1532. [CrossRef]

23. Warwick, J.; Stoker, C.; Meyer, T. Radio Emission Associated With Rock Fracture: Possible Application to the Great Chilean Earthquake of May 22, 1960. *J. Geophys. Res.* **1982**, *87*, 2851–2859. [CrossRef]

24. Smith, B.; Johnston, M. A tectonomagnetic effect observed before a magnitude 5.2 earthquake near Hollister, California. *J. Geophys. Res.* **1976**, *81*, 3556–3560. [CrossRef]

25. Zhao, Y.; Qian, F. Geoelectric precursors to strong earthquakes in China. *Tectonophysics* **1994**, *233*, 99–113.

26. Wallace, R.; Teng, T. Prediction of the Sungpan–Pingwu earthquakes, 1976. *Bull. Seismol. Soc. Am.* **1980**, *70*, 1199–1223. [CrossRef]

27. Gokhberg, M.; Morgounov, V.; Yoshino, T.; Tomizawa, I. Experimental Measurement of Electromagnetic Emissions Possibly Related to Earthquakes in Japan. *J. Geophys. Res.* **1982**, *87*, 7824–7828. [CrossRef]

28. Gershenzon, N.; Gokhberg, M. On the origin of electrotelluric disturbances prior to an earthquake in Kalamata, Greece. *Tectonophysics* **1993**, *224*, 169–174. [CrossRef]

29. Molchanov, O.A.; Kopytenko, Y.A.; Voronov, P.M.; Kopytenko, E.A.; Matiashvili, T.G.; Fraser-Smith, A.C.; Bernardi, A. Results of ULF magnetic field measurements near the epicenters of the Spitak (Ms = 6.9) and Loma–Prieta (Ms = 7.1) earthquakes: Comparative analysis. *Geophys. Res. Lett.* **1992**, *19*, 1495–1498. [CrossRef]

30. Kopytenko, Y.; Matiashviali, T.; Voronov, P.; Kopytenko, E.; Molchanov, O. Detection of ultra–low–frequency emissions connected with the Spitak earthquake and its aftershock activity, based on geomagnetic pulsations data at Dusheti and Vardzia observatories. *Phys. Earth Planet. Inter.* **1993**, *77*, 85–95.

31. Fraser–Smith, A.; Bernardi, A.; McGill, P.; Ladd, M.; Helliwell, R.; Villard, O. Low–frequency magnetic field measurements near the epicenter of the Ms 7.1 Loma Prieta earthquake. *Geophys. Res. Lett.* **1990**, *17*, 1465–1468. [CrossRef]

32. Serebryakova, O.N.; Bilichenko, S.V.; Chmyrev, V.M.; Parrot, M.; Rauch, J.L.; Lefeuvre, F.; Pokhotelov, O.A. Electromagnetic ELF radiation from earthquake regions as observed by low altitude satellites. *Geophys. Res. Lett.* **1992**, *19*, 91–94. [CrossRef]

33. Dea, J.; Hansen, P.; Boerner, W. Long–term EMF background noise measurements, the existence of window regions and applications to earthquake precursor emission studies. *Phys. Earth Planet. Inter.* **1993**, *77*, 109–125. [CrossRef]

34. Shalimov, S.; Gokhberg, M. Lithosphere–ionosphere coupling mechanism and its application to the earthquake in Iran on June 20, 1990. A review of ionospheric measurements and basic assumptions. *Phys. Earth Planet. Inter.* **1998**, *105*, 211–218. [CrossRef]

35. Ondoh, T. Ionospheric disturbances associated with great earthquake of Hokkaido southwest coast, Japan of July 12, 1993. *Phys. Earth Planet. Inter.* **1998**, *105*, 261–269. [CrossRef]

36. Hayakawa, M.; Kawate, R.; Molchanov, O.; Yumoto, K. Results of ultra–low frequency magnetic field measurements during the Guam earthquake of 8 August 1993. *Geophys. Res. Lett.* **1996**, *23*, 241–244. [CrossRef]

37. Hayakawa, M.; Ito, T. Fractal analysis of ULF geomagnetic data associated with the Guam earthquake on August 8, 1993. *Geophys. Res. Lett.* **1996**, *26*, 2797–2800. [CrossRef]

38. Smirnova, N.; Hayakawa, M. Fractal characteristics of the ground observed ULF emissions in relation to geomagnetic and seismic activities. *J. Atmos. Sol.–Terr. Phys.* **2007**, *69*, 1833–1841. [CrossRef]

39. Hayakawa, M.; Ida, Y.; Gotoh, K. Multifractal analysis for the ULF geomagnetic data during the Guam earthquake. Electromagnetic Compatibility and Electromagnetic Ecology. In Proceedings of the IEEE 6th Int Symposium, St. Petersburg, Russia, 21–24 June 2005; pp. 239–243.

40. Fujinawa, Y.; Takahashi, K. Electromagnetic radiations associated with major earthquakes. *Phys. Earth Planet. Inter.* **1998**, *105*, 249–259. [CrossRef]

41. Enomoto, Y.; Tsutsumi, A.; Fujinawa, Y.; Kasahara, M.; Hashimoto, H. Candidate precursors: Pulse–like geoelectric signals possibly related to recent seismic activity in Japan. *Geophys. J. Int.* **1997**, *131*, 485–494. [CrossRef]

42. Maeda, K.; Tokimasa, N. Decametric radiation at the time of the Hyogoken Nanbu earthquake near Kobe in 1995. *Geophys. Res. Lett.* **1996**, *23*, 2433–2436. [CrossRef]

43. Bernard, P.; Pinettes, P.; Hatzidimitriou, P.M.; Scordilis, E.M.; Veis, G.; Milas, P. From Precursors to prediction: A few recent cases from Greece. *Geophys. J. Int.* **1997**, *131*, 467–477. [CrossRef]

44. Contoyiannis, Y.; Diakonos, F.; Kapiris, P.; Peratzakis, A.; Eftaxias, K. Intermittent dynamics of critical pre–seismic electromagnetic fluctuations. *Phys. Chem. Earth* **2004**, *29*, 397–408. [CrossRef]

45. Varotsos, P.; Sarlis, N.; Eftaxias, K.; Lazaridou, M.; Bogris, N.; Makris, J.; Abdulla, A.; Kapiris, P. Prediction of the 6.6 Grevena–Kozani Earthquake of May 13, 1995. *Phys. Chem. Earth* **1999**, *24*, 115–121. [CrossRef]

46. Varotsos, P.; Sarlis, N.; Skordas, E. Electric Fields that "Arrive" before the Time Derivative of the Magnetic Field prior to Major Earthquakes. *Phys. Rev. Lett.* **2003**, *91*, 148501–1/148501–4. [CrossRef]

47. Varotsos, P.; Sarlis, N.; Skordas, E.; Lazaridou, M. Electric pulses some minutes before earthquake occurrences. *Appl Phys Lett.* **2007**, *90*, 064104–1/064104–3. [CrossRef]

48. Hayakawa, M.; Ohta, K.; Maekawa, S.; Yamauchi, T.; Ida, Y.; Gotoh, T.; Yonaiguchi, N.; Sasaki, H.; Nakamura, T. Electromagnetic precursors to the 2004 Mid Niigata Prefecture earthquake. *Phys. Chem. Earth* **2006**, *31*, 351–364. [CrossRef]

49. Yonaiguchi, N.; Ida, Y.; Hayakawa, M.; Masuda, S. Fractal analysis for VHF electromagnetic noises and the identification of preseismic signature of an earthquake. *J. Atmos. Sol.–Terr. Phys.* **2007**, *69*, 1825–1832. [CrossRef]

50. Biagi, P.; Ermini, A.; Kingsley, S. Disturbances in LF Radio Signals and the Umbria–Marche (Italy) Seismic Sequence in 1997–1998. *Phys. Chem. Earth* **2001**, *26*, 755–759. [CrossRef]

51. Fino, J.M.S.; Caneva, A.; Vargas, C.A.; Ochoa, L.H. Electrical and magnetic data time series' observations as an approach to identify the seismic activity of non–anthropic origin. *Earth Sci. Res. J.* **2021**, *25*, 297–307. [CrossRef]

52. Vargas, C.A.; Gómez, J.S.; Gómez, J.J.; Solano, J.M.; Caneva, A. Comment on seismic electric signals (SES) and earthquakes: A review of an updated VAN method and competing hypotheses for SES generation and earthquake triggering by Daniel S. Helman. *Phys. Earth Planet. Inter.* **2021**, *313*, 106676. [CrossRef]

53. Varotsos, P.A.; Sarlis, N.V.; Skordas, E.S.; Tanaka, H.K.; Lazaridou, M.S. Entropy of Seismic Electric Signals: Analysis in the Natural Time under Time Reversal. *Phys. Rev. E* **2006**, *73*, 031114. [CrossRef]

54. Varotsos, P.A.; Sarlis, N.V.; Skordas, E.S. Detrended fluctuation analysis of the magnetic and electric field variations that precede rupture. *Chaos Interdiscip. J. Nonlinear Sci.* **2009**, *19*, 02311. [CrossRef] [PubMed]

55. Uyeda, S.; Nagao, T.; Kamogawa, M. Short–term earthquake prediction: Current status of seismo–electromagnetics. *Tectonophysics* **2009**, *470*, 205–213. [CrossRef]

56. Cicerone, R.; Ebel, J.; Britton, J. A systematic compilation of earthquake precursors. *Tectonophysics* **2009**, *476*, 371–396. [CrossRef]

57. Zhang, X.M.; Zeren, Z.M.; Shen, X.H. Analysis on variation of electric field spectrum at cut–off frequency before strong earthquakes: Taking 2006 Tonga Mw8.0 earthquake as an example. *Acta Seismol. Sin.* **2011**, *33*, 451–460.

58. Sarlis, N.V.; Skordas, E.S.; Varotsos, P.A.; Nagao, T.; Kamogawa, M.; Uyeda, S. Spatiotemporal variations of seismicity before major earthquakes in the Japanese area and their relation with the epicentral locations. *Proc. Natl. Acad. Sci. USA* **2015**, *112*, 986–989. [CrossRef] [PubMed]

59. Stanica, D.; Stanica, D.A. Anomalous pre–seismic behavior of the electromagnetic normalized functions related to the intermediate depth earthquakes occurred in Vrancea zone, Romania. *Nat. Hazards Earth Syst. Sci.* **2011**, *11*, 3151–3156. [CrossRef]

60. Petraki, E.; Nikolopoulos, D.; Chaldeos, Y.; Coulouras, G.; Nomicos, C. Fractal evolution of MHz electromagnetic signals prior to earthquakes: Results collected in Greece during 2009. *Geomat. Nat. Hazards Risk* **2016**, *7*, 550–564. [CrossRef]

61. Koulakov, I.; Vargas, C.A. Evolution of the magma conduit beneath the Galeras volcano inferred from repeated seismic tomography. *Geophys. Res. Lett.* **2018**, *45*, 7514–7522. [CrossRef]

62. Ugalde, A.; Carcolé, E.; Vargas, C.A. S–wave attenuation characteristics in the Galeras volcanic complex (south western Colombia). *Phys. Earth Planet. Inter.* **2010**, *181*, 73–81. [CrossRef]

63. Carcolé, E.; Ugalde, A.; Vargas, C.A. Three–dimensional spatial distribution of scatterers in Galeras volcano, Colombia. *Geophys. Res. Lett.* **2006**, *33*, L08307. [CrossRef]

Article

Space–Time Variations of the Apparent Resistivity Associated with Seismic Activity by Using 1D-Magnetotelluric (MT) Data in the Central Part of Colombia (South America)

Carlos A. Vargas, J. Sebastian Gomez, Juan J. Gomez, Juan M. Solano and Alexander Caneva *

Department of Geosciences, Universidad Nacional de Colombia at Bogota, Bogota 111321, Colombia
* Correspondence: acanevar@unal.edu.co

Abstract: In this work, we apply multi-temporal 1D-magnetotelluric (MT) surveys to estimate the space–time variations of the apparent resistivity ρ_a and correlate these changes with seismic activity in the central part of Colombia (South America). We use the time series of the Earth's natural electric and magnetic fields registered at two MT stations of the National University of Colombia Seismological Network (RSUNAL), located in the Eastern Andean Cordillera, in the central part of Colombia, over several days. Assuming that large earthquakes may generate these types of anomalies, we identified positive results for the Mesetas earthquake (Mw6.0, Lon = 74.184° W, Lat = 3.462° N, H = 13 km-depth, 24 December 2019, UTC 19:03:55), with anomalies registered eight hours before the mainshock. The depth at which the resistivity anomaly was identified coincides with the depth of the earthquake hypocenter. The origin of these anomalies may be associated with the migration of fluids due to the change in the stress regime before, during, and after the earthquake. We hypothesize that before the occurrence of an earthquake, the stress field generates pore pressure gradients, promoting alterations in fluid migration that change the resistivity of the upper crust.

Keywords: apparent resistivity; correlation; earthquakes; magnetotellurics; electromagnetic signals; electromagnetic anomalies

Citation: Vargas, C.A.; Gomez, J.S.; Gomez, J.J.; Solano, J.M.; Caneva, A. Space–Time Variations of the Apparent Resistivity Associated with Seismic Activity by Using 1D-Magnetotelluric (MT) Data in the Central Part of Colombia (South America). *Appl. Sci.* **2023**, *13*, 1737. https://doi.org/10.3390/app13031737

Academic Editors: Eleftheria E. Papadimitriou and Alexey Zavyalov

Received: 20 December 2022
Revised: 19 January 2023
Accepted: 27 January 2023
Published: 29 January 2023

1. Introduction

Variations in the physical properties of the subsoil, such as electrical resistivity, caused by the occurrence of seismic activity have been studied in recent decades. Authors such as Du et al. [1,2] established a non-linear relationship between the magnitude of the earthquakes and the amplitude–duration of the recorded anomaly in the apparent resistivity ρ_a. Additionally, various electromagnetic (EM) anomalies have been reported, both in the ionosphere and subsurface, for large-magnitude seismic events [3,4]. These mentioned studies use the maximum entropy method (MEM) and spectral density to analyze their signals.

This study seeks to estimate the variation, both in time and in space (depth), of the ρ_a of the subsoil using seismic events of magnitude > 4.5 in Colombia, using the magnetotelluric (MT) method. In the MT method, deepened and applied to geophysical exploration by Cagniard [5], relationships are obtained that allow the electrical resistivity and its distribution in the subsoil to be inferred from the relationship between the fluctuations in the orthogonal components of the natural electric and magnetic fields. These equations were coupled in an algorithm developed in Python to obtain ρ_a variations as a function of time from the time series of the registered electric and magnetic fields. The method allows us to reach depths from tens of meters to hundreds of kilometers [6], considering that this depends on the used time window. By choosing different time windows, we can achieve a depth ranging between ~100 m and ~100 km. It is possible to reach greater depths, for which it would be sufficient to vary the time window used.

2. Materials and Methods

2.1. Electromagnetic Signals of Seismic Origin

The scientific community has studied the possible relationship between electromagnetic phenomena and seismic events since the second half of the 20th century. There is more and more evidence about a connection between the generation of electromagnetic signals and the occurrence of seismic events (in addition to the well-known variations of the Earth's magnetic field associated with the incidence of the solar wind). This field of study requires combining various disciplines, such as physics, geology, and even biology [7,8]. Anomalies in electromagnetic signals associated with strong seismic events have been identified, and physical mechanisms have been proposed to explain them [9–13].

The study of seismic–electromagnetic phenomena has acquired greater importance since the 1980s [7]. Anomalies were found in the behavior of electromagnetic signals before some strong earthquakes, shedding new light on the study of seismic precursors. Several authors have proposed different mechanisms for generating anomalous electromagnetic signals associated with seismic activity.

For example, (1) a mechanism that could explain the anomalies is the displacement of the limits between cortical blocks of high and low conductivity [14]. Due to this, the electrical currents in the subsoil would be altered, also altering the induced magnetic fields. (2) Another mechanism could be the electrokinetic effect, which consists of the fact that, in the presence of fluids in the subsoil, the solid phase can be electrically charged, while the liquid phase acquires the opposite charge [13]. The latter, when flowing, generates an electric current that can induce magnetic fields. (3) On the other hand, the piezomagnetic effect occurs when the magnetization of ferromagnetic rocks changes when subjected to stress, giving rise to an alteration in the magnetic field in the form of very low-frequency signals [15]. (4) An alternate mechanism could be associated with the formation of microfractures in the rock, in the presence of fluids. As the pressure increases before the seismic event, the rocks dehydrate and lose conductivity. These ruptures can also cause the separation of ionic bonds, generating a variation in the potential difference, which can generate electromagnetic signals [13,15,16].

Zhang and Shen [17] have proposed four stages before large-magnitude earthquakes based on the Wenchuan event, M8.0, on 12 May 2008. The first stage would consist of the accumulation of mechanical stresses. The pressure would close the subsurface microfractures, altering their ρ_a. In the final phase of this stage, electromagnetic signals would be recorded. In the case of the Wenchuan event, this stage lasted for more than two years. The second stage would consist of a blockade of the subsoil. At this point, the microfractures would have closed entirely, and tectonic activity (and therefore seismicity) would have been reduced to a minimum, avoiding the generation of electromagnetic anomalies. The third stage would consist of an "unblocking" and an expansion of the subsoil. The temperature would rise, generating electromagnetic signals in the infrared. Microfractures would re-form, and small seismic events would occur with very low-frequency electromagnetic emissions. The fourth stage would take place hours or days before the big event and would give rise to anomalies in the electric field of the ionosphere. Although the authors base their model on the observations associated with this specific event, they could be considered for the study of other events.

Solar activity can also be a source of electromagnetic signals, for which it is necessary to distinguish between these and those possibly associated with seismic activity [18]. Likewise, anthropic activity can generate electromagnetic noise, especially in areas where there are electrical power distribution systems [19]. Other problems associated with the identification of very low-frequency anomalies are their low intensity, their similarity with very low-frequency signals that are not of seismic origin, as well as the difficulty in identifying the focus of the earthquake to which the anomalies may be associated, before the earthquake occurs [14].

2.2. Magnetotelluric Sounding (MT)

The magnetotelluric method studies the penetration and propagation of electromagnetic waves inside the Earth, associated with the action of electrical storms and/or the incidence of the solar wind on the Earth. The method is based on measurements of the natural electric fields on the Earth's surface (by means of 2 perpendicular electric dipoles, E_x and E_y) and magnetic (using 3 coils or triaxial magnetometers, B_x, B_y, and B_z).

Among the advantages of the magnetotelluric method is the fact that it is based on using a natural electromagnetic source. The Earth's surface partially reflects the fluctuating electromagnetic fields originating in the ionosphere, and the ionosphere reflects the returning fields again due to its conductive characteristics. This happens repeatedly, so that the fields eventually have a strong vertical component and can be considered as vertically propagating plane waves, characterized by covering a broad spectrum of frequencies. These fields penetrate the ground and induce telluric (electric) currents, which generate secondary magnetic fields. The telluric currents, detected by two pairs of electrodes that make up each pair of dipoles, are oriented in the N–S and E–W directions. The three components of the magnetic fields are measured: the vertical component and two horizontal components, parallel to each one of the electrical components [6,20].

This method provides information about resistivity (conductivity) values for much greater depths than artificial source induction methods. Using long-period signals in the range from 10 to 1000 s, the MT method is relevant to the investigation of the crustal and upper mantle structure [20].

The two electrical and three magnetic components of an MT sounding are recorded continuously during long observation intervals, which can last hours or days. The registered magnetic fields have external contributions from the ionosphere and internal ones related to the distribution of the induced currents. Although an MT sounding can be carried out in a range of infrasound frequencies ($f \approx 10$–100 Hz), its main application is the determination of electrical conductivity (resistivity) values at great depths, using very low frequencies ($f \approx 1$ Hz) [6,20].

2.3. Magnetotelluric Fundamentals

The theoretical principles of the method lie in Maxwell's equations, which describe the behavior of electric and magnetic fields, as well as their interaction.

Based on the assumptions of this method [6,21] and the electromagnetic induction phenomena described by Ampere's and Faraday's laws, we define *skin depth* (δ). This concept describes the distance, in-depth, which the electromagnetic wave has traveled when its amplitude has decreased by a factor of $1/e$, assuming a homogeneous half-space. This value depends on the resistivity of the medium (ρ) and period (T) of the electromagnetic wave. Using SI units, the *skin depth* is approximated as:

$$\delta \approx 500\sqrt{\rho T}, \tag{1}$$

Remembering that period (T), frequency (f), and angular frequency (ω) are related by $T = 1/f = 2\pi/\omega$, we can use this last equation to analyze the behavior of the resistivity (ρ) as a function of the period (T), and thus obtain a first approximation of the depth of investigation. This depth (δ) is used to approximate the magnetotelluric data.

Transfer Function and Apparent Resistivity ρ_a

The transfer function term involves an Earth model describing a linear system with a predictable input and output. Here, the transfer function C of Schmucker–Weidelt [6,22,23] was used, which depends on the frequency and can be calculated from the field measurements in their orthogonal directions—in other words, by measuring the electric field in the E–W component (E_x-HQE) and the magnetic field in the N–S component (B_y-HFN), or with their equivalents (E_y-HQN, B_x-HFE), as follows:

$$C(\omega) = \frac{E_x}{i\omega B_y} = -\frac{E_y}{i\omega B_x} \tag{2}$$

Then, we can calculate the ρ_a, defined as the resistivity's average of a homogeneous half-space. Like C, the ρ_a is also expressed as a function of frequency [6]:

$$\rho_a(\omega) = |C(\omega)|^2 \mu_0 \omega, \tag{3}$$

where μ_0 is the magnetic permeability of free space.

In this study, we use the apparent resistivity ρ_a. However, it is essential to note that for MT soundings, being C-complex, it is also important to obtain the phase parameter, which, like these previous functions, will also depend on the frequency:

$$\phi_{1-D}(\omega) = \tan^{-1}\left(E_x(\omega)/B_y(\omega)\right) \tag{4}$$

Note that these MT parameters are usually plotted as a function of the period (T), as seen in the Results and Discussion Section.

2.4. K_p Index

According to NOAA's Space Weather Prediction Center [24], this index quantifies the impact on the Earth's horizontal magnetic field measured on the surface due to the geomagnetic activity, that is, the emission of charges and high solar radiation. This represents a significant source of noise for time series and data. The K_p index activity is classified as follows (Table 1):

Table 1. K_p index activity classification.

$K_p < 4$	Weak/Low solar activity
$K_p = 4$	Medium solar activity
$K_p > 4$	High solar activity

A K_p index > 4 (high solar activity) would imply a high disturbance in the terrestrial magnetic field and insufficient data reliability in the chosen time interval. This value was downloaded from the NOAA (National Oceanic and Atmospheric Administration) database on its website [24].

2.5. Mutiparameter Station

The USME station is a multi-parameter station that is part of the Seismological Network of the National University of Colombia [25]. Figure 1 shows the distribution of the network's stations in the central region of Colombia. The USME station has a broadband triaxial seismometer, triaxial magnetometer, and non-polarizable electrodes.

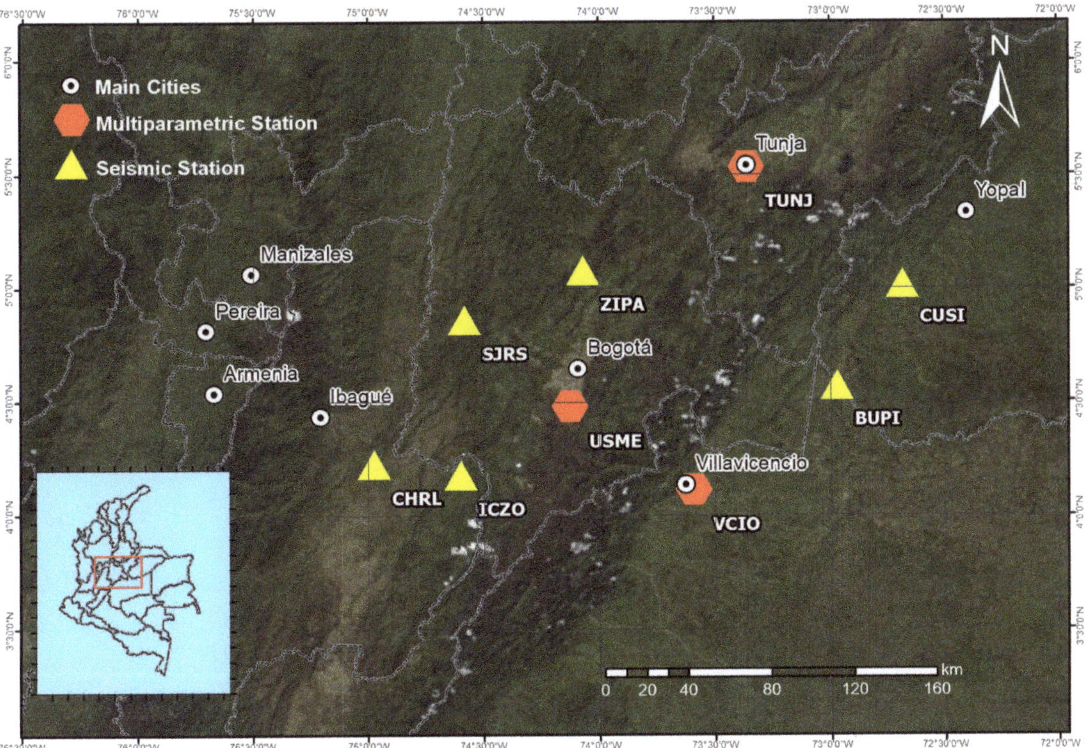

Figure 1. Distribution of the stations of the Geophysical Network of the National University of Colombia in the central region of Colombia.

2.5.1. Sensors

The sensors used in multi-parameter stations are [26]:

- Broadband Seismometer CME 4311: Electrochemical transducer designed for permanent or portable installation. It is a robust instrument that is easy to install. In addition, it does not require maintenance, mass blocking, or leveling. This sensor offers an effective solution for installations with a noise level close to the low-noise model with a response between 1/60 and 50 Hz [27].
- Non-polarizable electrode for burial Tinker & Rasor DB-A: Copper and copper sulfate ($Cu/CuSO_4$) electrode, non-polarizable, which allows direct exposure of copper sulfate over a large contact area. It has a shelf life of up to 10 years. Its structure is made of PVC/ABS and has a low freezing point and high evaporation point, which makes it robust for most environments. Dimensions: 7 cm in diameter, 12.2 cm high, and a weight of 896 g [28].
- Magnetometer Bartington Mag648L: Low-power, low-noise triaxial magnetometer with ± 60 µT range. It has vehicular, perimeter security, and ground measurement applications [29].

Figure 2 shows the instrumental deployment corresponding to the USME station.

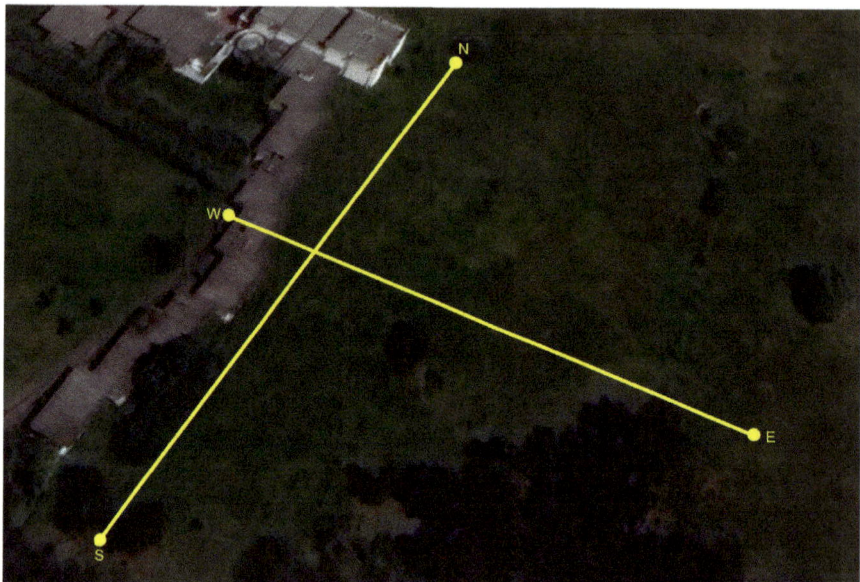

Figure 2. Schematic representation of the location of the orthogonal dipoles, corresponding to the USME station.

2.5.2. Acquisition Hardware

The digitizer is based on a Raspberry Pi 3 CPU with 1.2 GHz processing and 2 GB of RAM, with an independent video chip. Different peripherals are integrated into it via GPIO or USB. The 24-bit Analog–Digital Converter (ADC) is connected via serial communication protocol, having eight single channels to sample up to 500 sps that can be expanded to 16 channels. To achieve time synchronization with an error of fewer than 10 μs, we use a GPS connected via asynchronous serial communication, with the possibility of an external high-gain antenna to improve the view of the sky. Data transmission is via 3G modem. The power supply is obtained using a 5 V DC source at 2.5 A that is provided using a solar panel. The maximum consumption is around 10 W. The entire system is installed in a hard box.

2.6. Data Processing

The MT data processing in this study begins by converting a time series of magnetic and electrical signals to their frequency domain (using the Fourier transform (FT)), a stacking process over a 1-h time window (it is desired to see changes in the resistivity behavior each hour), to later operate their respective components, calculate the Schmucker–Weidelt transfer function from Equation (2), and finally obtain their associated values of ρ_a. This procedure was performed by developing our algorithm code in Python, which can be obtained by consulting the authors. The processing was performed following all the mathematical guidelines given in [6].

2.6.1. Time Windows

To estimate the space–time ρ_a variation, particular time intervals should be defined depending on the changes to be seen. Therefore, we define three principal time windows: *total window*, *partial window*, and *calculation window*.

- Total window: It is the complete time interval on which to work (e.g., you want to analyze six days along which there was seismic activity: total window = 6 days);
- Partial Window: It is the time interval in which you want to see changes in the ρ_a of the rocks (e.g., you want to see changes every hour: partial window = 1 h);

- Calculation window: It is the time interval over which we will calculate the Fourier Transform (FT) and, therefore, the variable used as the period (*T*) in the MT method equations (e.g., 10 s, 100 s, or 1000 s, depending on the depth of penetration to be obtained).

In this study:

- Total window: 6 days;
- Partial window: 1 h;
- Calculation window: 1000 s.

2.6.2. Apparent Resistivity Dispersion—Apparent Resistivity Curve

The ρ_a parameter (ρ_a), using Equations (2) and (3), will result in several dispersion points, as observed in Figure 3. Note how the figure is plotted as a function of the period ($T = 1/f$) for ease of analysis (the longer the period, the greater the *skin depth*). After obtaining all the associated ρ_a, a statistical data treatment is carried out for values of frequencies (periods) chosen in such a way that a reliable resistivity curve is obtained.

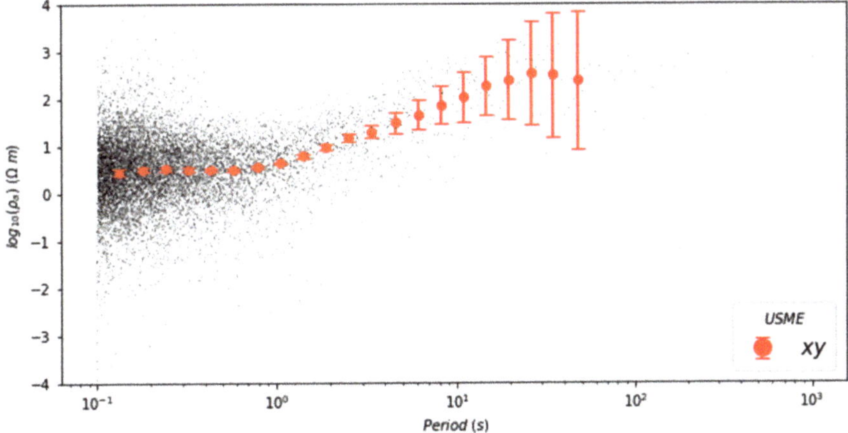

Figure 3. Apparent resistivity ρ_a dispersion as a function of the period (*T*). The gray dots represent the raw data computed from Equation (3). The red dots represent the fitted curve used for the apparent resistivities of the partial time window ('xy' means HQE-HFN).

To choose the period points to be evaluated, we consider the conditions stipulated in [6]:

- The chosen periods (or frequencies) must be equally spaced on a logarithmic scale;
- Ideally, it would help to have between 6–10 frequencies per magnitude. More frequencies are unnecessary, and fewer may result in unreliable values.

Points are selected as follows:

$$
\begin{aligned}
f_1 &= f_{max}, \\
f_2 &= f_{max}/\sqrt{1.8}, \\
f_3 &= f_{max}/1.8, \\
&\vdots \\
f_k &= f_{max}/1.8^{\sqrt{(k-1)}},
\end{aligned}
\tag{5}
$$

where $f_{max} = 10$ Hz (sample rate = 20 Hz).

After obtaining the desired frequencies, we observe that the dispersion of the data for each point will have a Gaussian behavior, as seen in Figure 4. This allows us to find a single ρ_a value associated with a single frequency/period and the corresponding standard deviation. Figure 3 shows an example of a curve obtained from fitting our dispersion data.

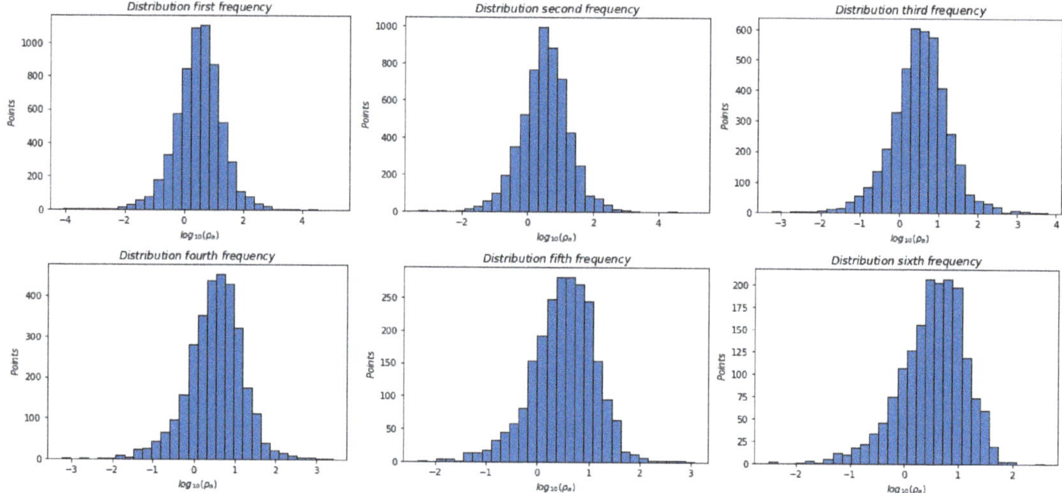

Figure 4. Distribution of the first six frequencies chosen to represent the apparent resistivity curve.

Now that we have the ρ_a curve for each hour, an ρ_a map is made in the total window (days), thus allowing precise observation of the space–time changes where the target seismic event occurred.

2.6.3. Apparent-Resistivity ρ_a Map

The ρ_a map (Figure 5), which is obtained from our algorithm code, can be seen as a matrix of values of N rows and M columns (Equation (6)) where each column represents a 1-h time window, and each row represents each frequency/period point (N = 'Number of point periods' and M = 'Number of partial windows'). Each column corresponds to the apparent resistivity-fitted curve (Figure 3); the only difference is the color assigned to each resistivity value (Figure 5).

$$
\begin{matrix}
\rho a_{(0,0)} & \rho a_{(0,1)} & \cdots & \rho a_{(0,M-1)} \\
\rho a_{(1,0)} & \rho a_{(1,1)} & \cdots & \rho a_{(1,M-1)} \\
& & \vdots & \\
\rho a_{(N-1,0)} & \rho a_{(N-1,1)} & \cdots & \rho a_{(N-1,M-1)}
\end{matrix}
\tag{6}
$$

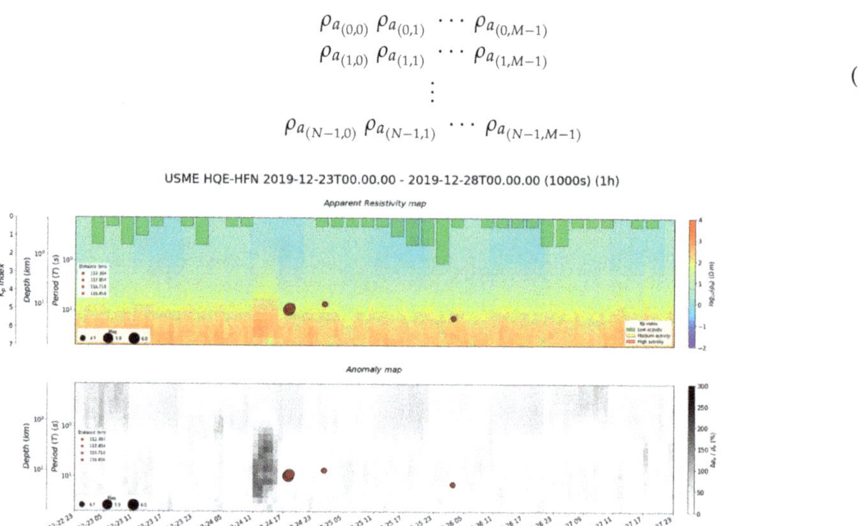

Figure 5. Anomaly registered before the 6.0 Mw Mesetas earthquake calculating the apparent resistivity from USME HQE-HFN channels. (Format of digitized data on the time axis: year-month-day hour).

As a second option, moving on to the interpretation of the ρ_a map, this can also be seen as a depth profile. Each period in each column means a different and greater depth; let us remember the relation of Equation (1), which tells us that the longer the period, the greater the *skin depth* (keeping in mind that this parameter also depends on the resistivity of the medium).

2.6.4. Anomaly Map

As the ρ_a color map can show some anomalous behaviors at specific depths (space–time variations), sometimes this phenomenon cannot be seen clearly. Therefore, it is necessary to create an anomaly map that represents the same time interval and depth ranges.

The values of the anomaly map (shown in the lower part of Figure 5) are the result of finding how anomalous each value of the ρ_a map is, that is, quantifying how much each one varies according to the average of the total window. First, the average value of the window is found, which will be a column of N rows resulting from the average resistivity for each period:

$$
\begin{array}{cccccc}
\rho_{a_{(0,0)}} & \rho_{a_{(0,1)}} & \cdots & \rho_{a_{(0,M-1)}} & \sum_{i=0}^{M-1} \rho_{a_{(0,i)}}/M & \overline{\rho_{a_0}} \\
\rho_{a_{(1,0)}} & \rho_{a_{(1,1)}} & \cdots & \rho_{a_{(1,M-1)}} & \sum_{i=0}^{M-1} \rho_{a_{(1,i)}}/M & \overline{\rho_{a_1}} \\
\vdots & & \Rightarrow & \vdots & = & \vdots
\end{array}
\tag{7}
$$

Starting from the matrix, which represents the apparent resistivity map (left part), then ($\Rightarrow$) we average each row (middle part), which is equivalent (=) to an average value corresponding to each row (period); we get:

$$
\rho_{a_{(N-1,0)}} \; \rho_{a_{(N-1,1)}} \; \cdots \; \rho_{a_{(N-1,M-1)}} \quad \sum_{i=0}^{M-1} \rho_{a_{(N-1,i)}}/M \; \overline{\rho_{a_{N-1}}},
$$

thus, obtaining an average ρ_a value for each period.

Finally, each value of the initial apparent resistivity matrix $(\rho_{a_{(i,j)}})$ is operated with its corresponding average $(\overline{\rho_{a_i}})$ as follows:

$$
Anomaly_{(i,j)} = A_{(i,j)} = \frac{\rho_{a_{(i,j)}} - \overline{\rho_{a_i}}}{\rho_{a_i}} \cdot 100,
\tag{8}
$$

to obtain a percentage of the anomaly. This value means how different each ρ_a value $(\rho_{a_{(i,j)}})$ is from the average $(\overline{\rho_{a_i}})$ of the total window, e.g., getting an anomaly value of 300% means that we have a ρ_a value three times greater than the average.

By obtaining all values of the anomalies, we will be able to plot a map with the same dimensions as the ρ_a map. This map is presented in grayscale (Figure 5) and allows us to see more clearly the area where an anomalous resistivity value occurs according to the average value of the total window.

3. Results and Discussion

Considering several variables such as solar activity, ρ_a, and associated anomalies, the results of this study provide visual information for interpreting in a simpler and more complete way. Figure 5 shows the space–time changes of the ρ_a for the USME station during the time interval when the Mesetas earthquake occurred (Mw6.0, Lon = 74.184° W, Lat = 3.462° N, H = 13 km-depth, 24 December 2019, UTC 19:03:55) and its associated aftershocks.

Solar activity (K_p index): Average low solar activity (<2). A behavior without significant variation is observed throughout the time window, except for a small peak of 2.5 on 26 December.

Apparent resistivity: The variation of the ρ_a depending on the depth H can be classified into three ranges of periods (Table 2):

Table 2. Variations of the ρ_a depending on the depth H.

Period T (s)	Depths H (km)	Apparent Resistivity ρ_a ($\Omega \cdot$m)
$T \lesssim 2$	$0.2\, H \lesssim 3$	$10^0 \lesssim \rho \lesssim 10^1$
$2 \lesssim T \lesssim 15$	$3 \lesssim H \lesssim 30$	$\rho \sim 10^1$
$15 \lesssim T$	$30 \lesssim H \lesssim 80$	$10^2 \lesssim \rho \lesssim 10^3$

Associated anomalies: There is a clear anomaly presented 11–16 h before 24 January (Table 3):

Table 3. ρ_a anomaly depending on the depth H.

Period T (s)	Depths H (km)	Anomaly (%)
$1 \lesssim T \lesssim 30$	$1.5 \lesssim H \lesssim 40$	81–224

Figure 6 shows the ρ_a curves found for a partial window before, during, and after the anomaly found (black: ρ_a curve before the anomaly, red: ρ_a curve during the anomaly, blue: ρ_a curve after the anomaly). A slight increase can be observed in the periods (depths) involved in the red curve, which means that at this specific time interval, there was a slight increase in the ρ_a of the subsurface between the periods shown in Figure 5. Geologically speaking, and because resistivity in rocks largely depends on their saturation fluids, this anomaly can be related to a momentary migration of fluids in the rock due to stresses produced in the mainshock, and then a return to initial equilibrium.

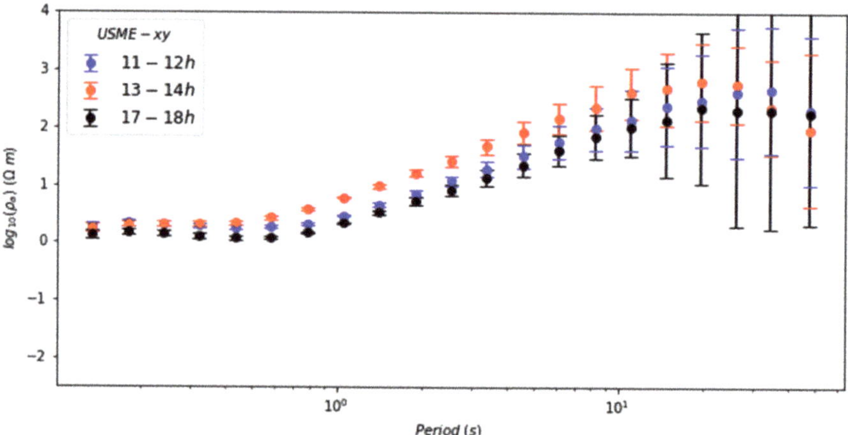

Figure 6. Comparison between three apparent resistivity curves, each corresponding to a different hour interval of December 24 (UTC). The red curve represents the time when the anomaly is observed.

The same procedure was done for the TUNJ station (also RSUNAL station) to rule out the idea that the found anomaly simply meant instrumental noise at the station. The results are illustrated in Figure 7. In this case, we see more anomalies, which are interpreted as external noise due to their periodicity. It is observed that there is also a disturbance of lesser magnitude in a depth range (periods) close to the depth of the seismic event, even if the event occurred at a greater distance from the station (~250 km).

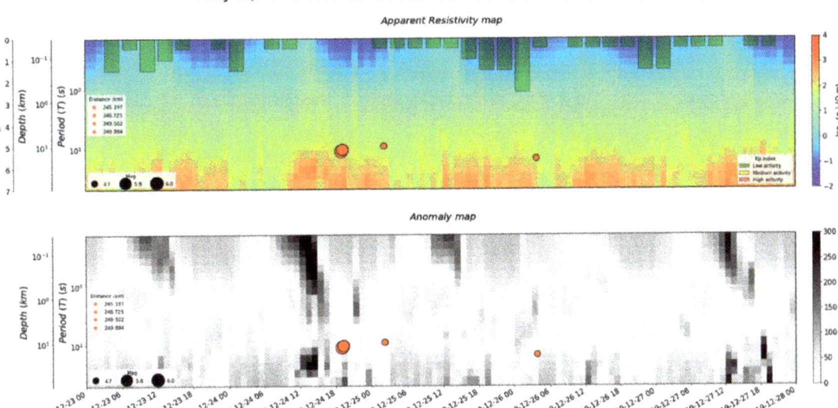

Figure 7. Anomalies registered at TUNJ station for the same Mesetas events.

4. Conclusions

The particularities identified in the case of the earthquakes in Mesetas yield suggestive results of an ρ_a anomaly in the subsoil eight hours before the first recorded seismic event (6.0 Mw), a duration 4–5 h from the start of the anomaly and within a range of periods between 1–30 s (depths between 1.5–40 km). It is also shown that this disturbance does not correspond to instrumental noise since it also exists and is registered at the TUNJ station. The origin of these anomalies may be associated with the migration of fluids due to the change in the stress regime before, during, and after the earthquake. We hypothesize that before the occurrence of an earthquake, the stress field generates pore pressure gradients, promoting alterations in fluid migration that change the resistivity of the upper crust.

The anomalies detected and observed at the USME multi-parameter station do not show a direct and causal relationship with seismic events of less than 5.8 Mw or distances greater than 120 km. However, since the notable anomalies in the designed algorithm must exceed 100% (double the average value), it is not ruled out that there are patterns of changes in the ρ_a of the subsoil of lower percentages related to these events. For these cases, we have used a minor anomaly scale.

Author Contributions: Conceptualization, C.A.V., J.S.G., J.J.G., J.M.S. and A.C.; Methodology, C.A.V., J.S.G. and J.J.G.; Software, J.S.G. and J.J.G.; Validation, J.S.G. and J.J.G.; Formal analysis, C.A.V., J.S.G. and A.C.; Writing, C.A.V., J.S.G. and A.C.; All authors have read and agreed to the published version of the manuscript.

Funding: Colombia Móvil–TIGO (Colombia) contributed with the data transmission, through mobile network, from remote sites to the data processing center, located at the National University of Colombia at Bogota.

Institutional Review Board Statement: Not applicable.

Informed Consent Statement: Not applicable.

Data Availability Statement: Once the paper is approved, the dataset will be openly available.

Acknowledgments: The authors want to express their gratitude to the Universidad Nacional Colombia at Bogotá and Universidad Antonio Nariño at Bogotá. Without their support and mutual collaboration, the permanent operation of the multi-parameter network would not be possible. The authors thank the Special Issue Editors (Guest Editors), for their patience, as well as two anonymous reviewers, for their valuable comments and suggestions.

Conflicts of Interest: The authors declare no conflict of interest.

References

1. Du, X.-B.; Xue, S.-Z.; Hao, Z.; Zhang, S.-Z. On the relation of moderate-short term anomaly of earth resistivity to earthquake. *Acta Seism. Sin.* **2000**, *13*, 393–403. [CrossRef]
2. Du, X.-B.; Li, N.; Ye, Q.; Ma, Z.-H.; Yan, R. A Possible Reason for the Anisotropic Changes in Apparent Resistivity Near the Focal Region of Strong Earthquake. *Chin. J. Geophys.* **2007**, *50*, 1555–1565. [CrossRef]
3. Fan, Y.-Y.; Du, X.-B.; Jacques, Z.; Tan, D.-C.; An, Z.-H.; Chen, J.-Y.; Zheng, G.-L.; Liu, J.; Xie, T. The Electromagnetic Phenomena Before theMs8.0 Wenchuan Earthquake. *Chin. J. Geophys.* **2010**, *53*, 997–1010. [CrossRef]
4. Liu, J.; Du, X.B.; Zlotnicki, J.; Fan, Y.Y.; An, Z.H.; Xie, T.; Zheng, G.L.; Tan, D.C.; Chen, J.Y. The changes of the ground and ionosphere electric/magnetic fields before several great earthquakes. *Chin. J. Geophys.* **2011**, *54*, 2885–2897. [CrossRef]
5. Cagniard, L. Basic Theory of the Magneto Telluric Method of Geophysical Prospecting. *Geophysics* **1953**, *18*, 605–635. [CrossRef]
6. Simpson, F.; Bahr, K. *Practical Magnetotellurics*; Cambridge University Press: Cambridge, UK, 2005. [CrossRef]
7. Uyeda, S.; Nagao, T.; Kamogawa, M. Short-term earthquake prediction: Current status of seismo-electromagnetics. *Tectonophysics* **2008**, *470*, 205–213. [CrossRef]
8. Uyeda, S.; Nagao, T.; Kamogawa, M. Earthquake Precursors and Prediction. In *Encyclopedia of Solid Earth Geophysics*; Gupta, H.K., Ed.; Encyclopedia of Earth Sciences Series; Springer: Dordrecht, The Netherlands, 2011.
9. Sarlis, N.V.; Skordas, E.S.; Varotsos, P.A. Natural Time Analysis: Results Related to Two Earthquakes in Greece during 2019. In Proceedings of the 2nd International Electronic Conference on Geosciences, Online, 8–15 June 2019; Available online: https://iecg2019.sciforum.net/ (accessed on 26 January 2023).
10. Cicerone, R.D.; Ebel, J.E.; Britton, J. A systematic compilation of earthquake precursors. *Tectonophysics* **2009**, *476*, 371–396. [CrossRef]
11. Varotsos, P.A.; Sarlis, N.V.; Skordas, E.S.; Lazaridou, M.S. Electric pulses some minutes before earthquake occurrences. *Appl. Phys. Lett.* **2007**, *90*, 064104. [CrossRef]
12. Varotsos, P.; Sarlis, N.; Skordas, E.; Lazaridou, M. Additional evidence on some relationship between Seismic Electric Signals (SES) and earthqueake focal mechanism. *Tectonophysics* **2006**, *412*, 279–288. [CrossRef]
13. Johnston, M.J.S. Review of Electric and Magnetic Fields Accompanying Seismic and Volcanic Activity. *Surv. Geophys.* **1997**, *18*, 441–476. [CrossRef]
14. Shrivastava, A. Are pre-seismic ULF electromagnetic emissions considered as a reliable diagnostics for earthquake prediction? *Curr. Sci.* **2014**, *107*, 596–600.
15. Petraki, E.; Nikolopoulos, D.; Nomicos, C.; Stonham, J.; Cantzos, D.; Yannakopoulos, P.; Kottou, S. Electromagnetic pre-earthquake precursors: Mechanisms, data and models-a review. *J. Earth Sci. Clim. Chang.* **2015**, *6*, 250.
16. Stanica, D.; Stanica, D.A. Anomalous pre-seismic behavior of the electromagnetic normalized functions related to the intermediate depth earthquakes occurred in Vrancea zone, Romania. *Nat. Hazards Earth Syst. Sci.* **2011**, *11*, 3151–3156. [CrossRef]
17. Zhang, X.; Shen, X. Electromagnetic Anomalies around the Wenchuan earthquake and their relationship with earthquake preparation. *Int. J. Geophys.* **2011**, *2011*, 904132. [CrossRef]
18. Koulouras, G.; Balasis, G.; Kiourktsidis, I.; Nannos, E.; Kontakos, K.; Stonham, J.; Ruzhin, Y.; Eftaxias, K.; Cavouras, D.; Nomicos, C. Discrimination between pre-seismic electromagnetic anomalies and solar activity effects. *Phys. Scr.* **2009**, *79*, 045901. [CrossRef]
19. Febriani, F.; Han, P.; Yoshino, C.; Hattori, K.; Nurdiyanto, B.; Effendi, N.; Maulana, I.; Gaffar, E.Z. Ultra low frequency (ULF) electromagnetic anomalies associated with large earthquakes in Java Island, Indonesia by using wavelet transform and detrended fluctuation analysis. *Nat. Hazards Earth Syst. Sci.* **2014**, *14*, 789–798. [CrossRef]
20. Lowrie, W. *Fundamentals of Geophysics*, 2nd ed.; Cambridge University Press: New York, NY, USA, 2007; 393p.
21. Chave, A.; Jones, A. *The Magnetotelluric Method, Theory and Practice*; Cambridge University Press: Cambridge, UK, 2012. [CrossRef]
22. Weidelt, P. The inverse problem of geomagnetic induction. *Z. Geophys.* **1972**, *38*, 257–289. [CrossRef]
23. Schmucker, U. Regional induction studies: A review of methods and results. *Phys. Earth Planet. Inter.* **1973**, *7*, 365–378. [CrossRef]
24. NOAA (National Oceanic and Atmospheric Administration). Available online: https://www.swpc.noaa.gov/products/planetary-k-index. (accessed on 19 December 2022).
25. Vargas, C.A.; Caneva, A.; Monsalve, H.; Salcedo, E.; Mora, H. Geophysical Networks in Colombia. *Seism. Res. Lett.* **2018**, *89*, 440–445. [CrossRef]
26. Fino, J.M.S.; Caneva, A.; Jiménez, C.A.V.; Ochoa, L.H. Electrical and magnetic data time series' observations as an approach to identify the seismic activity of non-anthropic origin. *Earth Sci. Res. J.* **2021**, *25*, 297–307. [CrossRef]
27. CME. Broadband Seismometer Model CME-4111. Available online: http://r-sensors.ru/en/products/digital_seismometers_eng/cme-4211nd-eng/ (accessed on 19 December 2022).
28. Tinker-Rasor. Direct Burial—Tinker & Rasor. 2021. Available online: https://www.tinker-rasor.com/direct-burial (accessed on 19 December 2022).
29. Bartington Instruments. Mag648 & Mag649 Low Power Three-Axis Magnetic Field Sensors. Available online: https://www.bartington.com/products/low-power-magnetometers/mag648-649/ (accessed on 19 December 2022).

Article

Statistical and Comparative Analysis of Multi-Channel Infrared Anomalies before Earthquakes in China and the Surrounding Area

Yingbo Yue [1,2], Fuchun Chen [1,*] and Guilin Chen [1]

[1] Key Laboratory of Infrared System Detection and Imaging Technology, Shanghai Institute of Technical Physics, Chinese Academy of Sciences, Shanghai 200083, China
[2] University of Chinese Academy of Sciences, Beijing 100049, China
* Correspondence: fuchun.chen@mail.sitp.ac.cn

Abstract: Abundant infrared remote sensing images and advanced information processing technologies are used to predict earthquakes. However, most studies only use single long-wave infrared data or its products, and the accuracy of prediction is not high enough. To solve this problem, this paper proposes a statistical method based on connected domain recognition to analyze multi-channel anomalies. We extract pre-seismic anomalies from multi-channel infrared remote sensing images using the relative power spectrum, then calculate positive predictive values, true positive rates and probability gains in different channels. The results show that the probability gain of the single-channel prediction method is extremely low. The positive predictive value of four-channel anomalies is 41.94%, which is higher than that of single-channel anomalies with the same distance threshold of 200 km. The probability gain of the multi-channel method is 2.38, while that of the single-channel method using the data of any channel is no more than 1.26. This study shows the advantages of the multi-channel method to predict earthquakes and indicates that it is feasible to use multi-channel infrared remote sensing images to improve the accuracy of earthquake prediction.

Keywords: earthquake prediction; infrared remote sensing; multi-channel; pre-seismic anomaly; relative power spectrum; connected region

Citation: Yue, Y.; Chen, F.; Chen, G. Statistical and Comparative Analysis of Multi-Channel Infrared Anomalies before Earthquakes in China and the Surrounding Area. *Appl. Sci.* **2022**, *12*, 7958. https://doi.org/10.3390/app12167958

Academic Editors: Eleftheria E. Papadimitriou and Alexey Zavyalov

Received: 8 July 2022
Accepted: 5 August 2022
Published: 9 August 2022

Publisher's Note: MDPI stays neutral with regard to jurisdictional claims in published maps and institutional affiliations.

1. Introduction

Earthquake prediction is a complex and challenging theme. Since scientists discovered pre-seismic infrared abnormal phenomena in the 1980s [1], scholars around the world have undertaken various relevant research. Qiang et al. put forward a relatively reasonable theoretical mechanism, which indicates that the main causes of the abnormal temperature increase before the earthquake are gases released from the earth's crust and the change of the electric field, based on various observations and experiments [2]. The change of water content in the earth's surface soils is also able to cause infrared anomalies before earthquakes [3]. Some experiments proved that the infrared radiation of the rock changes when it is pressed by stress [4]. Their studies provide theoretical support for the development of earthquake prediction using infrared radiation data.

Infrared remote sensing images have a definite advantage of wide-field and continuous observation over ground-based observation and are widely applied to earthquake prediction [5–8]. Most researchers only analyzed single-type data, mainly using long-wave infrared radiation or its products, such as land surface temperature (LST) and outgoing long-wave radiation (OLR) [9–11]. In some case studies, the pre-seismic infrared anomaly is discovered from remote sensing images as a common phenomenon [12–14]. There are only a few studies about the middle-wave infrared anomalies before earthquakes, whose trend is similar to the anomalies in the long-wave channel [5,15]. Other earthquake case studies found that there are abnormal changes in water vapor content in the atmosphere before and after the event [16,17]. Most infrared data used to predict earthquakes are from

NOAA satellites, Terra/Aqua satellites and Fengyun satellites [18,19]. Wei et al. found the infrared anomalies before the Ms8.0 earthquake in Sichuan, China, using single-channel images from the FY-2C satellite [20]. Ouzounov et al. found the anomalies before several strong earthquakes in Xinjiang, China, using OLR data from NOAA satellites and the FY-2D satellite [21]. Zhong et al. found the infrared anomaly associated with the 2017 M6.5 Jiuzhaigou earthquake from the data of two Fengyun satellites (FY-2E and FY-2G) [22].

Researchers have proposed different methods that were successfully used to extract pre-seismic anomalies, such as robust satellite techniques [23], interquartile, wavelet transform, Kalman filter methods [24], power spectrum [25] and some artificial intelligent methods [26]. Although abundant remote sensing data and advanced information processing technologies are used to predict earthquakes, there has not been any stable and valid algorithm to eliminate the influence of non-seismic factors, such as seasonal changes, weather conditions and human activities [18]. This is because of the complexity and variability of the earth system and the space environment. As a result, most methods merely work well in a few cases, lacking statistical evidence. Some statistical results in a small region show that there is an infrared anomaly before most earthquakes, but they did not analyze the proportion of the anomalies followed by an earthquake. In a study of 20 earthquakes in the Tibet region, the anomalies of the brightness temperature appeared before 17 earthquakes, and that of long-wave radiation appeared before 16 earthquakes [27]. The accuracy of earthquake prediction is unable to be estimated without PPV. Some statistical studies show a low positive predictive value (PPV). The statistical results based on the robust satellite techniques indicate the true positive rate (TPR) is high but the PPV of pre-seismic infrared anomalies, which is 25.9% in Sichuan province, is too low to put into practice [28]. The accuracy of earthquake prediction in the statistics by Jiao and Shan is 6.01% [29]. The PPV calculated by Filizzola et al. is 7.61% [23]. This means that most prediction results are wrong. Some studies do not even support the feasibility of earthquake prediction based on infrared remote sensing data [30]. Although some researchers obtained the PPV of 76.1% and the TPR of 67.1%, the spatiotemporal occupation of anomalies is high at 43.4%, which may cause a large prediction range and excessive public panic [31].

To explore a more valid algorithm for earthquake prediction using the data of infrared remote sensing, both data and research methods are improved. For one thing, multi-channel infrared images are used in this paper, and the four channels could provide more information about the state of the earth's surface and atmosphere. For another thing, the statistical method based on connected region recognition is proposed to analyze the correlation between infrared anomalies and earthquakes, which could recognize spatiotemporal characteristics of anomalies in the long-term and wide-region studies. In this paper, the relative power spectrum method is respectively used to extract the anomalies from data of every channel. Both statistical analysis and case study are used to compare the prediction performance of the data from any single channel. Finally, four-channel anomalies are analyzed statistically. The results show that multi-channel anomalies could provide larger PPV and probability gain. It proves the potential of multi-channel infrared data in earthquake prediction and shows that it is possible to identify anomalies associated with earthquakes using multi-dimensional or multi-source data. The statistical method based on connected region recognition could be used to analyze pre-seismic anomalies from most kinds of remote sensing data. It lays the foundation for more data to be used in earthquake prediction.

2. Materials and Methods

The full research method is shown in Figure 1. Section 2.1 introduces the data and research range (time and region). The wavelet decomposition and relative power spectrum are used to extract anomalies from the time series on every pixel, which is detailed in Section 2.2. Then, the statistical method based on connected region recognition is proposed to analyze the correlation between anomalies and earthquakes, which is detailed in Section 2.3.

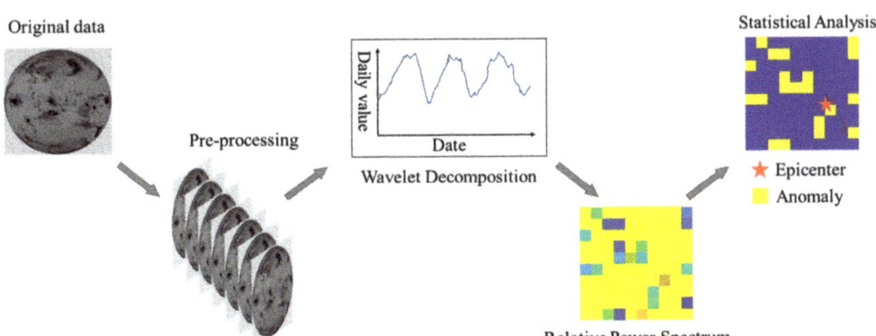

Figure 1. The flow of the research method.

2.1. Data and Study Area

Feng Yun-2G is the fifth operational satellite of the first generation of geostationary meteorological satellites launched by China on 31 December 2014, which can observe the parameters of the earth's surface and atmosphere widely and continuously. The main remote sensor is the visible and infrared spin scanning radiometer with 5 channels, including one visible channel (VIS), one medium-wave infrared channel (IR4), one water vapor channel (IR3) and two long-wave infrared channels (IR1 and IR2). The detailed parameters of each channel are shown in Table 1.

Table 1. The specific information of each channel.

Channel	Spectral Interval	Spatial Resolution
VIS	0.55–0.75 μm	1.25 km (5 km)
IR1	10.3–11.3 μm	5 km
IR2	11.5–12.5 μm	5 km
IR3	6.3–7.6 μm	5 km
IR4	3.5–4.0 μm	5 km

In this paper, the original data are the images from four infrared channels of the Feng Yun-2G satellite, which are provided by the National Satellite Meteorological Center. (http://satellite.nsmc.org.cn/PortalSite/Data/Satellite.aspx, accessed on 25 June 2022). One full-disk data file is generated every half hour or one hour. The file structure is shown in Figure 2, including file attributes and scientific datasets. Every file includes the geographic location of the image center, cloud classification, five-channel images and calibration tables and is saved as the hierarchical data format (HDF). The red digits are the number of the datasets, while the white digits are the size of one dataset. Moreover, the center also provides a lookup table so that the users could calculate the longitude and latitude of every pixel in the image.

Due to the high transmittance of the earth's atmosphere in the atmospheric windows, the radiation value from two long wave channels and one middle wave channel mainly depends on the temperature and the emissivity of the earth's surface. The emissivity of the same target is usually constant. The peak wavelength of most objects on the earth's surface is in the long-wave infrared channel, so the brightness temperature data from the long-wave infrared channel could show the temperature change trend of most objects on the earth's surface [32]. The data from the medium-wave infrared channel show the temperature change of the high-temperature targets and are affected by the reflection of solar radiation in the daytime, so data at night are selected in this paper. The water vapor channel is in the strong absorption band of water vapor, which is one of the main infrared absorption gases in the earth's atmosphere [33].

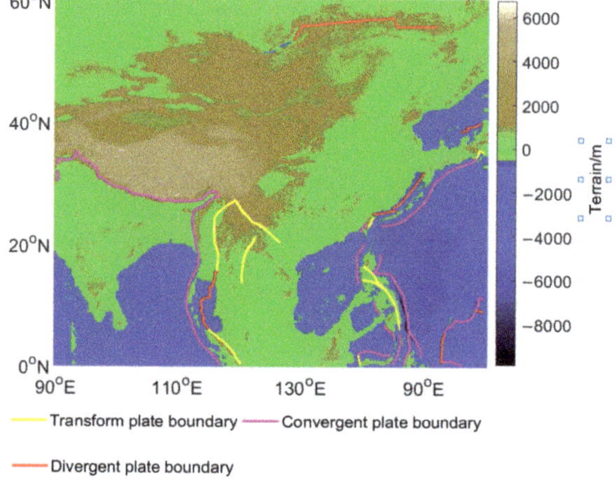

Figure 2. The structure of the data file.

China is located between the Pacific seismic zone and Euro–Asia seismic zone, so earthquakes happen frequently there, especially in Qinghai–Tibet Plateau, Yunnan–Guizhou Plateau and Taiwan. In China, the terrain is high in the west and low in the east, as shown in Figure 3. The lines in different colors are the different types of plate boundaries. The topography data are provided by the University of Califonia San Diego. (http://topex.ucsd.edu/marine_topo/mar_topo.html, accessed on 23 July 2022). China has a wide territory and rich soil resources, mainly including 15 types of soil. The variation of water content in the soil during the earthquake preparation period may also lead to anomalies in infrared remote sensing images [3]. The study area of this paper is between 0° N to 60° N and 70° E to 140° E, including China and the surrounding area. All images in this region from June 2015 to December 2020 are applied to extract abnormal signals. There are 358 earthquakes with a magnitude over five in the area from 31 January 2016 to 1 January 2021. The strongest earthquake in the study range was the earthquake with a magnitude of 7.3 that happened in Kyushu, Japan on 16 April 2016. According to the theoretical model proposed by Dobrovolsky, the relationship between the radius of earthquake preparation region R and the earthquake magnitude M is shown in Equation (1) [34].

$$R = 10^{0.43M},\tag{1}$$

The earthquake magnitude in this study is between 5 and 7.3, so the radius of the influenced region varies from 141 to 1377 km.

Figure 3. The terrain in the study area.

2.2. Anomaly Extraction

The relative power spectrum is a common method to extract seismic information from infrared remote sensing images and is used in some case studies with significant results. The process is shown in Figure 4 and was programmed using Python in the Spyder platform. The algorithm is introduced in detail as follows [35,36]:

(1) Pre-processing: We could not gather complete earth surface radiation data because of cloud coverage and the limitation of the remote sensing system. Some pre-processing is necessary before extracting anomalies. Infrared radiation could not go through clouds, so the infrared images in the region covered by clouds reveal the temperature of the top of the clouds, which is far lower than that of the earth's surface. After obtaining brightness temperatures by the look-up table method, we remove the invalid values that are 1.5-times the standard deviation below the average to decrease the effect of clouds [27]. Then, the average temperature every day is calculated. Finally, the spatiotemporal data become continuous through the nearest-neighbor interpolation method.

(2) Wavelet decomposition: The infrared radiation on the earth's surface is also affected by seasonal changes, weather conditions, geological activities, human activities, and so on. High-frequency information and low-frequency information are separated by wavelet decomposition and wavelet reconstruction. The low-pass part of seventh-order wavelet decomposition retains long-period information and is regarded as the background field. The low-pass part of second-order wavelet decomposition eliminates high-frequency information and is used to attenuate the influence of weather and human factors. The low-pass part of the second-order wavelet decomposition is subtracted by that of the seventh-order wavelet decomposition to eliminate the background field and high-frequency components.

(3) Power spectrum: The power spectrum reflects the change of signal power with frequency. We take 64 days as the window length and 1 day as the step length to calculate the time–frequency distribution of the power spectrum in each window [37].

(4) Characteristic frequency: We calculate the relative amplitude for each pixel and frequency, and select the frequency with the largest amplitude change as the characteristic frequency to obtain the spatiotemporal data of the relative power spectrum at the characteristic frequency.

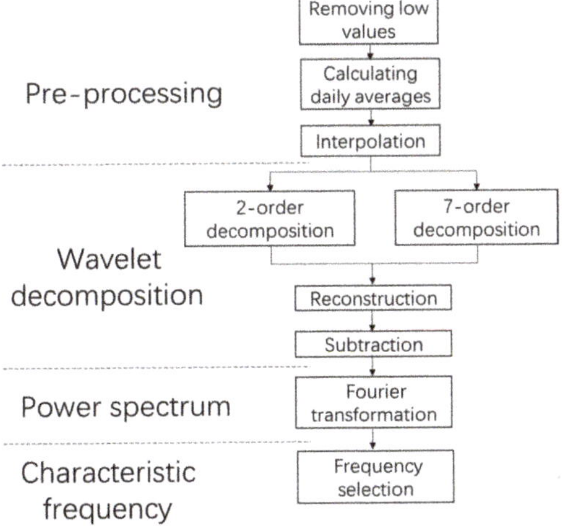

Figure 4. Flow chart of anomaly extraction algorithm.

In this paper, the algorithm is used to process the data from every channel of FY-2G, respectively, to compare the difference between channels and improve prediction performance using multi-channel data.

2.3. Statistical Method

Accuracy of earthquake prediction is the basis of practical application, because wrong predictions may cause serious economic losses and public panic. The PPV of infrared abnormal signals can evaluate the accuracy of its application in earthquake prediction, which is defined as the proportion of abnormal signals related to an earthquake in total anomalies, as shown in Equation (2) [28].

$$\text{PPV} = \frac{N_0}{N_a}, \tag{2}$$

where N_0 is the number of abnormal signals related to an earthquake, and N_a is the number of total abnormal signals in the research area. To calculate the PPV, we propose a method based on the connected domain to identify the spatiotemporal occupancy range of abnormal signals. We deem the point where the relative power spectrum is more than five as abnormal. The connected region of abnormal points is regarded as the spatial–temporal range of a single abnormal phenomenon, and abnormal regions with relatively small areas or short-term are ignored to eliminate the infrared radiation increase caused by human activities. In this paper, an anomaly that covers less than 500 pixels or lasts less than 3 days is ignored. In previous studies, anomalies appeared on average one to three weeks before the earthquake [38]. As a result, the anomaly is regarded to be related to the earthquake, if the distance between the abnormal region and the epicenter is less than the distance threshold (T_d) and the anomaly appears within one month before the earthquake.

The TPR is the proportion of earthquakes with pre-seismic infrared abnormalities in total earthquakes, which is defined as Equation (3) [28].

$$\text{TPR} = \frac{N_1}{N_e}, \tag{3}$$

where N_1 is the number of earthquakes with any pre-seismic anomaly, and N_e is the number of total earthquakes in the research area. It is used to assess the universality of infrared anomalies before earthquakes.

It provides higher PPV and TPR to increase the distance threshold, but the spatial accuracy of prediction would decrease. Therefore, earthquake prediction needs to predict more earthquakes successfully with lower spatiotemporal occupancy. The probability gain is the ratio of TPR to spatiotemporal occupancy, shown as Equation (4) [39,40]. It can be regarded as a criterion for selecting the optimal distance threshold.

$$\text{Gain} = \frac{\text{TPR}}{\tau}, \tag{4}$$

where TPR is the true positive rate, τ is the fraction of space–time occupied by the predicted range, and is associated with the distance threshold. Matlab platform was used for statistical analysis and results display.

3. Results and Discussion

3.1. Channels Comparison

Some previous research found the epicenter of an impending earthquake may be far from the anomaly [41–45], so both the abnormal region and the region around the anomaly should be considered as the predicted region. The possibility that there is any upcoming earthquake in the predicted region is larger if the area of the predicted region is larger. PPVs with different distance thresholds in four channels are calculated and shown in Figure 5. There are a total of 619 anomalies from 2016 to 2020, including 155 in the IR1 channel, 161 in the IR2 channel, 132 in the IR3 channel, and 171 in the IR4 channel. PPVs vary obviously with the change of the distance threshold. The difference between the PPVs in any two channels is less than 0.08 for the same distance threshold. It means that there is not much difference between the PPVs of anomalies from different channels.

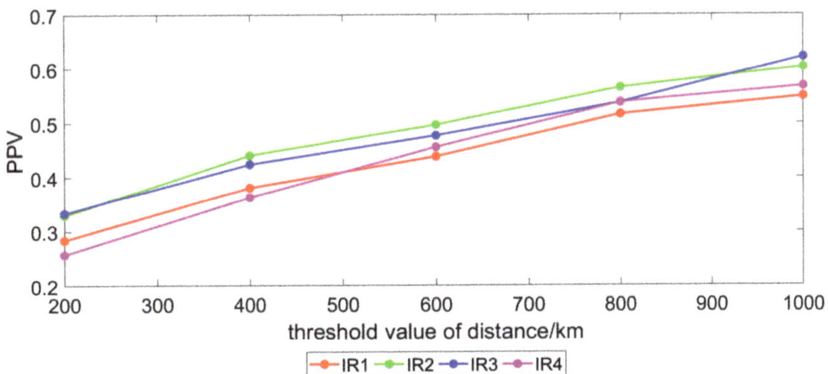

Figure 5. Positive predictive values in different channels at different distance thresholds.

Figure 6 shows the TPRs with different distance thresholds in four channels. The TPRs in the two long-wave infrared channels are similar. The TPR in the IR4 channel is the highest among all channels whatever the distance threshold is, while that in the IR1 channel is the lowest. The difference between the TPRs in the IR2 channel and the IR3 channel becomes large with the increase in the distance threshold. The long-wave infrared channels (IR1 and IR2) behave worse on PPV and TPR, although they were earlier used for earthquake prediction.

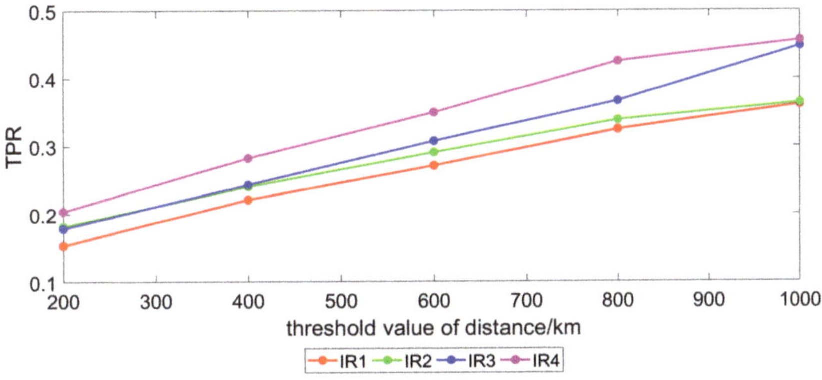

Figure 6. True positive rates in different channels at different distance thresholds.

The infrared anomalies of 358 earthquakes in the study area were analyzed statistically. As shown in Figure 7, the horizontal axis is a four-digit number whose four digits represent the state of four channels (IR1, IR2, IR3, IR4) successively, where 1 means there was an anomaly before the earthquake, and 0 means there was no anomaly before the earthquake. For example, 1111 means there were abnormalities in all four channels. The state of the anomalies in the two long-wave channels is similar. There were only 26 earthquakes with four-channel anomalies within 400 km around the epicenter, and 18 of them were within 200 km.

As shown in Figure 8, the Gain in each channel varies slightly with the distance threshold. The probability gain cannot be improved by changing the distance threshold.

The gains in the IR3 channel are highest among those in all four channels for the same distance threshold and only a little higher than that in other channels.

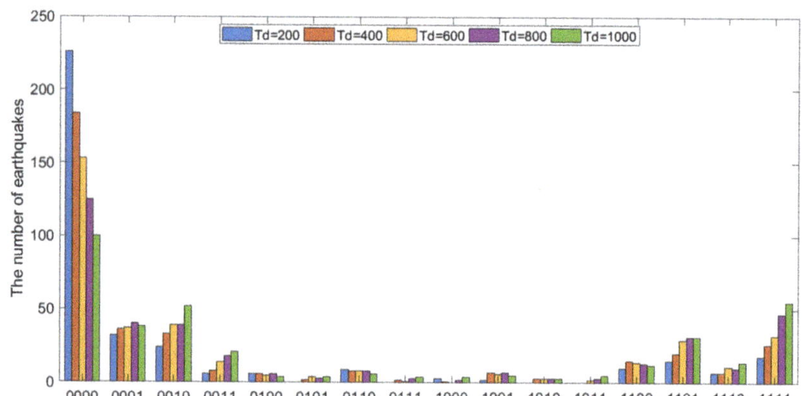

Figure 7. Infrared anomalies before earthquakes.

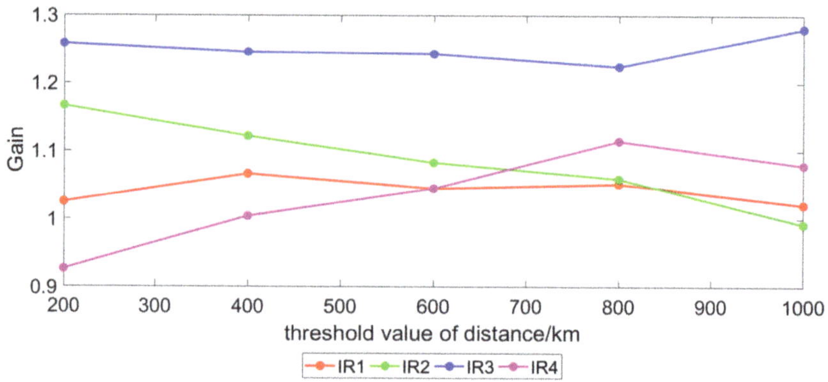

Figure 8. Probability gains in different channels at different distance thresholds.

3.2. Case Study

Considering regional differences, three seismic cases in different provinces were taken as examples. Detailed seismic information is shown in Table 2. The magnitude of the Sichuan earthquake was the largest. All three earthquakes occurred at shallow depths.

Table 2. Seismic information.

Province	Date	Magnitude (M)	Latitude (°N)	Longitude (°E)	Depth (km)
Sichuan	8 August 2017	7.0	33.20	103.82	20
Taiwan	16 December 2018	5.2	23.71	121.80	26
Tibet	19 July 2019	5.6	27.67	92.89	10

3.2.1. Sichuan Earthquake

There was a strong earthquake with a magnitude of 7.0 in Sichuan at 21:19:46 on 8 August 2017 (LT). The epicenter was at 103.82° E, 33.20° N. The depth was 20 km. At

07:27:52 on the next day (LT), an earthquake with a magnitude of 6.6 occurred in Xinjiang. The epicenter was at 82.89° E, 44.27° N. The depth was 11 km.

The information on pre-seismic anomalies in different channels is shown in Table 3. In Table 3, the start time and the end time are the numbers of days that the start time and end time of the anomaly relative to the earthquake time. Negative numbers are before the earthquake and positive numbers are after the earthquake. The anomalies of the two long-wave channels appeared 25 days before the earthquake, as shown in Figure 9. On that day, there were discrete high-value points north of the epicenter in the images of the four channels, and the anomaly area of the long-wave infrared channels was larger, so it was identified earlier. As shown in Figure 10, there were high-amplitude and large-area anomalies in four channels on the 19 days before the earthquake. On that day, IR1, IR2 and IR4 obtained their maximum abnormal value. The anomalies were located between the epicenters of the two earthquakes and more close to that of the Sichuan earthquake. Figure 11 shows the largest coverage area of anomalies in the earthquake preparation region. During the abnormal period, the abnormal area of the four channels went through two times of increase and decrease. The abnormal area in the IR3 channel was less than 500 pixels 19 days before the earthquake, while that in the IR4 channel was also too small during the first time of increase and decrease. They are both ignored in Table 3. Anomalies in all channels lasted until 18 days before the earthquake.

Table 3. The spatial–temporal characteristics of anomalies before the Sichuan earthquake.

Channel	Start Time (Days)	End Time (Days)	Maximum	Time of Maximum (Days)	Maximum Cover Area (Pixes)	Distance (km)
IR1	−25	−18	28.9379	−19	2275	244.1042
IR2	−25	−18	29.9185	−19	2569	244.1042
IR3	−24	−22	27.2232	−24	765	267.2322
	−20	−18	26.1566	−19	1447	482.0594
IR4	−21	−18	25.7784	−19	1146	567.8146

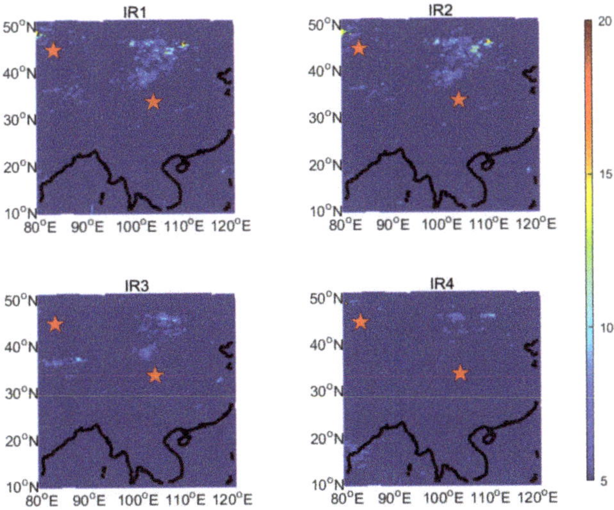

Figure 9. Power spectrum images on 14 July 2017.

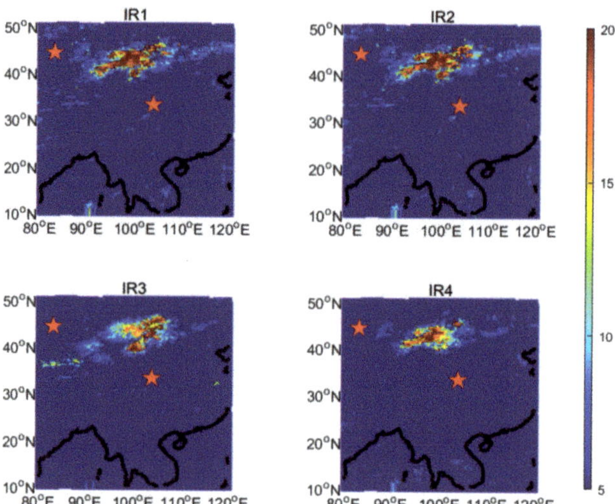

Figure 10. Power spectrum images on 20 July 2017.

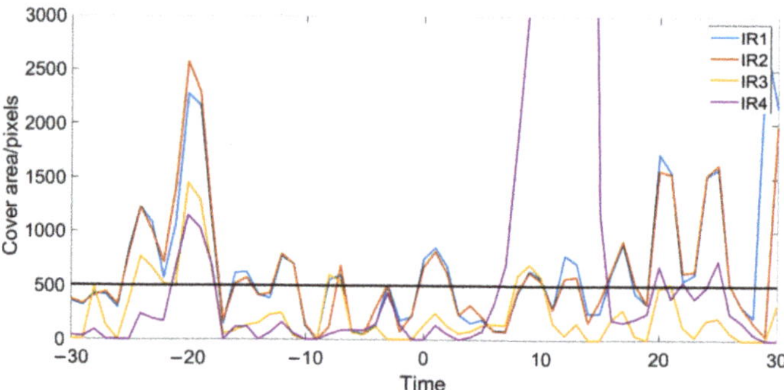

Figure 11. The cover area of the anomalies before and after the Sichuan earthquake.

3.2.2. Taiwan Earthquake

An earthquake with a magnitude of 5.2 happened in Taiwan on 16 December 2018, 05:21:05 (LT). The location of the epicenter was 23.71° N, 121.80° E, and the focal depth was 26 km. On the same day, an earthquake with a magnitude of 5.7 broke out at 12:46:07 (LT) in Sichuan. The location of the epicenter was 28.24° N, 104.95° E, and the focal depth was 12 km.

As shown in Table 4, the IR3 channel appeared abnormal at first, and anomalies in other channels appeared simultaneously and disappeared simultaneously, which were a little later than that in the IR3 channel. In images from the IR3 channel, a high-value region arose on 6 December 2018, shown in Figure 12. In the IR3 channel image, there was a weak anomaly between the two epicenters. Relatively, there were more obvious anomalies in the high latitude region, but they were far from the two epicenters, which is not significant for the prediction of this earthquake. It became an isolated region on 8 December 2018, while the anomalies in the other three channels just started, as shown in Figure 13. The anomalous positions of the two long-wave channels and the medium-wave channel were close, but the anomaly area of the medium-wave channel was smaller. The anomaly area of the IR3 channel was the largest, and it has obvious location deviation from the other three channels. On 9 December 2018, the anomalies in IR1 and IR3 obtained their maximums, as shown in Figure 14. In the abnormal region of the IR3 channel image, the southern part is closer to the epicenter of Taiwan, in which anomaly intensity was significantly higher than that in the northern part. Similar to the first case, the anomaly occurred between the epicenters of the two earthquakes. The abnormal region was mainly on the land and distributed along the coastline. The epicenter of the Taiwan earthquake was at the edge of the anomaly. The anomaly in the IR3 channel was closest to the epicenter. There was only 1.7 km between the anomaly and the epicenter. The anomalies in four channels all disappeared completely on 12 December 2018. According to Figure 15, the anomalies went through appearance, increase, decrease and disappearance before the earthquake. The anomaly in the IR4 channel intersected with that in other channels, but the distance from the epicenter was far beyond the radius of the earthquake preparation region calculated theoretically.

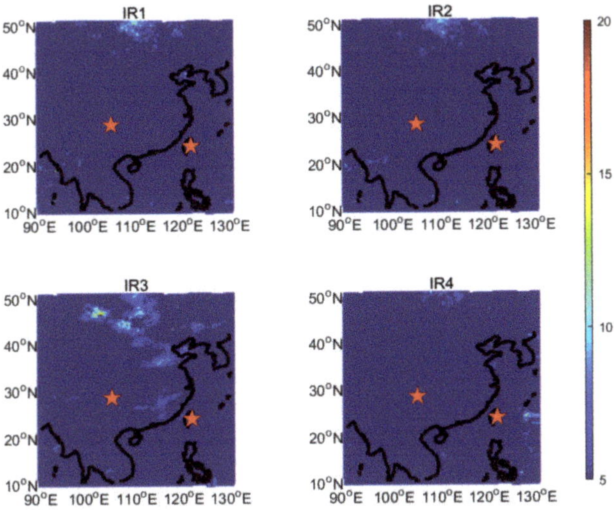

Figure 12. Power spectrum images on 6 December 2018.

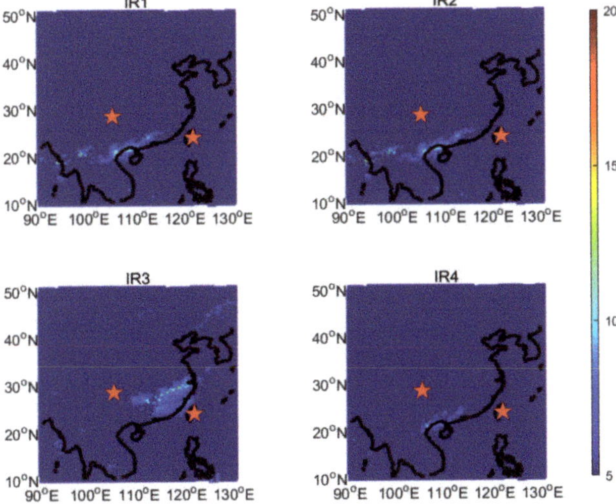

Figure 13. Power spectrum images on 8 December 2018.

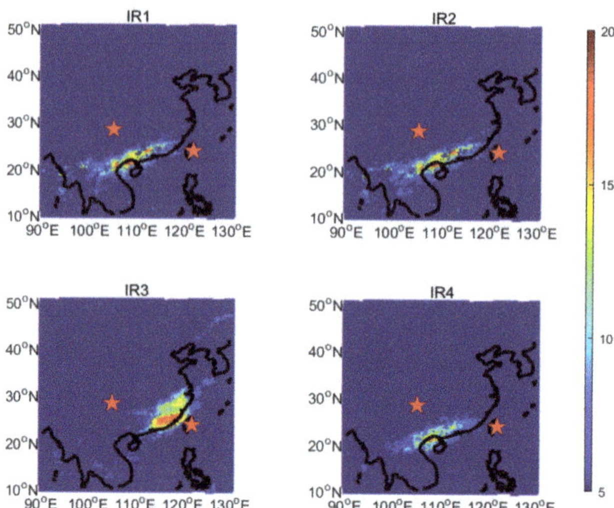

Figure 14. Power spectrum images on 9 December 2018.

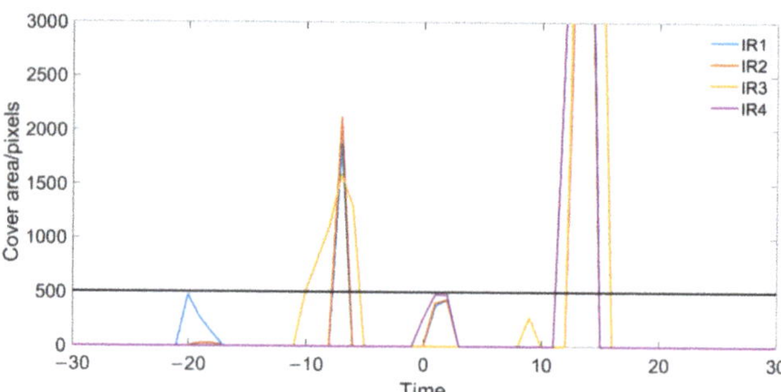

Figure 15. The cover area of the anomalies before and after Taiwan earthquake.

Table 4. The spatial–temporal characteristics of anomalies before the Taiwan earthquake.

Channel	Start Time (Days)	End Time (Days)	Maximum	Time of Maximum (Days)	Maximum Cover Area (Pixes)	Distance (km)
IR1	−8	−5	22.1271	−7	2068	134.1929
IR2	−8	−5	21.1125	−6	2202	134.1929
IR3	−10	−6	19.0396	−7	1587	1.6510
IR4	−8	−5	23.6295	−6	1625	240.8767

3.2.3. Tibet Earthquake

Tibet is located on the Qinghai–Tibet Plateau with high altitudes and complex terrain. There was an earthquake with a magnitude of 5.6 at 17:22:14 on 19 July 2019. The epicenter was located at 27.67° N, 92.89° E. The depth was 10 km. Anomalies before the earthquake are shown in Table 5. From 30 days to 20 days before the earthquake, only the IR4 channel showed abnormalities. Anomalies in the IR3 and IR4 channels appeared on the 11th day before the earthquake. IR1 and IR2 showed abnormalities later. The occurrence time and disappearance time of the anomalies in the four channels were relatively close. The relative powers spectrum on 9 July 2019 is shown in Figure 16. Anomalies in all four channels were close in location. The distance between the anomalies and the epicenter is 11.5 km. The cover area of the anomalies before and after the earthquake is shown in Figure 17. The variation trend of the cover area of anomalies in the four channels is similar during the four-channel abnormal period. For this seismic case, the anomalies of the four channels show obvious similarities in spatiotemporal characteristics and the evolution process.

Table 5. The spatial–temporal characteristics of anomalies before the Tibet earthquake.

Channel	Start Time (Days)	End Time (Days)	Maximum	Time of Maximum (Days)	Maximum Cover Area (Pixels)	Distance (km)
IR1	−10	−6	25.8302	−10	2366	11.4747
IR2	−10	−6	21.7490	−7	2569	11.4747
IR3	−11	−5	17.1277	−9	3147	11.4747
IR4	−30	−20	36.5649	−25	7125	184.2958
	−11	−5	35.2557	−9	3237	11.4747

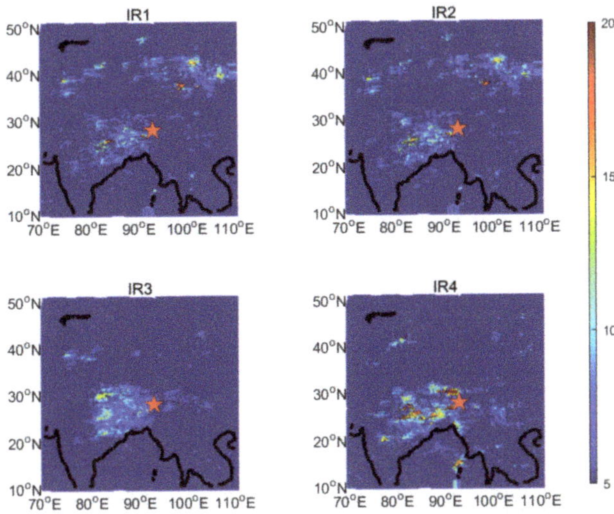

Figure 16. Power spectrum images on 7 July 2019.

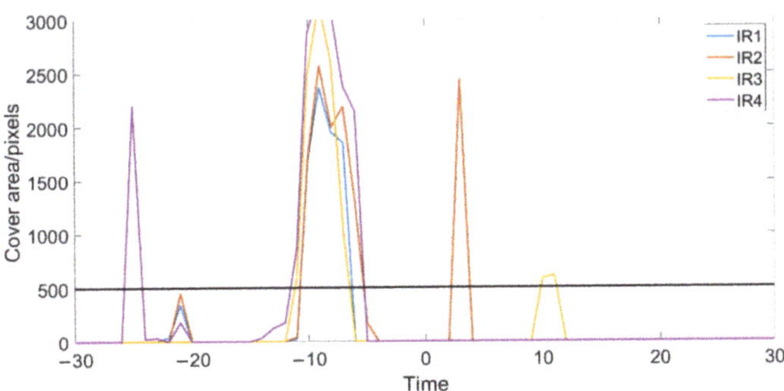

Figure 17. The cover area of the anomalies before and after Tibet earthquake.

3.3. Multi-Channel Anomalies

To improve the accuracy of earthquake prediction, we scan the relative power spectrum data in the four channels and identify the region of four-channel anomalies, which means the region where all of the four channels are abnormal at the same time. The statistical result is shown in Table 6. A larger distance threshold can improve PPV and TPR, but can not improve the probability gain. The probability gain is the highest when the distance threshold is 200 km. Figure 18 shows the comparison between the PPV, TPR and gain of multi-channel methods with that of the single-channel method with the distance threshold of 200 km. The PPV and probability gain of the multi-channel method are higher than that of any single-channel method. In particular, the probability gain is roughly doubled, although the TPR is very small. Only 19 earthquakes are associated with the multi-channel anomalies because the multi-channel method only considers simultaneous anomalies of four channels, and earthquakes with anomalies of no more than three channels will be missed. The locations of the epicenters of these 19 earthquakes are shown in Figure 19. Four-channel infrared anomalies were found before the earthquakes in different regions, which indicates the phenomenon is not only suitable for a specific region. The distribution of anomaly time is shown in Figure 20. Eight earthquakes occurred one to two weeks after the anomalies appeared. Only one earthquake occurred three to four weeks after the anomaly appeared. Figure 21 is the Molchan diagram of the different methods. The spatiotemporal occupancy of the four-channel anomaly is much smaller than that of the single-channel anomaly. Although the miss rate is so high, enough small space–time occupations could provide a high value of probability gain.

Table 6. Statistical result of multi-channel anomalies.

Distance Threshold	200 km	400 km	600 km	800 km	1000 km
PPV	0.4194	0.4516	0.4839	0.5484	0.6129
TPR	0.0531	0.0670	0.0754	0.1061	0.1117
Gain	2.3833	1.9869	1.6286	1.7731	1.5110

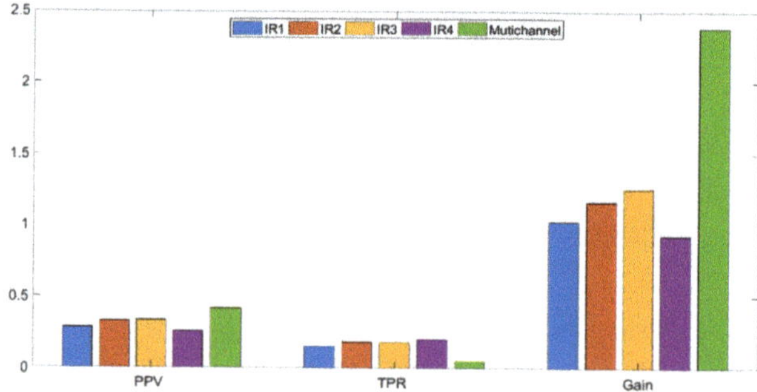

Figure 18. Comparison of single-channel method and multi-channel method.

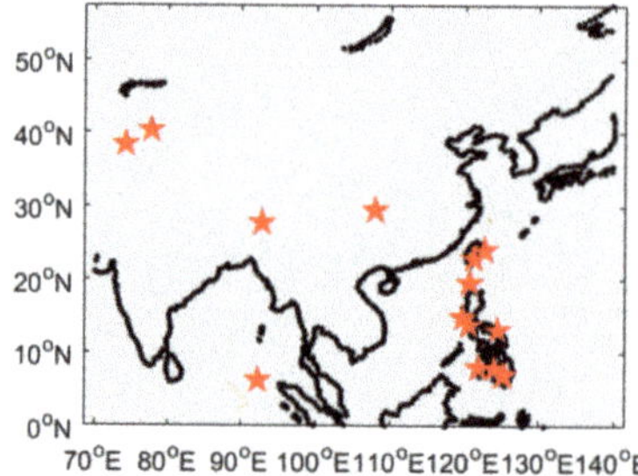

Figure 19. Epicentres of earthquakes following four-channel anomaly.

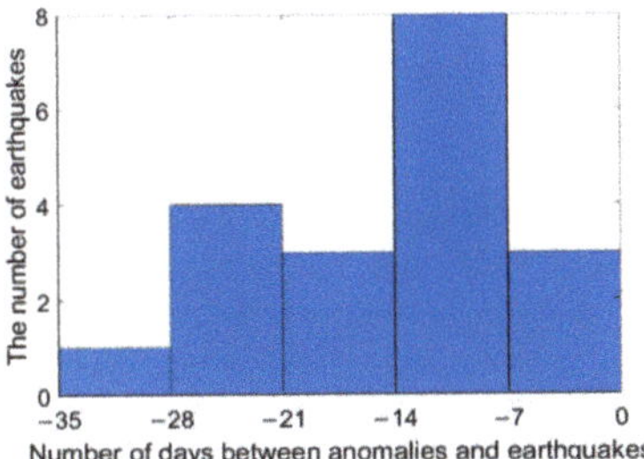

Figure 20. The number of days between anomalies and earthquakes.

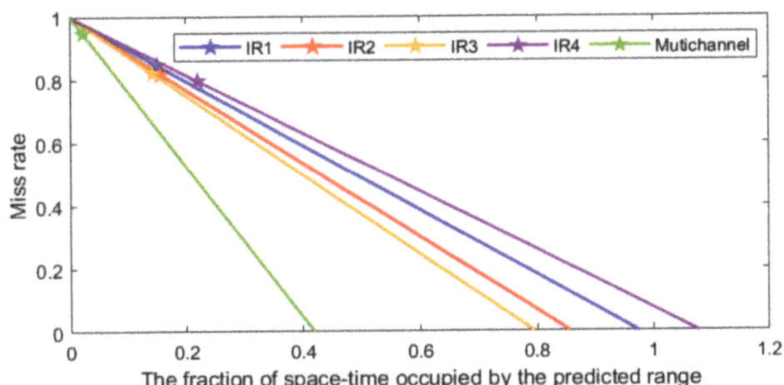

Figure 21. Molchan diagram of different methods.

To compare the multi-channel method with other methods, the authors, data, study region, study period, earthquake magnitude and their results in previous studies are shown in Table 7 [23,28,29,31,46]. The anomaly time means the number of days from anomaly to earthquake. The multi-channel method could obtain the highest probability gain, although the TPR is the lowest. This means that this method reduces the uncertainty of earthquake prediction more obviously than other methods. The PPV of the multi-channel method is also the highest in the studies on earthquakes of magnitude five and above. It means that the multi-channel method has higher accuracy in earthquake prediction than previous methods for earthquakes with magnitude five or above.

Table 7. The statistical results in previous studies.

Authors	Data	Region	Period	Magnitude	PPV	TPR	Gain	Anomaly Time
Ching-Chou Fu et al. [46]	OLR [1]	Taiwan, China	2009–2019	≥6.0	None [5]	77%	None	<25 days
Ying Zhang et al. [28]	ST [2]	Sichuan, China	2002–2018	≥5.0	25.9%	10.8%	<1.5	<30 days
Ying Zhang et al. [31]	OLR	China	2007–2017	≥4.0	76.1%	58.1%	1.34	<50days
Carolina Filizzola et al. [23]	TIR [3]	Turkey	2004–2015	≥4.0 ≥5.0	32.9% None	≈12% [6] ≈20%	1.3 2.2	<30 days
Zhong-Hu Jiao et al. [29]	ST	Global	2010–2018	≥5.0	7.61%	98.4%	None	<120 days
Yingbo Yue et al.	BT [4]	China	2016–2020	≥5.0	41.9%	5.3%	2.38	<30 days

[1] OLR means outgoing long-wave radiation; [2] ST means land surface temperature; [3] TIR means thermal infrared radiation; [4] BT means brightness temperature; [5] None means that the parameter was not mentioned in the study; [6] ≈ means the value was estimated according to the point in the images.

4. Conclusions

Based on the relative power spectrum method, we propose a statistical method based on connected domain recognition to calculate the PPVs, TPRs and probability gains in different channels. The results show that the PPV and TPR could be improved by increasing the distance threshold. The probability gain is low and its change with distance threshold is not obvious. In addition, we also statistically analyzed the multi-channel infrared anomalies before 358 earthquakes. There is at least one channel anomaly within one kilometer of the epicenter within one month before 36.87% of the earthquakes, but there are only 26 earthquakes with four-channel anomalies within 400 km of the epicenter and 18 earthquakes with four-channel anomalies within 200 km of the epicenter.

In the study of three earthquake cases, four-channel anomalies appeared and disappeared before the earthquake. The epicenter is at or some distance from the edge of the anomaly. Due to the low PPV and probability gain of the earthquake prediction method using single-channel data, multi-channel infrared remote sensing images are used for earthquake prediction. The PPV of four-channel anomalies is 41.94%. This is higher than that of single-channel anomalies at the same distance threshold of 200 km. Meanwhile, the method causes a lower TPR. Significantly, the spatial–temporal occupancy of four-channel anomalies is very low, and the probability gain is doubled.

This study shows the difference between pre-earthquake anomalies in multi-channel infrared remote sensing images and indicates that multi-channel infrared remote sensing images may have more advantages in the PPV and the probability gain of earthquake prediction than single-channel data. In earthquake prediction, the PPV, which indicates the reliability of the algorithm, is more important than the TPR. This is because the accurate prediction of a single earthquake can also save a lot of life and property. However, it is still difficult to use the four-channel infrared data to obtain high enough accuracy for the practical application of earthquake prediction. The purpose of this paper is to show the advantages of multi-channel data over single-channel data. The results could be compared with other types of pre-seismic anomalies to study the mechanism of anomalies during the earthquake preparation period and explore a better method for earthquake prediction.

Earthquake prediction needs a lot of remote sensing data and ground-based observation data. In future studies, we can improve the performance of earthquake prediction with the following aspects:

a. FY-2 contains eight satellites. By assimilating data from multiple satellites, it is possible to capture longer observations and count more earthquakes;

b. The second generation of Fengyun geostationary meteorological satellites (FY-4) carried the advanced geostationary radiation imager and the geostationary interferometric infrared sounder. The former has 14 infrared channels, while the latter can detect the temperature and humidity of the vertical atmosphere. The data may provide more information for earthquake prediction;

c. The new prediction method that combines remote sensing data and surface observation data should be explored.

The anomaly extraction algorithm in this paper can be used for time series analysis of other remote sensing data, and the statistical method can be used for other types of wide-field and long-time spatial–temporal data. Some parameters may need to be adjusted for using other data.

Author Contributions: Conceptualization, G.C.; methodology, F.C.; investigation, Y.Y.; writing—original draft preparation, Y.Y.; writing—review and editing, F.C.; supervision, G.C.; funding acquisition, F.C. All authors have read and agreed to the published version of the manuscript.

Funding: This research received no external funding.

Institutional Review Board Statement: Not applicable.

Informed Consent Statement: Not applicable.

Data Availability Statement: Not applicable.

Acknowledgments: Thanks to the National Satellite Meteorological Center of China for providing the data of the FY-2G satellite.

Conflicts of Interest: The authors declare no conflict of interest.

References

1. Gornyi, V.I.; Sal'Man, A.G.; Tronin, A.A.; Shilin, B.V. The terrestrial outgoing infrared radiation as an indicator of seismic activity. *Dokl. Akad. Nauk. Sssr* **1988**, *301*, 67–69.
2. Qiang, Z.J.; Kong, L.C.; Zheng, L.Z.; Guo, M.H.; Wang, G.P.; Zhao, Y. An experimental study on temperature increasing mechanism of satellitic thermo-infrared. *Acta Seismol. Sin.* **1997**, *10*, 247–252. [CrossRef]
3. Guo, Z.Q.; Qiang, S.Q.; Wang, C.; Liu, Z.; Gao, X.; Zhang, W.G.; Yu, Y.; Zhang, H.; Qiu, J.H. The mechanism of earthquake's thermal infrared radiation precursory on remote sensing images. In Proceedings of the IEEE International Geoscience and Remote Sensing Symposium (IGARSS 2002)/24th Canadian Symposium on Remote Sensing, Toronto, ON, Canada, 24–28 June 2002; pp. 2036–2038.
4. Sun, H.; Ma, L.; Liu, W.; Spearing, A.; Han, J.; Fu, Y. The response mechanism of acoustic and thermal effect when stress causes rock damage. *Appl. Acoust.* **2021**, *180*, 108093. [CrossRef]
5. Ouzounov, D.; Freund, F. Mid-infrared emission prior to strong earthquakes analyzed by remote sensing data. *Adv. Space Res.* **2004**, *33*, 268–273. [CrossRef]
6. Qiang, Z.J.; Li, L.Z.; Dian, C.G.; Xu, M.; Ge, F.S. Use remote sensing technique to predict earthquakes. In Proceedings of the Optical Remote Sensing for Industry and Environmental Monitoring, Beijing, China, 15–17 September 1998; Volume 3504, pp. 342–345.
7. Tronin, A.A. Remote sensing and earthquakes: A review. *Phys. Chem. Earth Parts A/B/C* **2006**, *31*, 138–142. [CrossRef]
8. Wu, L.X.; Liu, S.J.; Wu, Y.H. The experiment evidences for tectonic earthquake forecasting based on anomaly analysis on satellite infrared image. In Proceedings of the 2006 IEEE International Geoscience and Remote Sensing Symposium, Denver, CO, USA, 31 July–4 August 2006; p. 2158. [CrossRef]
9. Ouzounov, D.; Liu, D.; Chunli, K.; Cervone, G.; Kafatos, M.; Taylor, P. Outgoing long wave radiation variability from IR satellite data prior to major earthquakes. *Tectonophysics* **2007**, *431*, 211–220. [CrossRef]
10. Song, D.; Xie, R.; Zang, L.; Yin, J.; Qin, K.; Shan, X.; Cui, J.; Wang, B. A new algorithm for the characterization of thermal infrared anomalies in tectonic activities. *Remote Sens.* **2018**, *10*, 1941. [CrossRef]
11. Su, B.; Li, H.; Ma, W.; Zhao, J.; Yao, Q.; Cui, J.; Yue, C.; Kang, C. The outgoing longwave radiation analysis of medium and strong earthquakes. *IEEE J. Sel. Top. Appl. Earth Obs. Remote Sens.* **2021**, *14*, 6962–6973. [CrossRef]
12. Bellaoui, M.; Hassini, A.; Bouchouicha, K. Pre-seismic anomalies in remotely sensed land surface temperature measurements: The case study of 2003 Boumerdes earthquake. *Adv. Space Res.* **2017**, *59*, 2645–2657. [CrossRef]
13. Khalili, M.; Alavi, P.; Seyed, K.; Abdollahi, E.; Seyed, S. Using Robust Satellite Technique (RST) to determine thermal anomalies before a strong earthquake: A case study of the Saravan earthquake (16 April 2013, MW = 7.8, Iran). *J. Asian Earth Sci.* **2019**, *173*, 70–78. [CrossRef]
14. Jing, F.; Shen, X.H.; Kang, C.L.; Xiong, P. Variations of multi-parameter observations in atmosphere related to earthquake. *Nat. Hazards Earth Syst. Sci.* **2013**, *13*, 27–33. [CrossRef]
15. Guo, X.; Zhang, Y.S.; Wei, C.X.; Zhong, M.J. Anomalies of middle infrared brightness before 2008 Yutian Ms7.3 and 2010 Yushu Ms7.1 earthquakes. *Acta Seismol. Sin.* **2014**, *36*, 175–183. [CrossRef]
16. Dey, S.; Sarkar, S.; Singh, R. Anomalous changes in column water vapor after Gujarat earthquake. *Adv. Space Res.* **2004**, *33*, 274–278. [CrossRef]
17. Liu, S.; Cui, L.; Wu, L.; Wang, Z. Analysis on the water vapor anomaly before Wenchuan earthquake based on MODIS data. In Proceedings of the 2009 IEEE International Geoscience and Remote Sensing Symposium, Cape Town, South Africa, 12–17 July 2009; pp. 412–415. [CrossRef]
18. Zhao, X.; Pan, S.; Sun, Z.; Guo, H.; Zhang, L.; Feng, K. Advances of satellite remote sensing technology in earthquake prediction. *Nat. Hazards Rev.* **2021**, *22*, 03120001. [CrossRef]
19. Sekertekin, A.; Inyurt, S.; Yaprak, S. Pre-seismic ionospheric anomalies and spatio-temporal analyses of MODIS Land surface temperature and aerosols associated with 24 September 2013 Pakistan Earthquake. *J. Atmos. Sol. Terr. Phys.* **2020**, *200*, 105218. [CrossRef]
20. Wei, L.; Guo, J.; Liu, J.; Lu, Z.; Li, H.; Cai, H. Satellite thermal infrared earthquake precursor to the Wenchuan Ms 8.0 earthquake in Sichuan, China, and its analysis on geo-dynamics. *Acta Geol. Sin. Engl. Ed.* **2009**, *83*, 767–775. [CrossRef]
21. Ouzounov, D.; Pulinets, S.; Sun, K.; Shen, X.; Kafatos, M. Atmosphere response to pre-earthquake processes revealed by satellite and ground observations. Case study for few strong earthquakes in Xinjiang, China (2008–2014). *Ann. Geophys.* **2020**, *63*, 552. [CrossRef]
22. Zhong, M.; Shan, X.; Zhang, X.; Qu, C.; Guo, X.; Jiao, Z. Thermal infrared and ionospheric anomalies of the 2017 MW6.5 jiuzhaigou earthquake. *Remote Sens.* **2020**, *12*, 2843. [CrossRef]
23. Filizzola, C.; Corrado, A.; Genzano, N.; Lisi, M.; Pergola, N.; Colonna, R.; Tramutoli, V. RST Analysis of anomalous TIR sequences in Relation with earthquakes occurred in Turkey in the period 2004–2015. *Remote Sens.* **2022**, *14*, 381. [CrossRef]

24. Saradjian, M.R.; Akhoondzadeh, M. Thermal anomalies detection before strong earthquakes (M > 6.0) using interquartile, wavelet and Kalman filter methods. *Nat. Hazards Earth Syst. Sci.* **2011**, *11*, 1099–1108. [CrossRef]

25. Wei, C.X.; Zhang, Y.S.; Guo, X.; Hui, S.X.; Qin, M.Z.; Zhang, Y. Thermal Infrared Anomalies of Several Strong Earthquake Remote sensing observations of pre-earthquake thermal anomalies. *Sci. World J.* **2013**, *2013*, 208407. [CrossRef]

26. Akhoondzadeh, M. A comparison of classical and intelligent methods to detect potential thermal anomalies before the 11 August 2012 Varzeghan, Iran, earthquake (Mw = 6.4). *Nat. Hazards Earth Syst. Sci.* **2013**, *13*, 1077–1083. [CrossRef]

27. Lu, X.; Meng, Q.Y.; Gu, X.F.; Zhang, X.D.; Xie, T.; Geng, F. Thermal infrared anomalies associated with multi-year earthquakes in the Tibet region based on China's FY-2E satellite data. *Adv. Space Res.* **2016**, *58*, 989–1001. [CrossRef]

28. Zhang, Y.; Meng, Q. A statistical analysis of TIR anomalies extracted by RSTs in relation to an earthquake in the Sichuan area using MODIS LST data. *Nat. Hazards Earth Syst. Sci.* **2019**, *19*, 535–549. [CrossRef]

29. Jiao, Z.H.; Shan, X.J. Statistical framework for the evaluation of earthquake forecasting: A case study based on satellite surface temperature anomalies. *J. Asian Earth Sci.* **2021**, *211*, 104710. [CrossRef]

30. Pavlidou, E.; van der Meijde, M.; van der Werff, H.; Hecker, C. Time series analysis of land surface temperatures in 20 earthquake cases worldwide. *Remote Sens.* **2018**, *11*, 61. [CrossRef]

31. Zhang, Y.; Meng, Q.Y.; Ouillon, G.; Sornette, D.; Ma, W.Y.; Zhang, L.L.; Zhao, J.; Qi, Y.; Geng, F. Spatially variable model for extracting TIR anomalies before earthquakes: Application to Chinese Mainland. *Remote Sens. Environ.* **2021**, *267*, 112720. [CrossRef]

32. Richards, A.; Johnson, G. Atmospheric effects on infrared imaging systems. In Proceedings of the SPIE European Symposium on Optics and Photonics in Security and Defence, Bruges, Belgium, 26–29 September 2005; pp. 1–14. [CrossRef]

33. Kirkham, M.B. Chapter 25—Solar radiation, black bodies, heat budget, and radiation balance. In *Principles of Soil and Plant Water Relations*, 2nd ed.; Academic Press: Cambridge, MA, USA, 2014; pp. 453–472.

34. Dobrovolsky, I.P.; Zubkov, S.I.; Miachkin, V.I. Estimation of the size of earthquake preparation zones. *Pure Appl. Geophys.* **1979**, *117*, 1025–1044. [CrossRef]

35. Wei, C.; Lu, X.; Zhang, Y.; Guo, Y.; Wang, Y. A time-frequency analysis of the thermal radiation background anomalies caused by large earthquakes: A case study of the Wenchuan 8.0 earthquake. *Adv. Space Res.* **2019**, *65*, 435–445. [CrossRef]

36. Xie, T.; Ma, W. Possible thermal brightness temperature anomalies associated with the Lushan (China) M S7.0 earthquake on 20 April 2013. *Earthq. Sci.* **2015**, *28*, 37–47. [CrossRef]

37. Zhang, X.; Zhang, Y.; Tian, X.; Zhang, Q.; Tian, J. Tracking of thermal infrared anomaly before one strong earthquake-in the case of Ms6.2 earthquake in Zadoi, Qinghai on 17 October 2016. *J. Phys. Conf. Ser.* **2017**, *910*, 012048. [CrossRef]

38. Bhardwaj, A.; Singh, S.; Sam, L.; Joshi, P.; Bhardwaj, A.; Martín-Torres, F.J.; Kumar, R. A review on remotely sensed land surface temperature anomaly as an earthquake precursor. *Int. J. Appl. Earth Obs. Geoinf.* **2017**, *63*, 158–166. [CrossRef]

39. Zechar, J.D.; Jordan, T.H. Testing alarm-based earthquake predictions. *Geophys. J. Int.* **2008**, *172*, 715–724. [CrossRef]

40. Eleftheriou, A.; Filizzola, C.; Genzano, N.; Lacava, T.; Lisi, M.; Paciello, R.; Pergola, N.; Vallianatos, F.; Tramutoli, V. Long-term RST analysis of anomalous TIR sequences in relation with earthquakes occurred in Greece in the period 2004–2013. *Pure Appl. Geophys.* **2015**, *173*, 285–303. [CrossRef]

41. Lu, G.J.; Zhang, H.; Li, H.; Wang, J.L. Infrared brightness temperature changes before moderate to strong earthquakes in north China. *Earthquake* **2018**, *38*, 96–106.

42. Lü, Q.-Q.; Ding, J.-H.; Cui, C.-Y. Probable satellite thermal infrared anomaly before the Zhangbei Ms = 6.2 earthquake on 10 January 1998. *Acta Seism. Sin.* **2000**, *13*, 203–209. [CrossRef]

43. Wang, Y.; Zhang, Y.; Wei, C. Comparative study on thermal infrared anomaly characteristics of several moderate and strong earthquakes in Yunnan province. *Imaging Sci. Photochem.* **2019**, *37*, 215–226.

44. Zhang, L.; Guo, X.; Zhang, X.; Tu, H. Anomaly of thermal infrared brightness temperature and basin effect before Jiuzhaigou MS7.0 earthquake in 2017. *Acta Seismol. Sin.* **2018**, *40*, 797–808.

45. Zhang, X.; Zhang, Y.S.; Guo, X.; Wei, C.X.; Zhang, L.F. Analysis of thermal infrared anomaly in the Nepal MS8.1 earthquake. *Earth Sci. Front.* **2017**, *24*, 227–233. [CrossRef]

46. Fu, C.-C.; Lee, L.-C.; Ouzounov, D.; Jan, J.-C. Earth's outgoing longwave radiation variability prior to M $\geq$ 6.0 earthquakes in the Taiwan area during 2009–2019. *Front. Earth Sci.* **2020**, *8*, 364. [CrossRef]

Article

Modeling Locations with Enhanced Earth's Crust Deformation during Earthquake Preparation near the Kamchatka Peninsula

Maksim Gapeev *,† and **Yuri Marapulets** †

Institute of Cosmophysical Research and Radio Wave Propagation FEB RAS, Kamchatka Region, Elizovskiy District, Mirnaya Str. 7, 684034 Paratunka, Russia
* Correspondence: owlptr@yandex.ru
† These authors contributed equally to this work.

Abstract: In seismically active regions of the Earth, to which the Kamchatka peninsula refers, pre-seismic anomalies are recorded in different geophysical fields. One of such fields is the acoustic emission of rocks, the anomalies of which are recorded 1–3 days before earthquakes at the distance of the first hundreds of kilometers from their epicenters. Results of joint acoustic-deformation measurements showed that growth of geoacoustic radiation intensity occurs during the increase in the level of deformations in rock masses by more than one order compared to the background values. Simulation studies of the areas with increased deformation are realized to understand the causes of anomalous acoustic-deformation disturbance occurrences before strong earthquakes. The model is based on the assumption that the Earth's crust in the first approximation can be considered as a homogeneous isotropic elastic half-space, and an earthquake source can be considered as a displacements along a rectangular fault plane. Based on these assumptions, deformation regions of Earth's crust were modeled during the preparations of two earthquakes with local magnitudes $M_L \approx 5$ occurred on the Kamchatka Peninsula in 2007 and 2009. The simulation results were compared for the first time with the data of a laser strainmeter-interferometer installed at the Karymshina observation site (52.83° N, 158.13° E). It was shown that, during the preparation of the both earthquakes, the Karymshina observation site was within the region of shear deformations $\approx 10^{-7}$, which exceeded the tidal ones by an order. On the whole, simulation results corresponded to the results of the natural observations. Construction of an adequate model for the generation of acoustic-deformation disturbances before strong earthquakes is topical for the development of an early notification system on the threat of catastrophic natural events.

Keywords: earthquake preparation; areas of increased shear deformations; mathematical simulation; rock deformation; acoustic emission of near-surface rocks

Citation: Gapeev, M.; Marapulets, Y. Modeling Locations with Enhanced Earth's Crust Deformation during Earthquake Preparation near the Kamchatka Peninsula. *Appl. Sci.* **2023**, *13*, 290. https://doi.org/10.3390/app13010290

Academic Editor: Marco Vona

Received: 8 December 2022
Revised: 22 December 2022
Accepted: 22 December 2022
Published: 26 December 2022

1. Introduction

It is generally accepted that mechanical processes play a leading role in the preparation of seismic events [1–3]. They cause increased stresses, leading to deformations of the Earth's crust around an earthquake source. These changes in the stress–strain state of rocks lead to the anomaly occurrences, classified as earthquake precursors, in various geophysical fields [4–9].

The increase in rock acoustic emission in kilohertz frequency range is one of the identified pre-seismic anomalous disturbances in geophysical fields. Such anomalies were observed in various seismically active regions of the world: in Armenia [10], in Italy [11], and in Russia on the Kamchatka peninsula, which is a part of the Circum-Pacific Orogenic Belt, also known as the "Ring of Fire" [12–14]. As a result of long-term studies of acoustic emission in Kamchatka, a high-frequency acoustic emission effect was revealed [15]. This effect consists in an increase of geoacoustic radiation intensity with an increase of rock deformation rate. This effect is determined by rock deformations at observation sites and manifests the most clearly in the kilohertz frequency range 1–3 days before earthquakes at a distance of the first hundreds of kilometers from their epicenters [16].

The peculiarity of the geoacoustic observations in Kamchatka is the application of broadband piezoceramic hydrophones installed in natural and artificial reservoirs to record the signals. The use of receivers of this type allowed us to expand the frequency range of registration to 0.1 Hz–11 kHz in comparison with standard geophones [16]. To confirm the deformation nature of acoustic emission anomalies, a laser strainmeter-interferometer of unequal shoulder type with a measuring base length of 18 m and sensitivity of 10–11 m was installed in 2005 at the Karymshina site (52.83° N, 158.13° E), located 41 km southwest of Petropavlovsk–Kamchatskiy [17]. Taking into account the peculiarities of its installation on the surface without optical waveguide, the calculated measurement accuracy of relative deformations was no less than 10^{-8} [18]. The data of joint acoustic-deformation observations at this site was studied in detail [13]. It is shown that with the deformation intensification at the observation site, the near-surface sedimentary rock deformation rate also increases simultaneously. These rocks have a polydisperse fluid-saturated porous structure of low strength. Such activation of deformations is accompanied by relative micro-displacements of rock fragments, their interaction and, as a consequence, generation of acoustic emission of increased intensity. Such effects are observed the most clearly at the final stage of earthquake preparation a few days before their onset. Two cases of high-frequency geoacoustic responses to the activation of deformation process before two earthquakes with local magnitudes $M_L \approx 5$ were detected and studied in detail [13]. These earthquakes occurred in 2007 and 2009 at the epicentral distances of about 150 km to the Karymshina site. It was reasonable to make model studies of the deformation fields that occurred during the seismic event preparation and to correlate them with the real deformation levels recorded at the observation site. Comparison of the results of the simulation and the natural experiment is topical to understand the causes of anomalous acoustic-deformation disturbances at the final stage of earthquake preparation.

2. Research Significance

Investigation of pre-seismic anomalies in geophysical fields is of high practical significance for the development of the methods of early notification on seismic hazard. Geophysical field intensity variations in seismically active regions are known to be associated with stress–strain state near earthquake sources during their preparation. In this case, pre-seismic anomalies are often recorded at the distances of hundreds of kilometers from preparing earthquake epicenters. That corresponds to the cases of acoustic emission of rock anomalies recorded in kilohertz frequency range. However, signals at such frequencies cannot propagate from preparing earthquake epicenters due to strong attenuation. Thus, they occur as the result of medium response at the recording site on the change of its stress-strain state. It appears that the changes of stress–strain state near earthquake sources propagate at the distances up to hundreds of kilometers. Undoubtedly, that requires theoretical, model and experimental confirmation. Such investigations of the fields of the Earth's crust stress and deformation are topical in order to obtain new knowledge on the physical processes occurring during earthquake source preparation.

The basic concepts of plate tectonics, theory of elasticity and rock plastic deformations were described by D. Turcotte and G. Shubert [19]. V. Nikolaevsky [20] formulated the basic concepts of deformation and destruction of fractured rocks under static and dynamic load. C. Sholz [1] described in detail the modern understanding of earthquake mechanics. The theory of earthquake-induced seismic wave propagation and modern method of observation data interpretation and processing were considered in detail by K. Aki and P. Richards [21]. Analytical solutions of the mathematical model, describing the Earth's crust deformations in medium elastic approximation under earthquake source effect, were considered in detail by Yu. Okada [22,23]. I. Dobrovolskiy [24] introduced the notion of precursor manifestation zone and showed that its size is determined by the energy of a preparing earthquake. I. Dobrovolskiy [25] also suggested limiting the dimensions of this zone by the boundaries, behind which anomalies in deformations do not exceed the deformation process background values of the order of 10^{-8}. A. Alekseev and his co-authors [26] connected the geophysical field anomalies, occurring in seismically active regions, with the appearance of zones of geo-environment nonlinear loosening (dilatancy). Dilatancy zones, which are formed in the vicinity of earthquake sources during the stresses close to destructive values for rocks, were modeled in the approximation of elastic half-space [26]. Three-dimensional block visco-elastic model of the south-eastern Tibetan Plateau was constructed [27]. Earth's crust model taking into account friable-plastic transitions in the Earth's crust was constructed. This model is based on Maxwell and Kelvin–Voigt models [28]. Visco-elastic fault-based model of crustal deformation was proposed for the 2023 Update to the "U.S. National Seismic Hazard Model" [29]. Earth's crust deformations model under earthquake source impact in the form of a rectangular plane was developed. In this model the Earth is a homogeneous elastic sphere [30]. Some complication of this model was made. It was generalized to the case of a layered elastic sphere [31]. The pre-seismic deformations before the Japanese earthquake occurred on 11 March 2011 in the Tohoku region were estimated [32]. An analysis of seismic anomalies associated with Ludian earthquake on 3 August 2014 was made [33]. Post-seismic deformations after Kokoxili earthquake occured on 14 November 2001 in the Northern Tibetan Plateau were estimated using satellite data [34]. A wide review of articles on the estimates of the Earth's crust deformations based on satellite data was made in 2018 [35].

Simulation of stress fields formed around Kamchatka earthquake sources was carried out earlier. The authors of the studies used different approaches to describe an earthquake source as a point source in the form of a single force [36] or a double force [37,38], as a distributed source in the form of a rectangular plane [39]. It was shown in all the cases that regions with deformations, exceeding the background values, occur around earthquake sources at the distances of hundreds of kilometers. Comparative modeling of deformation fields using these source models was carried out. It was shown that it is better to use the models of a distributed source in the form of a rectangular plane, as they describe more accurately the force action in the earthquake source [40].

The proposed paper continues those investigations. Based on the assumption that the Earth's crust can be considered as a uniform isotropic elastic half-space in first approximation and an earthquake source is a shift along a fracture rectangular plane, the authors modeled Earth's crust deformations, occurring around sources during the preparation of two earthquakes in Kamchatka. For the first time ever, the deformation values, obtained during the modeling, were compared with the data from a laser strainmeter-interferometer installed at the distance of several hundreds of kilometers from the earthquake sources.

3. Research Methods

The source of a tectonic earthquake is formed as a result of release of the stresses accumulated by elastic medium during tectonic deformation [21]. As a result of this release, a break of medium continuity appears. The accumulated elastic energy of deformation turns inelastic. According to this theory, an earthquake source can be described through a displacement along a fault plane [21]. It is notable that a displacement along a fault excites the same seismic waves as some system of forces distributed on the fault with zero total moment. In general, distribution of forces may have different form. However, in the case of an isotropic medium, it can always be chosen as a surface distribution of double pairs of forces.

In accordance with that, limiting ourselves to the consideration of a fault flat plane, the earthquake source model can be represented schematically as follows (Figure 1).

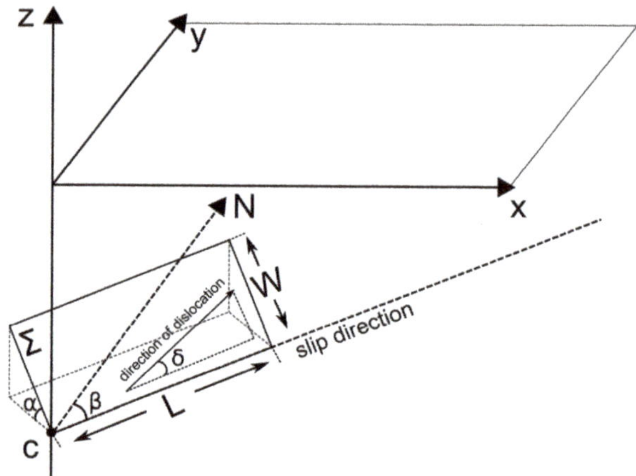

Figure 1. Schematic description of the earthquake source model. In the figure: α is the dip angle, β is the strike angle, δ is the angle of displacement direction, C is the hypocenter depth, L is the plane length, W is its width, N is the North direction (the axis is aligned with OY), and Σ is the fault plane with an equivalent distributed system of double forces with a moment.

Some parameters of this model (α, β, δ, C) can be accessed directly, for example, from the Harvard Catalog of Earthquake Mechanics CMT Catalog [41].

The linear dimensions of the fault plane, L (km) and W (km), as well as the displacement U (cm) magnitude can be estimated using the following correlation equations [42]:

$$
\begin{aligned}
lg(L) &= 0.75 \cdot M_W - 3.60, \\
lg(W) &= 0.75 \cdot M_W - 1.45, \\
lg(U) &= 0.75 \cdot M_W - 0.37,
\end{aligned}
\tag{1}
$$

where $M_W = 2/3(M_0 - 16.1)$ is the moment magnitude, M_0 is the scalar seismic moment.

The research objective is the simulation of stress and strain fields caused by energy accumulation during earthquake preparation. It is obvious that this energy is significantly greater than the released energy of elastic deformations at the times of earthquakes.

In a generalized form, the correlation Equations (1) are presented as follows:

$$
lg(N_E) = a \cdot M_W + b,
\tag{2}
$$

where a and b are some coefficients, N_E is the characteristic of an earthquake source, calculated taking into account the released energy of elastic deformations.

The efficiency coefficient of elastic deformation energy release is:

$$
\eta = \frac{E}{W},
\tag{3}
$$

where W is the total energy of elastic deformations in the area including the earthquake source before the fault activation.

In Equation (2), the moment magnitude is expressed in terms of the earthquake energy E, using the Gutenberg–Richter equation: $E = 10^{1.5 M_W + 5}$. Eliminating the logarithm, the following relation is obtained:

$$
N_E = \left(\frac{E}{10^5}\right)^{(2/3)a} \cdot 10^b.
\tag{4}
$$

In Equation (4), the earthquake energy E is replaced by the total energy of elastic deformations W, using Equation (3). The following relation is obtained:

$$N_W = \left(\frac{1}{\eta}\right)^{(2/3)a} \cdot N_E,$$ (5)

where N_W is the earthquake source characteristic calculated taking into account the total energy of elastic deformations. The coefficient $(1/\eta)^{(2/3)a}$ carries the meaning of an increasing coefficient for correlation Equation (1). This coefficient makes it possible to calculate the stress–strain state of rocks taking into account the total energy of elastic deformations during earthquake preparation.

There are various approaches to evaluate both effective released stress [43,44] and to evaluate the efficiency of elastic deformation energy release. For example, the most accurate approach to estimate the coefficient of efficiency of elastic deformation energy release η, which requires reconstruction of tectonic stress in a seismically active region, is described by Yu. Rebetsky [45]. I. Dobrovolsky [24] proposed a less accurate but a simpler variant to calculate the coefficient:

$$\eta = 10^{0.26 M_W} - 3.93.$$ (6)

Equation (6) was used in further calculations.

4. Simulation of Stress and Strain Fields

The following model for the formation of regions with increased deformation of the Earth's crust during earthquakes preparation is proposed. The Earth's crust is a homogeneous isotropic elastic half-space. The model of an earthquake source is a dislocation in the form of a rectangular plane with a constant displacement vector (Figure 1). The stress–strain state of the Earth's crust is determined by the accumulated elastic energy in the process of earthquake preparation. Zones of acoustic-deformation anomalies are the areas of daytime surface defined by the equation $z = 0$ with the level of relative deformations exceeding the tidal ones ($>10^{-8}$). Shear sources of acoustic emission prevail, since rock strength with respect to tangential stresses is less than to compression. Therefore, only shear deformations are taken into account in the simulation.

Using Mindlin's solutions [46,47], Yu. Okada [22,23] obtained compact analytical solutions for the displacement vector and its spatial derivatives in the case of three types of displacement: in the direction of strike, in the direction of dip and expansion.

The Navier equations underlying the model are linear. Therefore, the solution in case of an arbitrarily oriented displacement (not for expansion) can be obtained in the form of a linear combination of solutions for the displacement in the strike and dip directions:

$$U_{strike} = U \cdot cos(\delta),$$ (7)
$$U_{dip} = U \cdot sin(\delta),$$ (8)

where U_{strike} is the component of the displacement vector along the strike, U_{dip} is the component of the displacement vector along the dip, δ is the angle of displacement direction.

The Mercator projection was used to convert the geographical coordinates to Cartesian ones. An additional coordinate system was built for each earthquake to simplify the calculations. A system had a center at the earthquake epicenter and was oriented relative to the OZ axis of the original system by the angle of $\beta - 90°$. Thus, the axes OX and OY were parallel to the projections of the sides L and W of the displacement plane on the plane $z = 0$ (Figure 2).

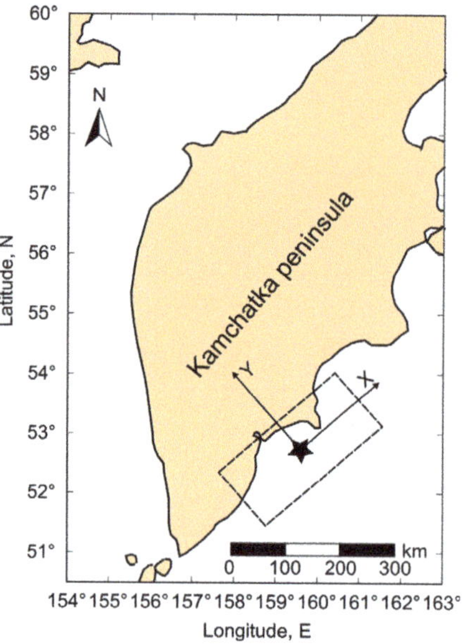

Figure 2. Schematic representation of an additional coordinate system centered at the earthquake epicenter (asterisk). The dotted line is the projection of the fault plane onto the Earth's surface ($z = 0$).

Two earthquakes, before which simultaneous anomalies of acoustic emission and rock deformations were observed at the Karymshina site, were simulated [13]. Earthquake No. 1 occurred on 2 May 2007, at 12:00:48.4 UT, the coordinates are 52.29° N, 160.55° E, the depth is 28 km, local magnitude M_L = 5.2. Earthquake No. 2 occurred on 8 October 2009, at 05:25:13.4 UT, the coordinates are 52.84° N, 160.15° E, the depth is 20 km, local magnitude M_L = 5.1. These earthquakes are not listed in the CMT catalog due to their low energy. The data were taken from the earthquakes catalog for Kamchatka and the Commander Islands [48]. Unfortunately, it is impossible to obtain the information on the orientation of the displacement plane from this catalog. This information is necessary for the computational experiment. To estimate it, the earthquakes, which occurred in the area of the earthquakes under the study and which were presented in the CMT catalog, were analyzed. The analysis was carried out under the assumption that for the Kamchatka subduction zone, the general directions of force impact from the sources of small-focus earthquakes, located in some small area, are quite constant. For this purpose, the entire observation period from 1 January 1976 to 30 June 2022, presented in the CMT catalog, was considered. Seismic events in the area, close to the simulated earthquakes (latitude interval: [52°, 53°], longitude interval: [160°, 161°]), were analyzed.

In total, nine small-focus earthquakes were represented in the CMT catalog in this area. More detailed information about these earthquakes, including information about the focal mechanism, is presented in Table 1. Epicenter locations are shown in Figure 3. Numbers of earthquakes in Figure 3 correspond to the ones in Table 1.

Table 1. Data on nine small-focus earthquakes that occurred during the period from 1 January 1976 to 30 June 2022 in the area under the study (latitude interval: [52°, 53°], longitude interval: [160°, 161°]).

No	Date, Time	Coordinates of the Epicenter	Depth, km	M_W	Strike Angle [1], β	Dip Angle [1], α	Angle of Displace-ment [1], δ	Focal Mechanism
1	1977/12/2, 12:57:22.6	52.32° N, 160.48° E	40.0	5.6	217	35	96	
2	1977/12/21, 16:39:40.9	52.60° N, 160.52° E	55.6	5.6	218	38	93	
3	1979/6/25, 18:45:57.2	52.68° N, 160.06° E	57.3	5.0	210	19	76	
4	1979/9/1, 17:54:59.9	52.86° N, 160.66° E	15.0	5.5	309	25	−150	
5	1980/1/23, 1:51:49.8	52.22° N, 160.69° E	20.3	5.8	213	26	86	
6	1980/1/23, 2:34:17.6	52.25° N, 160.79° E	15.0	5.7	192	21	57	
7	1980/1/23, 6:52:53.7	52.23° N, 160.84° E	19.6	5.5	216	28	90	
8	1980/1/23, 8:12:31.6	52.23° N, 160.65° E	15.0	5.6	219	21	92	
9	1980/1/23, 10:7:17.1	52.26° N, 160.57° E	17.2	5.2	205	22	77	

[1] Degree measures of the angle are given.

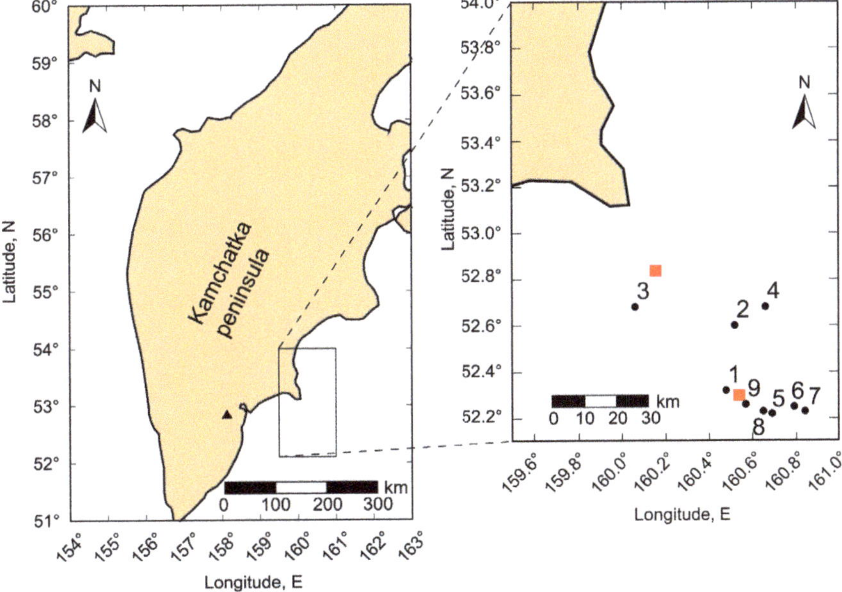

Figure 3. Map of Kamchatka peninsula with location of earthquake epicenters presented in the CMT catalog and occurred during the period from 1 January 1976 to 30 June 2022 (black circles) in the area under the study and location of simulated earthquake epicenters (red squares). Location of earthquake epicenters are shown in scaled part of map. The black triangle on the map indicates the location of the Karymshina observation site.

Only one of them (Earthquake No. 4 in Table 1) significantly differed in the orientation of the fault plane. All other earthquakes were very similar in these parameters. It was removed from the sample and the following statistical estimates of the orientation angles were obtained:

$$\bar{\alpha} = 211.25°, \; S(\alpha) = 8.48°, \tag{9}$$

$$\bar{\beta} = 26.25°, \; S(\beta) = 6.55°, \tag{10}$$

$$\bar{\delta} = 83.38°, \; S(\delta) = 12.08°, \tag{11}$$

where $\bar{\alpha}, \bar{\beta}, \bar{\delta}$ are average values of angles, $S(\alpha), S(\beta), S(\delta)$ are standard deviations.

The moment magnitude values are required for the application of correlation Equation (1). The relationship between the local magnitude M_L for Kamchatka earthquakes and the moment magnitude M_W is [49]:

$$M_L = M_W - 0.4. \tag{12}$$

The following parameters of the elastic half-space were taken: the shear modulus, $\mu = 3.675 \cdot 10^{10}$ N/m^2, the second Lame parameter, $\lambda = 3.675 \cdot 10^{10}$ N/m^2 [40]. The simulation was carried out on a grid with the dimensions of 8° in latitude and 8° in longitude with the step of 0.01°. Earthquake coordinates were the center of the grid.

5. The Results of Computational Experiment

5.1. Earthquake No. 1

Figure 4 shows the example of a simultaneous anomaly of acoustic emission and rock deformations recorded on 1 May 2007, 25 h before the earthquake that occurred on 2 May 2007, at 12:00 UT [13]. It is clear from Figure 4 that during the period from 1 to 9 o'clock, rather sharp compressions of rocks occurred, followed by the releases lasting from 1 to 5 min, which were accompanied by increases in the deformation rate and simultaneous increase in the emission level in kilohertz frequency range. The level of relative deformations during the compression reached the order of 10^{-7}, and the deformation rate increased to 10^{-8} s^{-1}.

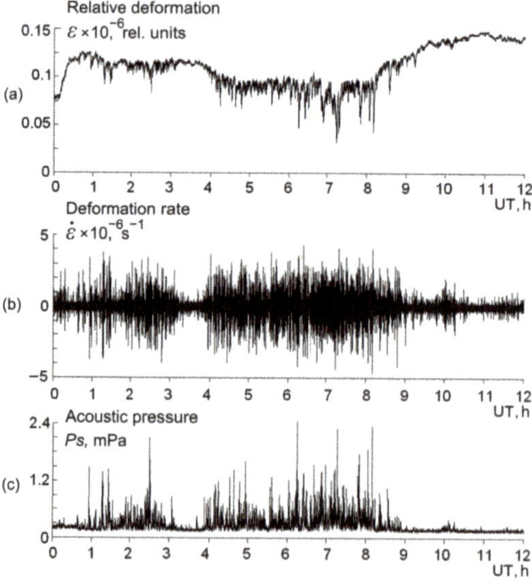

Figure 4. An example of a simultaneous acoustic-deformation anomaly before earthquake No. 1. (**a**) Variations of rock relative deformation ε, (**b**) variations of deformation rate $\dot{\varepsilon}$, (**c**) variations of acoustic pressure P_s, accumulated over 4 s in the frequency range of 2.0–6.5 kHz.

The simulation results for the zones of relative shear deformations that occurred during earthquake No. 1 preparation are presented in Figure 5.

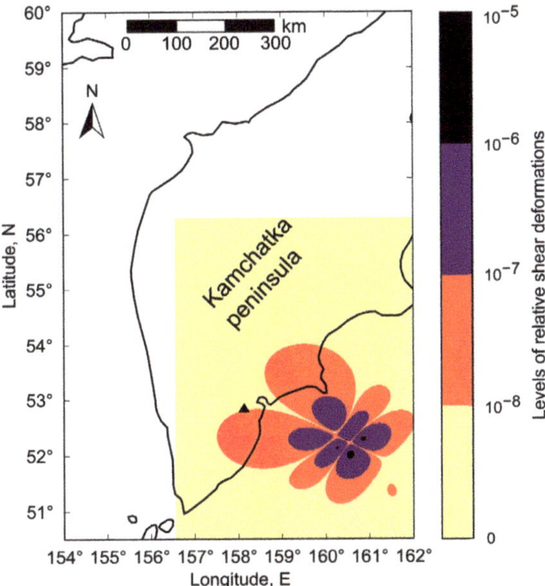

Figure 5. Zones of relative shear deformations on the Earth's surface $z = 0$ simulated for earthquake No. 1. The triangle on the map indicates the location of the Karymshina observation site.

It is clear from Figure 5 that the Karymshina observation site is on the boundary of the region of relative shear deformations of the order from 10^{-8} to 10^{-7}. That generally corresponds to the results of the natural experiment with a laser strainmeter-interferometer.

5.2. Earthquake No. 2

Figure 6 shows an example of a synchronous recording of acoustic emission and rock deformation from 6 October to 8 October 2009 before the earthquake that occurred on 8 October 2009, at 05:25 UT [13].

Figure 6a,b shows that, 35 h before the earthquake, there was a simultaneous anomaly of acoustic emission and rock deformation lasting for about 12 h. Figure 6c,d shows more detailed fragments of the record during the anomaly. For comparison, Figure 6e illustrates the subsequent calm period. The level of relative deformations during the anomaly was 10^{-7} and sometimes reached the order of 10^{-6}.

Figure 7 represents the simulation results for the zones of relative shear deformations occurring during this earthquake preparation.

It is clear from Figure 7 that the Karymshina site is in the region of relative shear deformations of the order from 10^{-8} to 10^{-7}, as in the case of earthquake No. 1, which corresponds to the results of the natural experiment using a laser strainmeter-interferometer.

In both cases presented, the calculated levels of relative shear deformation turned out to be slightly lower than the data of the natural experiments. In the case of earthquake No. 1, the Karymshina site is on the boundary of the deformation region from 10^{-8} to 10^{-7}, while according to deformation measurements, the relative deformation was about 10^{-7}. In the case of earthquake No. 2, the Karymshina site is in the area of deformations from 10^{-8} to 10^{-7}, while according to deformation measurements, the relative deformation was about 10^{-7} and sometimes reached the order of 10^{-6}.

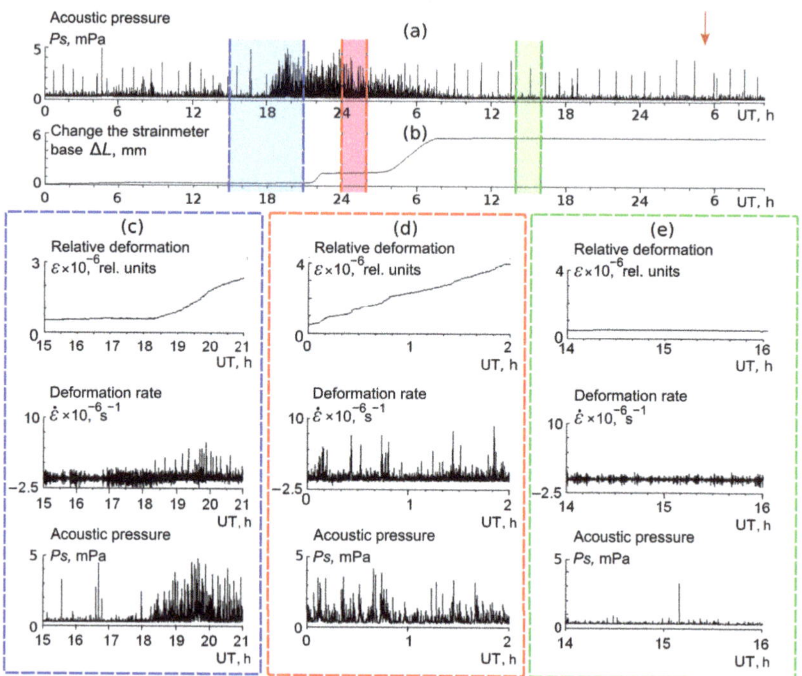

Figure 6. Record fragment of acoustic emission in different ranges and rock deformations from 00:00 on 6 October 2009, to 10:00 on 8 October 2009. (**a**) Variations of acoustic pressure P_s, accumulated over 4 s in the frequency range 2.0–6.5 kHz, (**b**) change in the strainmeter base ΔL. The red arrow shows the earthquake moment. At the bottom (**c–e**), enlarged fragments of the rock relative deformation ε, the deformation rate $\dot{\varepsilon}$, and sound pressure P_s, accumulated over 4 s in the frequency range of 2.0–6.5 kHz, are presented.

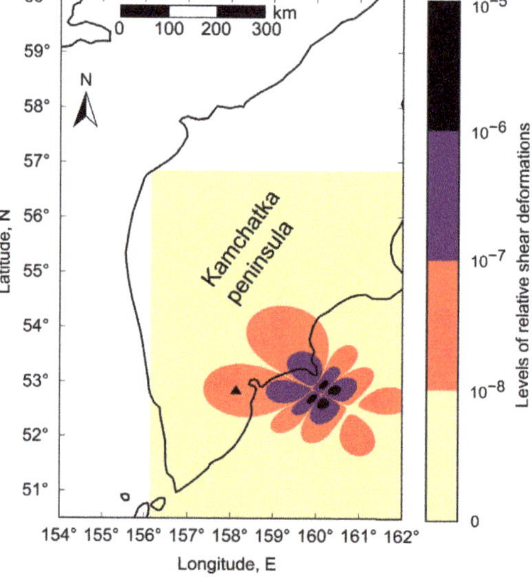

Figure 7. Zones of relative shear deformations on the Earth's surface $z = 0$ simulated for earthquake No. 2. The triangle on the map indicates the location of the Karymshina observation site.

6. Discussion

The suggested pre-seismic deformation model, based on the classical theory of elasticity and simplification of the Earth's crust model to isotropic elastic half-space, has its advantages and disadvantages. Such a model has analytical solutions that simplifies calculations, obviates the need for the estimate of numerical solution stability. For example, a more complicated model of a medium in the form of isotropic uniform elastic sphere requires numerical solution of differential equation system [30]. In that case, the differences in estimates turn to be significant mainly for deep earthquakes at the distance of about 5°, from their epicenters [30]. Both earthquakes considered in this paper are shallow. Their epicenters are at the distance of about 2°, from the Karymshina observation site. That makes it possible use the Earth's model in the form of elastic half-space, not taking into account its surface curvature.

One more disadvantage of the proposed model is the application of deformation statistical equations, which do not take into account the deformation rate variation. In this respect, rock mass deformation rate affects significantly the activation of acoustic emission before earthquakes [15]. The Earth's crust deformation rate makes it possible to take into account different visco-elastic models of a medium, in particular, the Maxwell visco-elastic model [50]. However, it is very difficult to apply such models to estimate the Earth's crust deformations during earthquake preparation. For example, it is impossible to determine the exact time of earthquake preparation and the function of force impact change in a source. Thus, application of deformation static equations is justified.

When modeling, the authors did not take into account plastic deformations and heterogeneous structure of the Earth's crust. In fact, the Karymshina site is located in the zone of different-rank tectonic faults that may result in the recording of the deformations with the levels exceeding the calculated ones. This fact, also known as "problem of far-distance effect of earthquake sources", was considered by the researchers before. For example, it was proposed, when modeling, to take into account the Earth's crust regions with anomalous regime of the stress state (fault zones, layers with high fluid pressure) [51]. It was shown that, when introducing such regions of postcritical deformation with inelastic properties into a model, the decrease in disturbed deformation level at a large distance is 10^4 times less than in case of medium elastic model. Moreover, during the simulation, only shear deformations, which prevail during acoustic radiation generation, were considered, whereas the strainmeter records rock deformation within its base, not taking into account its type.

However, the suggested model makes it possible to estimate the stress–strain state of the Earth's crust during earthquake preparation at different distances from their sources. The computational experiment, using the proposed model, showed that the deformation during earthquake preparation at the Karymshina observation site exceeded the tidal ones by an order. Overall, the simulation results corresponded to the deformations measured by the laser strainmeter-interferometer. That is the ground to state that the observed joint acoustic and deformation anomalies are associated with the process of earthquake preparation in Kamchatka.

7. Conclusions

In seismically active regions of the Earth, to which the Kamchatka peninsula refers, pre-seismic anomalies are recorded in different geophysical fields. One of such fields is the acoustic emission of rocks, the anomalies of which are recorded 1–3 days before earthquakes at the distance of the first hundreds of kilometers from their epicenters. The results of joint acoustic-deformation measurements showed that growth of geoacoustic radiation intensity occurs during the increase in the level of deformations in rock masses by more than one order compared to the background values. According to the assumption that the Earth's crust in the first approximation can be considered as a homogeneous isotropic elastic half-space, and an earthquake source can be considered as a displacements along a rectangular fault plane, the authors proposed to simulate the deformation of the

Earth's crust around the source of impending earthquake. Another assumption of the authors was that increased deformations occur at a distance of hundreds of kilometers from the epicenters, and that causes acoustic emission and rock deformations anomalies. The total energy of elastic deformations accumulated during the earthquake preparation was estimated for the simulation. It determines the stress–strain state of the Earth's crust around the epicenter and is significantly greater than the released energy of seismic waves. Based on these assumptions, deformation regions were modeled for two earthquakes with local magnitudes $M_L \approx 5$ occurred on the Kamchatka Peninsula in 2007 and 2009. Simultaneous anomalies of acoustic emission and deformation of near-surface rocks were recorded before these earthquakes.

The simulation results were compared for the first time with the data of the strainmeter-interferometer and geoacoustic system installed at the Karymshina observation site in Kamchatka. It was shown that during the preparation of both earthquakes, the Karymshina observation site was within the region of shear deformations $\approx 10^{-7}$, which exceeded the tidal ones by an order. On the whole, simulation results corresponded to the results of the natural observations. Comparison of the results of the simulation and of the natural analysis is topical for understanding the causes of anomalous acoustic-deformation disturbance occurrences at the final stage of earthquake preparation. Construction of an adequate model for generation of such disturbances is relevant for development an early notification system on the threat of earthquakes.

Author Contributions: Conceptualization, Y.M. and M.G.; methodology, Y.M.; software, M.G.; validation, Y.M. and M.G.; formal analysis, M.G.; investigation, Y.M. and M.G.; writing—original draft preparation, M.G.; writing—review and editing, Y.M.; visualization, M.G. All authors have read and agreed to the published version of the manuscript.

Funding: The research was carried out as a part of implementation of the Russian Federation state assignment AAAA-A21-121011290003-0.

Institutional Review Board Statement: Not applicable.

Informed Consent Statement: Not applicable.

Data Availability Statement: The data on nine small-focus earthquakes, which were used for estimated average dislocation parameters in the region (latitude interval: [52°, 53°], longitude interval: [160°, 161°]) are available at The Global Centroid-Moment-Tensor Catalog at https://www.globalcmt.org/cgi-bin/globalcmt-cgi-bin/CMT5/form?itype=ymd&yr=1976&mo=1&day=1&otype=ymd&oyr=2022&omo=6&oday=22&jyr=1976&jday=1&ojyr=1976&ojday=1&nday=1&lmw=0&umw=10&lms=0&ums=10&lmb=0&umb=10&llat=52&ulat=53&llon=160&ulon=161&lhd=0&uhd=90<s=-9999&uts=9999&lpe1=0&upe1=90&lpe2=0&upe2=90&list=0 (accessed on 27 June 2022). The data on two earthquakes, for which the simulation was performed, are available at Unified Information System of Seismological Data of the Kamchatka Branch of Geophysical Service Russian Academy of Science at http://sdis.emsd.ru/info/earthquakes/catalogue.php (accessed on 5 June 2022).

Conflicts of Interest: The authors declare no conflict of interest.

References

1. Sholz, C. *The Mechanics of Earthquakes and Faulting*, 3rd ed.; Cambridge University Press: Cambridge, UK, 2019; p. 512.
2. Kocharyan, G.G.; Kishkina, S.B. The physical mesomechanics of the earthquake source. *Phys. Mesomech.* **2021**, *24*, 343–356. [CrossRef]
3. Shah, M.; Abbas, A.; Adil, M.A.; Ashraf, U.; Oliveira-Junior, J.F.; Tariq, M.A.; Ahmed, J.; Ehsan, M.; Ali, A. Possible seismo-ionospheric anomalies associated with Mw > 5.0 earthquakes during 2000–2020 from GNSS TEC. *Adv. Space Res.* **2022**, *70*, 179–187. [CrossRef]
4. Cicerone, R.D.; Ebel, J.E.; Britton, J. A systematic compilation of earthquake precursors. *Tectonophysics* **2009**, *476*, 371–396. [CrossRef]
5. Marapulets, Y.; Rulenko, O. Joint anomalies of high-frequency geoacoustic emission and atmospheric electric field by the ground-atmosphere boundary in a seismically active region (Kamchatka). *Atmosphere* **2019**, *10*, 267. [CrossRef]
6. Riaz, M.S.; Bin, S.; Naeem, S.; Kai, W.; Xie, Z.; Gilani, S.M.; Ashraf, U. Over 100 years of faults interaction, stress accumulation, and creeping implications, on Chaman Fault System, Pakistan. *Int. J. Earth Sci.* **2019**, *108*, 1351–1359. [CrossRef]

7.	Ashraf, U.; Zhang, H.; Anees, A.; Mangi, H.N.; Ali, M.; Zhang, X.; Imraz, M.; Abbasi, S.S.; Abbas, A.; Ullah, Z.; et al. A Core Logging, Machine Learning and Geostatistical Modeling Interactive Approach for Subsurface Imaging of Lenticular Geobodies in a Clastic Depositional System, SE Pakistan. *Nat. Resour. Res.* **2021**, *30*, 2807–2830. [CrossRef]
8.	Ullah, J.; Luo, M.; Ashraf, U.; Pan, H.; Anees, A.; Li, D.; Ali, M.; Ali, J. Evaluation of the geothermal parameters to decipher the thermal structure of the upper crust of the Longmenshan fault zone derived from borehole data. *Geothermics* **2022**, *98*, 102268. [CrossRef]
9.	Kopylova, G.; Kasimova, V.; Lyubushin, A.; Boldina, S. Variability in the Statistical Properties of Continuous Seismic Records on a Network of Stations and Strong Earthquakes: A Case Study from the Kamchatka Peninsula, 2011–2021. *Appl. Sci.* **2022**, *12*, 8658. [CrossRef]
10.	Morgunov, V.A.; Lyuboshevskij, M.N.; Fabricius, V.Z. Geoacoustic precursor of the Spitak earthquake. *Vulkanol. Seismol.* **1991**, *4*, 104–106. (In Russian)
11.	Gregori, G.P.; Poscolieri, M.; Paparo, G.; DeSimone, S.; Rafanelli, C.; Ventrice, G. "Storms of crustal stress" and AE earthquake precursors. *Nat. Hazards Earth Syst. Sci.* **2010**, *10*, 319–337. [CrossRef]
12.	Kuptsov, A.V. Variations in the geoacoustic emission pattern related to earthquakes on Kamchatka. *Izv. Phys. Solid Earth* **2005**, *41*, 825–831.
13.	Marapulets, Y.; Shevtsov, B. *Mesoscale Acoustic Emission*; Dal'nauka: Vladivostok, Russia, 2012; p. 125. (In Russian)
14.	Rulenko, O.P.; Marapulets, Y.V.; Kuz'min, Y.D.; Solodchuk, A.A. Joint perturbation in geoacoustic emission, radon, thoron, and atmospheric electric field based on observations in Kamchatka. *Izv. Phys. Solid Earth* **2019**, *55*, 76–78. [CrossRef]
15.	Marapulets, Y.V. High-frequency acoustic emission effect. *Bull. KRASEC. Phys. Math. Sci.* **2015**, *10*, 39–48.
16.	Marapulets, Y.V.; Shevtsov, B.M.; Larionov, I.A.; Mishchenko, M.A.; Shcherbina, A.O.; Solodchuk, A.A. Geoacoustic emission response to deformation processes activation during earthquake preparation. *Russ. J. Pac. Geol.* **2012**, *6*, 457–464. [CrossRef]
17.	Dolgikh, G.I.; Shvets, V.A.; Chupin, V.A.; Yakovenko, S.V.; Kuptsov, A.V.; Larionov, I.A.; Marapulets, Y.V.; Shevtsov, B.M.; Shirokov, O.P. Deformation and acoustic precursors of earthquakes. *Dokl. Earth Sci.* **2007**, *413*, 281–285. [CrossRef]
18.	Larionov, I.A.; Marapulets, Y.V.; Shevtsov, B.M. Features of the Earth surface deformations in the Kamchatka peninsula and their relation to geoacoustic emission. *J. Geophys. Res. Solid Earth* **2014**, *5*, 1293–1300. [CrossRef]
19.	Turcotte, D.; Schubert, G. *Geodynamics*, 3rd ed.; Cambridge University Press: Cambridge, UK, 2014; p. 636.
20.	Nikolaevsky, V. *Mechanics of Porous and Fractured Media*; Nedra: Moscow, Russia, 1984; p. 232. (In Russian)
21.	Aki, K.; Richards, P. *Quantitative Seismology*, 2nd ed.; University Science Books: Cambridge, UK, 2002; p. 704.
22.	Okada, Y. Surface deformation due to shear and tensile faults in a half-space. *Bull. Seismol. Soc. Am.* **1985**, *75*, 1135–1154. [CrossRef]
23.	Okada, Y. Internal deformation due to shear and tensile faults in a half-space. *Bull. Seismol. Soc. Am.* **1992**, *82*, 1018–1040. [CrossRef]
24.	Dobrovolsky, I. *Mathematical Theory of Preparation and Prediction of a Tectonic Earthquake*; Fizmatlit: Moscow, Russia, 2009; p. 235. (In Russian)
25.	Dobrovolsky, I.P.; Zubkov, S.I.; Miachkin, V.I. Estimation of the size of earthquake preparation zones. *Pure Appl. Geophys.* **1979**, *117*, 1025–1044. [CrossRef]
26.	Alekseev, A.S.; Belonosov, A.S.; Petrenko, V.E. On the concept of multidisciplinary earthquake prediction using an integral predictor. *Vych. Seismol.* **2001**, *32*, 81–97. (In Russian)
27.	Li, Y.; Liu, M.; Li, Y.; Chen, L. Active crustal deformation in southeastern Tibetan Plateau: The kinematics and dynamics. *Earth Planet. Sci. Lett.* **2019**, *523*, 115708. [CrossRef]
28.	Dansereau, V.; Shapiro, N.; Campillo, M.; Weiss, J. A Burgers-Brittle model for the seismic-aseismic, brittle-ductile transition within the Earth crust. In Proceedings of the EGU General Assembly Conference Abstracts, vEGU21: Gather Online, Vienna, Austria, 19–30 April 2021; EGU21–7547.
29.	Pollitz, F. Viscoelastic Fault-Based Model of Crustal Deformation for the 2023 Update to the U.S. National Seismic Hazard Model. *Seismol. Res. Lett.* **2022**, *93*, 3087–3099. [CrossRef]
30.	Takagi, Y.; Okubo, S. Internal deformation caused by a point dislocation in a uniform elastic sphere. *Geophys. J. Int.* **2017**, *208*, 973–991. [CrossRef]
31.	Liu, T.; Fu, G.; She, Y.; Zhao, C. Co-seismic internal deformations in a spherical layered earth model. *Geophys. J. Int.* **2020**, *221*, 1515–1531. [CrossRef]
32.	Zhu, S. Inter-and pre-seismic deformations in the 2011 M_W 9.0 Tohoku-Oki earthquake: Implications for earthquake prediction. *Chin. J. Geophys.* **2020**, *63*, 427–439.
33.	Chen, C.H.; Su, X.; Cheng, K.C.; Meng, G.; Wen, S.; Han, P. Seismo-deformation anomalies associated with the M 6.1 Ludian earthquake on August 3, 2014. *Remote Sens.* **2020**, *12*, 1067. [CrossRef]
34.	Lv, X.; Shao, Y. Rheology of the Northern Tibetan Plateau Lithosphere Inferred from the Post-Seismic Deformation Resulting from the 2001 Mw 7.8 Kokoxili Earthquake. *Remote Sens.* **2022**, *14*, 1207. [CrossRef]
35.	Jiao, Z.H.; Zhao, J.; Shan, X. Pre-seismic anomalies from optical satellite observations: A review. *Nat. Hazards Earth Syst. Sci.* **2018**, *18*, 1013–1036. [CrossRef]
36.	Perezhogin, A.S.; Shevtsov, B.M.; Sagitova, R.N.; Vodinchar, G.M. Geoacoustic emission's zones modeling. *Math. Models Comput. Simul.* **2007**, *19*, 59–64.

37. Perezhogin, A.S. Geoacoustic emission zones in an elastic model of continuum. *Bull. KRASEC Earth Sci.* **2009**, *13*, 198–201. (In Russian)
38. Perezhogin, A.S.; Shevtsov, B.M. Models of an intense-deformed condition of rocks before earthquakes and their correlation with geo-acoustic emission. *Comput. Technol.* **2009**, *14*, 48–57. (In Russian)
39. Saltykov, V.A.; Kugaenko, Y.A. Development of near-surface dilatancy zones as a possible cause for seismic emission anomalies before strong earthquakes. *Russ. J. Pac. Geol.* **2012**, *6*, 86–95. [CrossRef]
40. Gapeev, M.I.; Marapulets, Y.V. Modeling of relative shear deformation zones before strong earthquakes in Kamchatka from 2018–2021. *Vestn. KRAUNC Fiz.-Mat. Nauki* **2021**, *37*, 53–66. (In Russian)
41. The Global Centroid-Moment-Tensor (CMT) Project. Available online: https://www.globalcmt.org/ (accessed on 27 June 2022).
42. Gusev, A.A.; Melnikova, V.N. Relations between magnitudes: Global and Kamchatka data. *J. Volcanol. Seismol.* **1990**, *6*, 55–63.
43. Kanamori, H. The radiated energy of the 2004 Sumatra-Andaman earthquake. In *Earthquakes: Radiated Energy and the Physics of Faulting (Geophysical Monograph Series)*; Abercrombie, R., McGarr, A., Kanomari, H., Di Toro, G., Eds.; American Geophysical Union: Washington, DC, USA, 2006; pp. 59–60.
44. Bath, M.; Duda, S. Earthquake volume, fault plane area, seismic energy, strain, deformation and related quantities. *Ann. Geophys.* **1964**, *17*, 353–368.
45. Rebetsky, Y. *Tectonic Stresses and Strength of Natural Massifs*; IKC "Akademkniga": Moscow, Russia, 2007; p. 406. (In Russian)
46. Mindlin, R. Force at a point in the interior of a Semi-Infinite solid. *J. Appl. Phys.* **1936**, *195*, 195–292. [CrossRef]
47. Mindlin, R.; Cheng, D. Nuclei of Strain in the Semi-Infinite Solid. *J. Appl. Phys.* **1950**, *21*, 926–930. [CrossRef]
48. Kamchatka Branch of Geophysical Service Russian Academy of Science. Available online: http://sdis.emsd.ru/info/earthquakes/catalogue.php (accessed on 5 June 2022).
49. Gusev, A.A.; Skorkina, A.A.; Pavlov, V.M.; Abubakirov, I.R. Obtaining mass estimates of regional moment magnitudes MW and establishing their relationship with ML3 for subduction Kamchatka earthquakes. In Proceedings of the Materials of the XX Regional Scientific Conference dedicated to the Day of Volcanologist, XX Regional Scientific Conference dedicated to the Day of Volcanologist, Petropavlovsk-Kamchatsky, Russia, 30–31 March 2017; Institute of Volcanology and Seismology FEB RAS: Petropavlovsk-Kamchatsky, Russia, 2017. (In Russian)
50. Mondal, D.; Debnath, P. An application of fractional calculus to geophysics: Effect of a strike-slip fault on displacement, stresses and strains in a fractional order Maxwell type visco-elastic half space. *Int. J. Appl. Math.* **2021** *34*, 873–888. [CrossRef]
51. Rebetsky, Y.L.; Lermontova, A.S. On the long-range Influence of earthquake rupture zones. *J. Volcanol. Seismol.* **2018**, *12*, 341–352. [CrossRef]

applied sciences

Article

Spatial Correlations of Global Seismic Noise Properties

Alexey Lyubushin

Institute of Physics of the Earth, Russian Academy of Sciences, Moscow 123242, Russia; lyubushin@yandex.ru

Abstract: A study of global seismic noise during 1997–2022 was carried out. A property of waveforms known as the Donoho–Johnston (DJ) index was used, which separates the values of the wavelet coefficients into "small" and "large". For each reference point in an auxiliary network of 50 points, a time series was calculated with a time step of one day for the median of the values at the five nearest stations. In a moving time window of 365 days, correlations between the index values at the reference points were calculated. A decrease in the average values of the DJ-index and an increase in correlations were interpreted as a sign of an increase in global seismic danger. After 2011, there was a sharp increase in the maximum distances between reference points with large correlations. The high amplitude of the response of the DJ-index to the length of the day for 2020–2022 could predict a strong earthquake in the second half of 2023. The purpose of this study was to improve the mathematical apparatus for assessing the current seismic hazard according to the properties of seismic noise.

Keywords: seismic noise; wavelet-based entropy; wavelet-based Donoho–Johnston index; correlations; day length

Citation: Lyubushin, A. Spatial Correlations of Global Seismic Noise Properties. *Appl. Sci.* **2023**, *12*, 6958. https://doi.org/10.3390/app13126958

Academic Editor: Roberto Zivieri

Received: 4 April 2023
Revised: 6 June 2023
Accepted: 7 June 2023
Published: 8 June 2023

1. Introduction

Information about seismic noise makes it possible to study the processes preceding strong earthquakes [1–3]. Atmosphere and ocean (cyclone movement and the impact of ocean waves on the shelf) are the main sources for the energy of seismic noise [4–10]. At the same time, processes inside the earth's crust are reflected in changes in the properties of seismic noise and studying these properties helps in investigating the structural features of the earth's crust [11–13].

The Donoho–Johnston index (DJ-index) of a random signal can be defined as the ratio of the number of the coefficients of orthogonal wavelet decomposition, which in absolute value exceed the threshold introduced in [14], to the total number of wavelet coefficients. The threshold that separates "large" wavelet coefficients from others was defined first in [14] and is used to shrink noise from signals and images using wavelets. It has turned out that the DJ-index has the ability to most clearly highlight the effects of the spatial correlation of seismic noise, and it outperforms other noise statistics (such as entropy and the width of the multi-fractal spectrum of the singularity) in terms of highlighting predictive effects [15–18]. This article presents a detailed study of the DJ-index of global seismic noise, both the change in time and space of the correlation properties of noise and its response to the irregularity of the Earth's rotation, as well as the possible use of this response to assess the current seismic hazard.

2. Initial Seismic-Noise Data

The data used were the vertical components of continuous seismic-noise records with a 1 s sampling step, which were downloaded via the address http://www.iris.edu/forms/webrequest/ (accessed on 1 January 2023) from 229 broadband seismic stations of three networks, http://www.iris.edu/mda/_GSN (accessed on 1 January 2023), http://www.iris.edu/mda/G (accessed on 1 January 2023), and http://www.iris.edu/mda/GE

(accessed on 1 January 2023), which belong to the Incorporated Research Institutions for Seismology.

The duration of the time interval of observations was 26 years, from 1 January 1997 to 31 December 2022. The data were resampled to 1 min time series by computing the mean values within adjacent time intervals of the length 60 s.

An auxiliary network of 50 reference points was introduced. The positions of these points were determined by cluster analysis of the positions of 229 seismic stations using the hierarchical "far neighbor" method, providing compact clusters [19]. The positions of the 229 seismic stations and 50 control points are shown in Figure 1. The numbering of the reference points was carried out in order as the latitude of the point decreases.

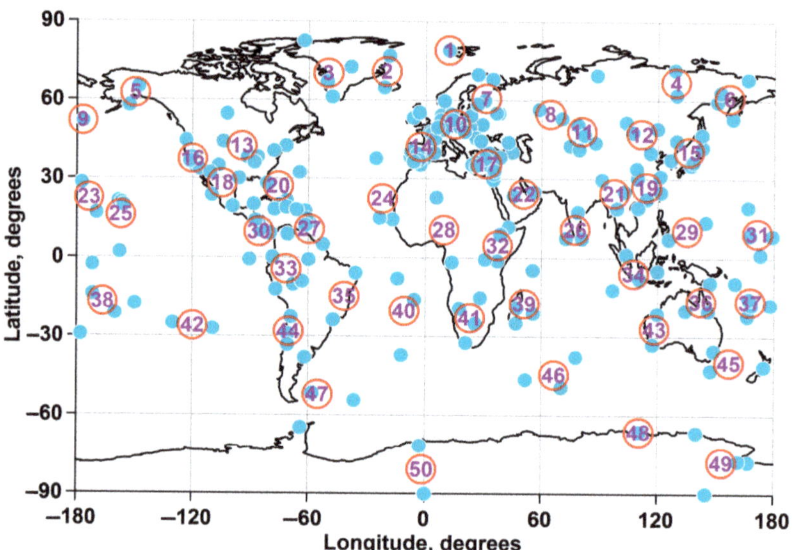

Figure 1. Positions of 229 broadband seismic stations (blue circles) and a network of 50 reference points (numbered red circles).

3. Wavelet-Based Measure of Time-Series Non-Stationarity

Let us consider a random signal $x(t)$ where $t = 1, \ldots, N$ is an integer discrete-time index. For a finite sample of the signal, the entropy [20] could be defined as:

$$En = -\sum_{k=1}^{N} p_k \cdot \log(p_k), \; p_k = c_k^2 / \sum_{j=1}^{N} c_j^2 \tag{1}$$

Here, c_k are the coefficients of orthogonal wavelet decomposition. Within a finite set of Daubechies wavelet bases [20] with the number of vanishing moments from 1 to 10, the optimal basis is chosen from the minimum of (1).

The DJ-index was introduced in the problem of noise reduction and compression of information by setting all "small" wavelet coefficients to zero and performing an inverse wavelet transform [14,20]. This operation of noise reduction is performed for the optimal wavelet basis which was found from the minimum of entropy. The problem is how to define the threshold which separates "large" wavelet coefficients from "small" ones. In [14], this problem was solved using a formula of the asymptotic probability of the maximum deviations of Gaussian white noise. The DJ threshold is given by the expression $T_{DJ} = \sigma\sqrt{2 \cdot \ln N}$, where σ is the standard deviation of wavelet coefficients of the first detail level of wavelet decomposition. The first level is the most high frequency and is considered as the most noisy. Thus, in order to estimate the noise standard deviation, it is possible to calculate the standard deviation of the first-level wavelet coefficients $c_k^{(1)}$. The robust estimate was proposed:

$$\sigma = med\left\{\left|c_k^{(1)}\right|,\ k = 1, \ldots, N/2\right\}/0.6745, \tag{2}$$

where $N/2$ is the number of wavelet coefficients on the first level, and *med* is the median. Formula (2) gives the relationship between the standard deviation and the median for the Gaussian random variable.

Finally, it is possible to define the dimensionless signal characteristic γ, $0 < \gamma < 1$, as the ratio of the number of the most informative wavelet coefficients, for which the inequality $|c_k| > T_{DJ}$ is satisfied, to the total number N of all wavelet coefficients. For stationary Gaussian white noise, which is a kind of ideal stationary random signal, the index $\gamma = 0$. That is why the DJ-index could be called a measure of non-stationarity.

For each reference point, the index value γ is calculated daily as a median of the values in the five nearest operational stations. Thus, continuous time series are obtained with a time sampling of one day at 50 reference points. Figure 2 shows the graphs of the index γ for 9 reference points (point numbers are indicated next to each graph).

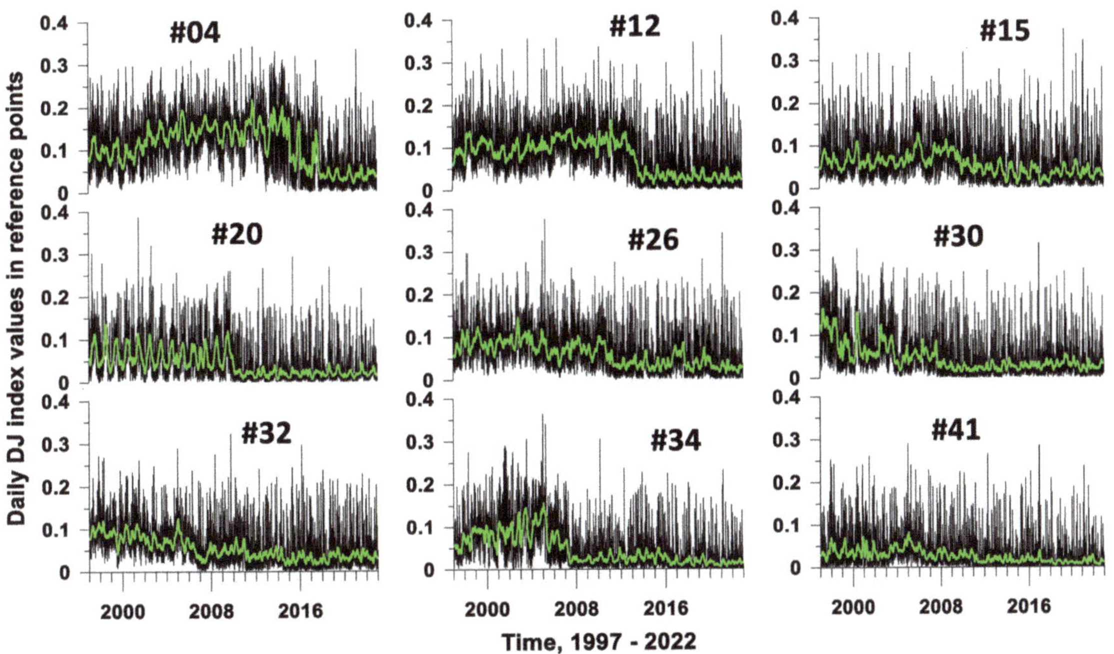

Figure 2. Graphs of daily index γ values for 9 reference points. The green lines represent the moving averages in a 57-day window.

Figure 3 shows a graph of average daily values of index γ calculated for all 50 reference points. The red lines represent the linear trends of the average values calculated for the time fragments 1997–2003, 2004–2015 and 2016–2022. The boundaries of time fragments were chosen so that the continuity of the broken line at the union of linear trends was approximately observed. For the time intervals 1997–2003 and 2016–2022, linear trends coincided with the average values and it can be seen that during the intermediate time interval 2004–2015 there had been a significant decrease in the global average γ. According to the results published in articles [15,17,18], a decrease in the average value of the index γ, which is a sign of the simplification of the statistical structure of noise (approximation of its properties to the properties of white noise), was associated with an increase in seismic hazard. Thus, the graph of average values in Figure 3 can be interpreted as a transition to a higher level of global seismic hazard during 2004–2015.

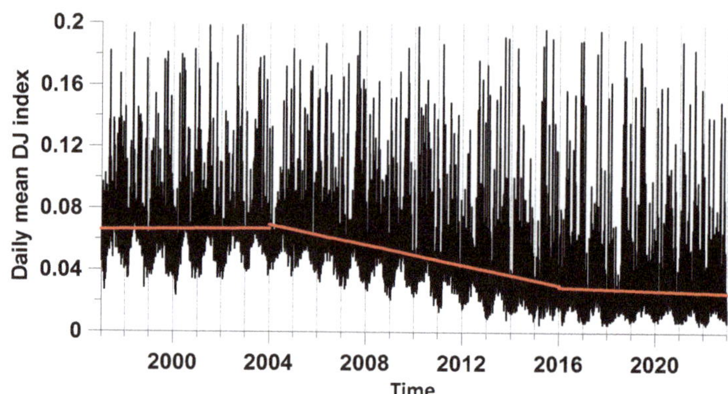

Figure 3. Graph of the average daily values of the DJ-index for all 50 reference points. Red lines present linear trends for 3 time intervals: 1997–2003, 2004–2015 and 2016–2022.

4. Probability Densities of Extreme Values of Seismic-Noise Properties

Further analysis of the properties of seismic noise was based on the identification of regions that are characterized by the most frequent occurrences of extreme values for certain seismic-noise statistics. To do this, it was necessary to estimate the probability densities of the distribution over the space of extreme values by their values in the network of reference points. The following process was used. Denote by ζ_k, $k = 1, \ldots, m = 50$ coordinates of the reference points. Let Z_k be the values of any property of the noise at the reference point with the number k. We will be interested in maps of the distribution of probability densities of extreme values (minimums or maxima) of values Z over the Earth's surface. To construct such maps, consider a regular grid of 50 nodes in latitude and 100 nodes in longitude covering the entire earth's surface. Next, maps will be built for various seismic-noise statistics Z, the values of which will be estimated in time windows of various lengths. For each such window, we define the time index t, which corresponds to its right end. For each time index t, find the reference point with the number k_t^*, at which the extremum of one or another property Z is realized, and let ζ_t^* be the coordinate of the reference point with the number k_t^*.

Let $\rho(\zeta_k, r_{ij})$ be the distance between the points ζ_k and the coordinates of the nodes of the regular grid r_{ij}. Then, the value of the probability density of extreme values for the quantity Z at the node r_{ij} of the regular grid can be calculated according to Gaussian distribution:

$$p(r_{ij}|t) = \exp(-\rho^2(\zeta_t^*, r_{ij})/(2h^2))/P(t) \tag{3}$$

Here, h is the distance, which corresponds to an arc of 15 angular degrees, which is approximately equal to 1700 km, and $P(t)$ is a divisor that provides the normalization condition so that the numerical integral of function (3) over the Earth's surface is equal to 1. Formula (3) gives an "elementary map" of the probability density of the property Z extrema distribution for each current window with a time index t. This makes it possible to calculate the average probability density for a given time interval from t_0 to t_1:

$$\overline{p}(r_{ij}|t_0, t_1) = \sum_{t=t_0}^{t_1} p(r_{ij}|t)/n(t_0, t_1) \tag{4}$$

where $n(t_0, t_1)$ is the number of time windows labeled from t_0 to t_1.

Another way to estimate the variability of the probability density of extreme properties of seismic noise is to construct histograms of the distribution of the numbers of control points, in which the statistics extrema are realized. The construction of such histograms in a moving time window, the length of which should include a sufficiently large number of initial time windows within which the noise properties are estimated, makes it possible to determine the reference points at which the minima or maxima of the seismic-noise properties Z are realized most often.

This method has the disadvantage that it does not give a direct distribution of the places of the most probable realization of the noise-property extrema in the form of a geographical map, although each reference point has geographical coordinates. On the other hand, this method makes it possible to compactly visualize the temporal dynamics and, in particular, to determine the time intervals when the places of concentration of extreme values for the noise statistics change sharply. To solve a similar problem of visualizing temporal dynamics using averaged probability densities (4), one would have to build a laborious and long sequence of maps for time fragments consisting of the same number of time windows.

Therefore, in the future, the probability densities of extreme values of noise properties will be presented simultaneously in two forms: as averaged densities according to Formula (4) for two time intervals, before and after 2011, and as a temporal histogram of the distribution of numbers of reference points in which extremes are realized. In this case, the number of base intervals (bins) of histograms is taken equal to the number of reference points, that is, 50, which makes it possible to visualize the dynamics of the emergence and disappearance of bursts of the probability of extreme values at each reference point.

Figure 4a,b show maps of the distribution of the minimum values of the DJ-index, calculated by Formulas (3) and (4) for windows with time stamps before and after 2011. The choice of 2011 as a boundary time mark is connected, as will be seen from the following presentation, with the fact that the correlation properties of seismic noise changed dramatically after two mega-earthquakes: on 27 February 2010, $M = 8.8$ in Chile, and on 11 March 2011, $M = 9.1$ in Japan. Previously, this change has been already detected in [3,18]. Increased attention to the areas where the minimum values γ are most often realized, as was already mentioned above when discussing Figure 3, is due to the fact that an increase in seismic hazard is associated with a simplification of the statistical noise structure and a decrease in γ. For regions that are quite densely covered with seismic networks, for example, Japan, this property of statistics γ makes it possible to estimate the location of a possible strong earthquake [17].

However, the global network of seismic stations is sparse and therefore the selection of places where the probability of occurrence of small values is increased does not necessarily coincide with the known places of the occurrence of strong earthquakes. In Figure 4a,b, the maxima of the probability-density distribution of the minimum values γ are located in the vicinity of reference points with numbers 23, 25 and 37, 38 in the western part of the Pacific Ocean, where the epicenters of strong earthquakes and volcanic manifestations are really concentrated, including the largest volcanic eruption, Hunga Tonga-Hunga Ha'apai on 15 January 2022 [21]. Note that in Figure 4c, which shows a diagram of the change in the histogram of the numbers of reference points, in which the minimum value γ was realized, one can see the switching of the histogram maximum from the reference point number 23 to the reference point number 38 just after 2011. As for the concentration of probabilities of minimum values γ in the vicinity of reference points with number 46 in the southern Indian Ocean (in the vicinity of Kerguelen Island) and in the vicinity of point 40 in the Atlantic Ocean (Saint Helena) after 2011, these features can be associated with a mantle-plume activation regime [22].

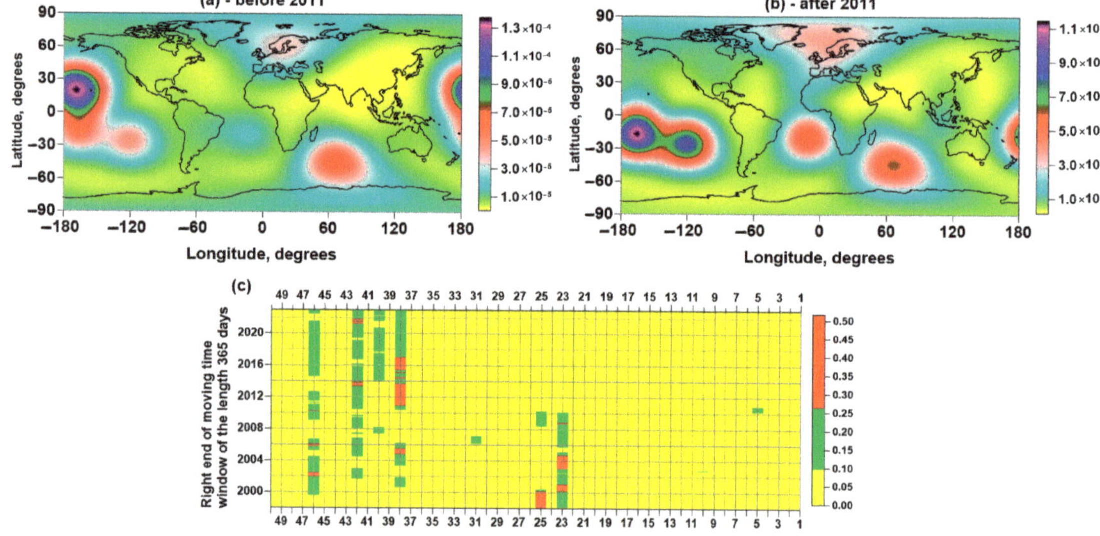

Figure 4. (**a,b**) The probability densities of the distribution of the minimum values of the seismic-noise index γ in a network of 50 control points before and after 2011; (**c**) histogram of numbers of control points, in which the minimum values of the index γ were realized in a moving time window 365 days long.

5. Estimation of Spatial Long-Range Correlations of Seismic-Noise Properties

The simplification of the statistical structure of seismic noise manifests itself in a decrease in the average value of the index γ (Figure 3), and it is also reflected in an increase in entropy (1) and a decrease in the width of the multi-fractal singularity spectrum support width ("loss of multifractality"). These changes in noise properties are indicators of the transition of a seismically active region (including the whole world) to a critical state [3,17].

Another indication of an increase in seismic hazard is an increase in the average correlation as well as the radius of strong correlations between noise properties in different parts of the system. In order to find out how the spatial scale of correlations changes with time, for each reference point we calculated the correlation coefficient between the index γ values at this point and at all other reference points. We performed these calculations in a sliding time window 365 days long with a shift of 3 days. Next, for each reference point, we calculated the average value of the correlation coefficient with the values of γ at all other reference points. Since the result of calculating such an average value depends on the time window, we obtained graphs of changes in average correlations for all reference points. Figure 5 shows examples of graphs of changes in correlations for nine reference points.

From the graphs in Figure 5, a general trend towards an increase in the average correlations for each reference point is noticeable. As a result of averaging all the graphs of the type shown in Figure 5 for all 50 reference points, we obtained a general graph of average correlations, as shown in Figure 6.

It can be noticed from the graph in Figure 6 that the time interval 2004–2016, in which the index γ decreases (Figure 3), corresponds to an increase in average correlations between the properties of noise at various reference points and that there is a transition from small correlations before 2004 to fairly large correlations after 2016.

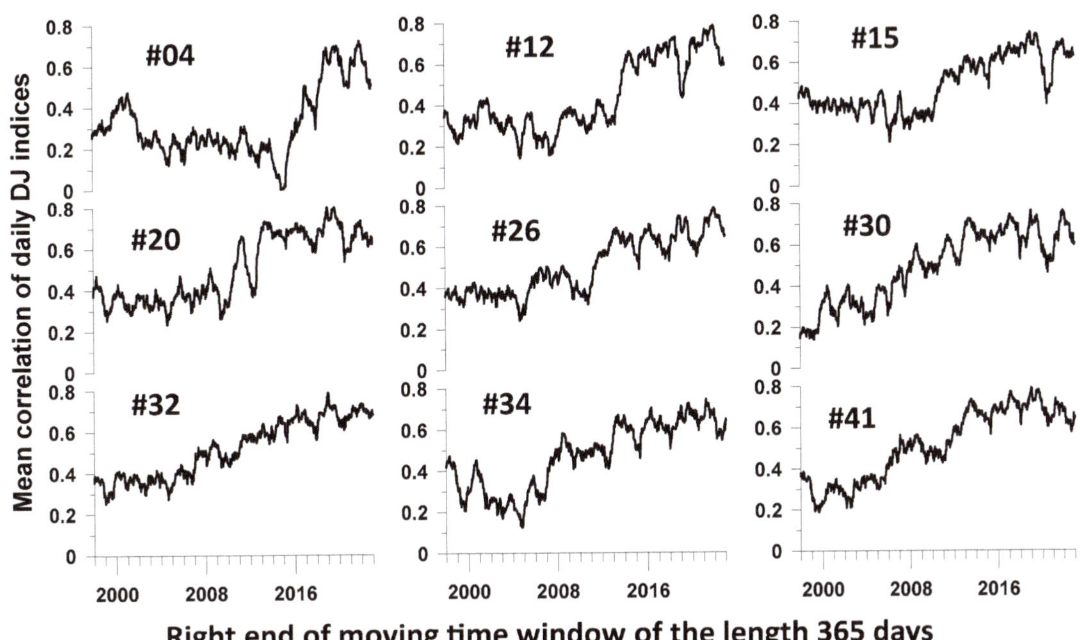

Figure 5. Graphs of the average correlation coefficients of the index γ for 9 reference points with values at other reference points in a moving time window of 365 days with a mutual shift of 3 days.

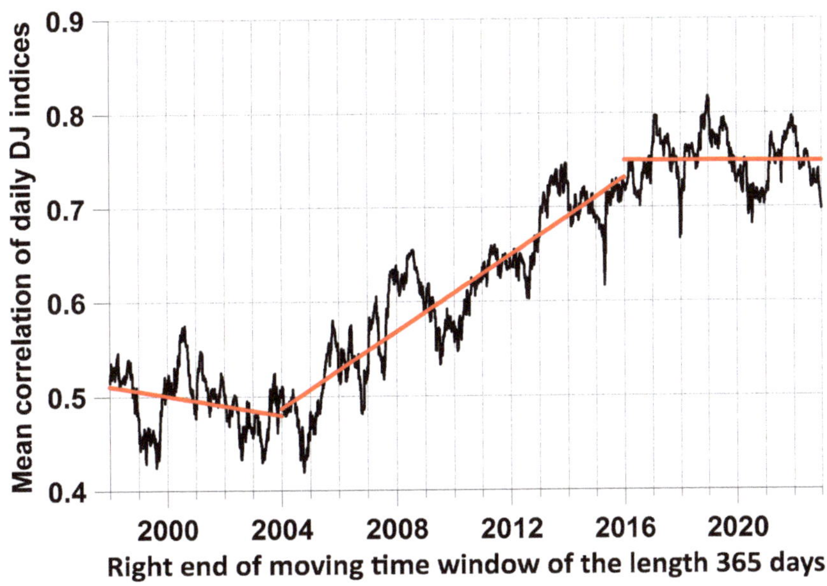

Figure 6. The average correlation coefficient between the values of the DJ-index γ at each control point with the values at other points. Red lines show linear trends for 3 intervals of time marks of the right ends of windows with a length of 365 days: 1998–2004, 2004–2016 and 2016–2023. On the last time fragment, the linear trend coincides with the average value.

In order to assess the scale of strong correlations between seismic-noise properties, for each reference point we defined another point that had a maximum correlation with this point, provided that this maximum exceeded the threshold of 0.8, and we calculated the distance between these points. Figure 7 shows the graphs of the changes in the average and maximum values of the distances between points for which the correlation maxima exceeded 0.8.

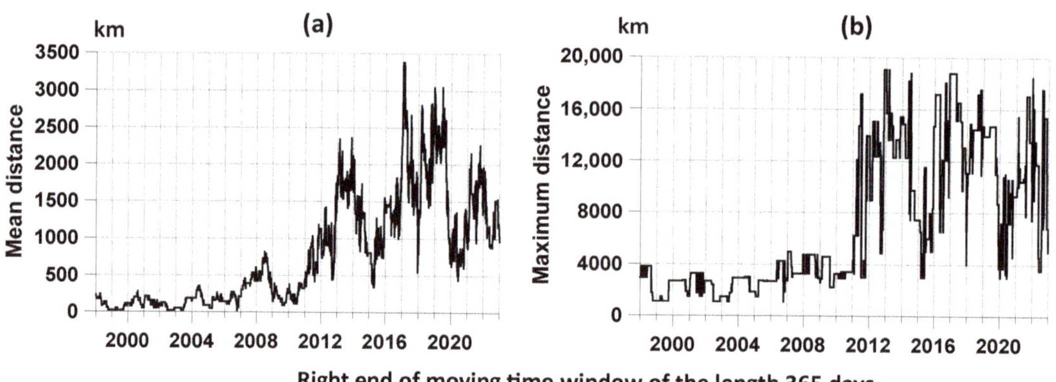

Figure 7. (**a**) The average distance between the reference points for which the maxima of correlations between the values of the DJ-index of seismic noise exceeded the threshold of 0.8; (**b**) the maximum distance between the reference points with the maximum correlation coefficient above the threshold of 0.8.

It can be seen from the graph in Figure 7b that after 2011 there was a sharp increase in the maximum distance at which strong maximum correlations occur. Thus, the growth of average correlations, which is visible in Figure 6, was supplemented by an explosive growth of the maximum scale of strong correlations.

It is of interest to identify those reference points in the vicinity of which strong correlations most often occur. As before, for this purpose we used estimates (3–4) of the probability density of the distribution over space of the maximum correlation values for the time intervals before and after 2011, as well as the distribution histograms of the numbers of control points in which the maximum correlation values occurred.

To select the length of the time window for calculating histograms, we recall that the correlation coefficients between the values at each reference point with the values of the DJ-index at other reference points are calculated in windows of 365 days with a shift of 3 days. Let us choose the length of the window for calculating histograms so that the dimensional length of the window is equal to five years. It is easy to calculate that this length will be 488 values of "short" time windows with a length of 365 days with an offset of 3 days. In this case, the histogram is calculated in a window of length $365 + 3 \cdot (488 - 1) = 1826$ days, which is approximately equal to five years, given that every fourth year is a leap year. As for the number of base intervals (bins) of histograms, in this case there will not be 50, as in Figure 4c, but 44, since for reference points with numbers 1–6 there are no realizations of correlation maxima.

Estimates of the position and temporal dynamics of places with an increased probability of maximum correlations are presented in Figure 8. Comparison of Figure 8a,b shows that the places of the concentration of maximum correlations have changed significantly since 2011. In particular, there has been a rapid shift of such a center from the Southwest Pacific to Central Eurasia, which can be seen in the histogram diagram in Figure 8c.

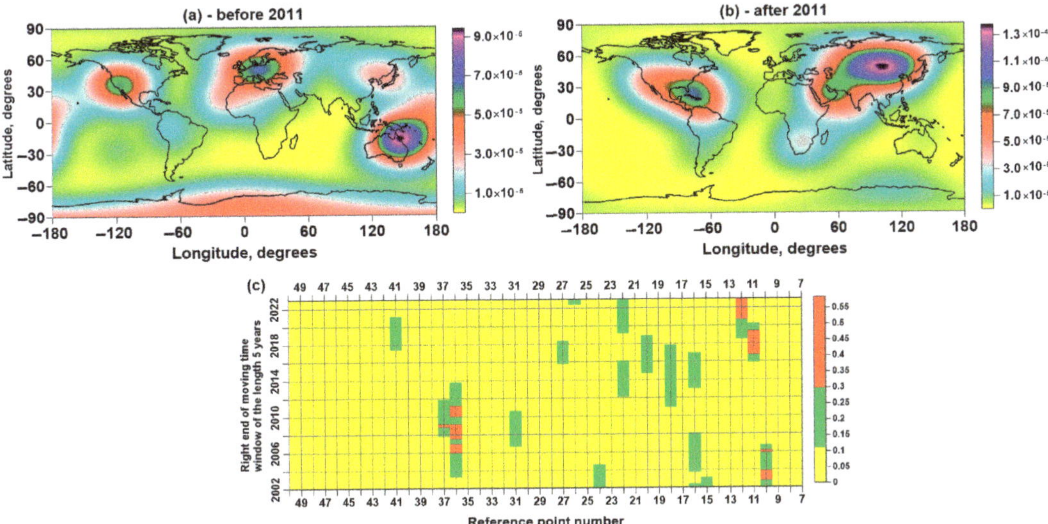

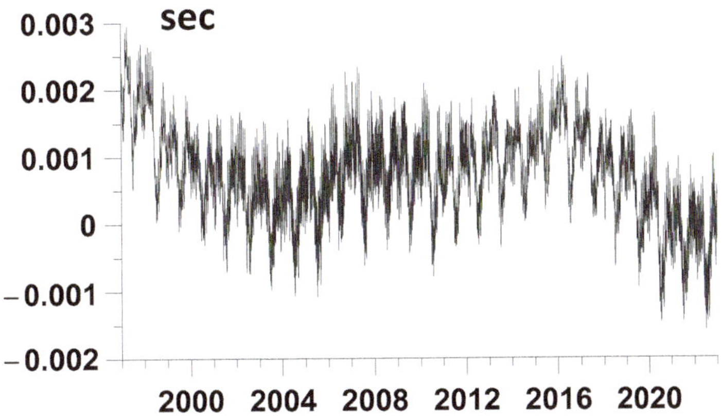

Figure 8. (**a**,**b**) The probability densities of the correlation maxima for index γ values at each reference point with index values at other reference points before and after 2011; (**c**) a histogram of the numbers of reference points, in which the correlation maxima for values of γ at each reference point with the values at other points in a moving time window of 5 years were realized.

6. Seismic-Noise Response to the Irregularity of the Earth's Rotation

The connection between the irregular rotation of the Earth and seismicity was investigated in [23]. The influence of strong earthquakes on the Earth's rotation was studied in [24,25]. Figure 9 shows a graph of the time series of the length of the day for the time interval 1997–2022.

Figure 9. Time series plot of the length of a day for the time interval 1997–2022. Day-length data (LOD) are available from the website at: https://hpiers.obspm.fr/iers/eop/eopc04/eopc04.1962-now (accessed on 1 January 2023).

Further, the LOD time series is used as a kind of "probing signal" for the properties of the seismic process [3,15–18]. To study the relationship between the properties of the seismic process and the irregularity of the Earth's rotation, the squared coherences maxima between the LOD and the properties of seismic noise at each reference point were estimated. The coherences were estimated in moving time windows of 365 days with a shift of 3 days using a vector (two-dimensional) fifth-order autoregressive model [15–18]. In what follows, these maximum coherences will be referred to as the seismic-noise responses to the LOD. Figure 8 shows examples of the LOD response graphs at nine reference points.

We define the integrated response of global seismic noise to the LOD as the average of the responses of the type shown in Figure 10 from all 50 reference points. Let us compare the behavior of the average response to the LOD with the amount of released seismic energy. To do this, we calculate the logarithm of the released energy as a result of all earthquakes in a moving time window 365 days long with a shift of 3 days. In Figure 11a,b, there are graphs of the logarithm of the released energy and the average response to the LOD. It is visually noticeable that bursts of the average maximum coherence in Figure 11b precede the energy spikes in Figure 11a. In order to quantitatively describe this delay, we calculated the correlation function between the curves in Figure 11a,b. The graph of this correlation function for time shifts of ±1200 days is shown in Figure 11c.

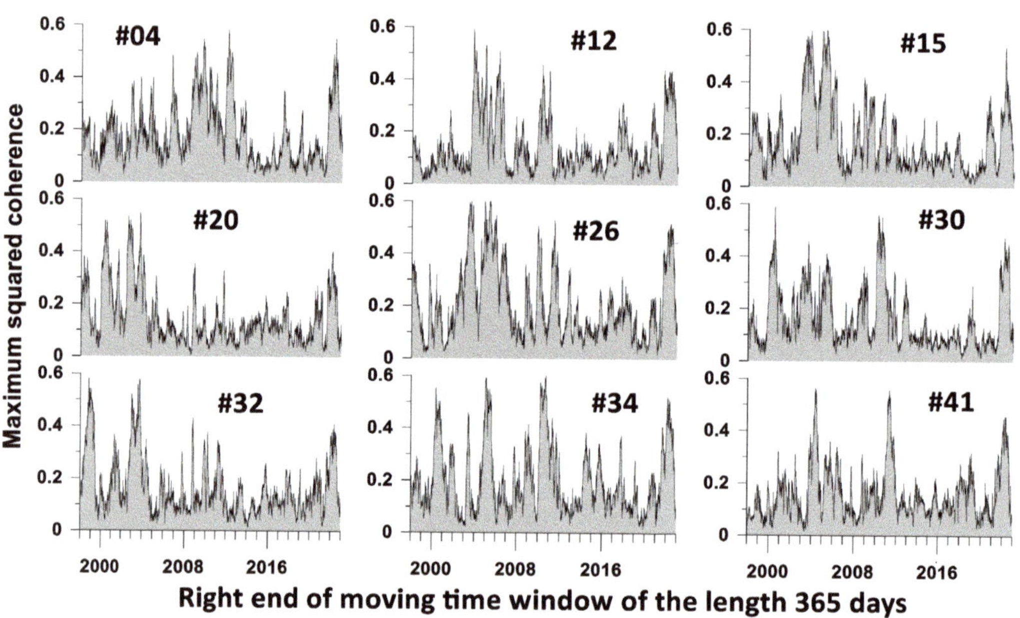

Figure 10. Plots of maximum squared coherence between index γ and LOD in a moving time window of 365 days with mutual shift of 3 days for 9 reference points.

The estimate of the correlation function in Figure 11c between the logarithm of the released seismic energy in Figure 11a and the average response of seismic noise to the LOD had a significant asymmetry and was shifted to the region of negative time shifts, which corresponded to the coherence maxima being ahead of the maxima of seismic energy emissions. The correlation maximum corresponded to a negative time shift of 534 days. This estimate of cross-correlations produced a foundation to propose that the burst shown in Figure 11b by the magenta line may precede a major earthquake with an average delay of 1.5 years.

The curve in Figure 11b can be split into three sections. The first two intervals with time marks of the right ends of windows before and after 2012 differ significantly from each other by their mean values presented by red lines.

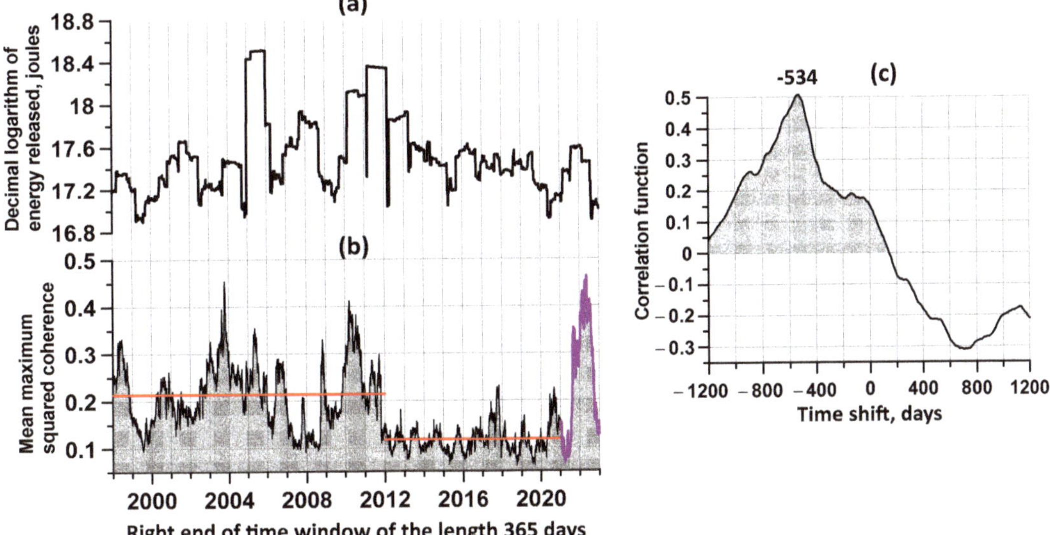

Figure 11. (**a**) Logarithm of the released seismic energy (in joules) in a sequence of time intervals 365-days long, taken with an offset of 3 days; (**b**) average values of maximum coherences between LOD and daily DJ-index values at 50 reference points; (**c**) the correlation function between the logarithm of the released seismic energy and the average value of the maximum coherence between the LOD and the seismic-noise DJ-index at 50 reference points.

The third time interval in Figure 11b is presented by the magenta line and refers to the processing of the most recent data which correspond to right-hand end windows after 2021, i.e., referring to the time span of 2020–2022. A short third segment was identified based on the results of data processing for 2021–2022 due to an unusually large spike in the response of noise properties to the LOD. The maximum value of the response in Figure 11b was reached at the time corresponding to the date 9 May 2022. The "naive" forecast, in which 534 days are added to this date, gives the most probable date for the future strong event as being 24 October 2023. A more realistic prediction would be to blur this date with some uncertainty interval. However, it is not yet possible to give such an interval due to the short history of using the seismic-noise response to the LOD as a predictor.

Similar to how it was done to identify the places of concentration of the maxima of spatial correlations of seismic noise in Figure 8, we can search for places where the maximum response of seismic-noise properties to the uneven rotation of the Earth is most likely. The results of such a search are shown in Figure 12. Comparison of Figure 12a,b shows that the maximum response to the LOD is concentrated in the subarctic regions.

In addition to the issue of the synchronization of seismic-noise properties, which are reflected in the graphs in Figures 5–7, we also studied the synchronization of seismic-noise responses to the irregularity of the Earth's rotation. Previously, this issue has been considered for the synchronization of seismic-noise responses to LODs in Japan and California [16].

For each reference point, we estimated the correlation of the response at this point with the responses at all other points. It should be taken into account that the correlated responses themselves were calculated in sliding time windows of 365 days with a shift of 3 days. These were "short" time windows. In this case, we needed to take a certain number of consecutive "short" windows and form a "long window" from them, such that it contained a sufficiently large number of estimates of the noise responses to the LOD in the "short" time windows. Next, 250 adjacent "short" windows were selected to form a "long" window. Correlation coefficients between the responses to the LOD at various reference points were calculated in a sliding "long" window with a shift of 1, that is, in a time window with a length of $365 + 3·(250 − 1) = 1112$ days, which is approximately 3 years. A shift of 1 count meant a real shift of 3 days.

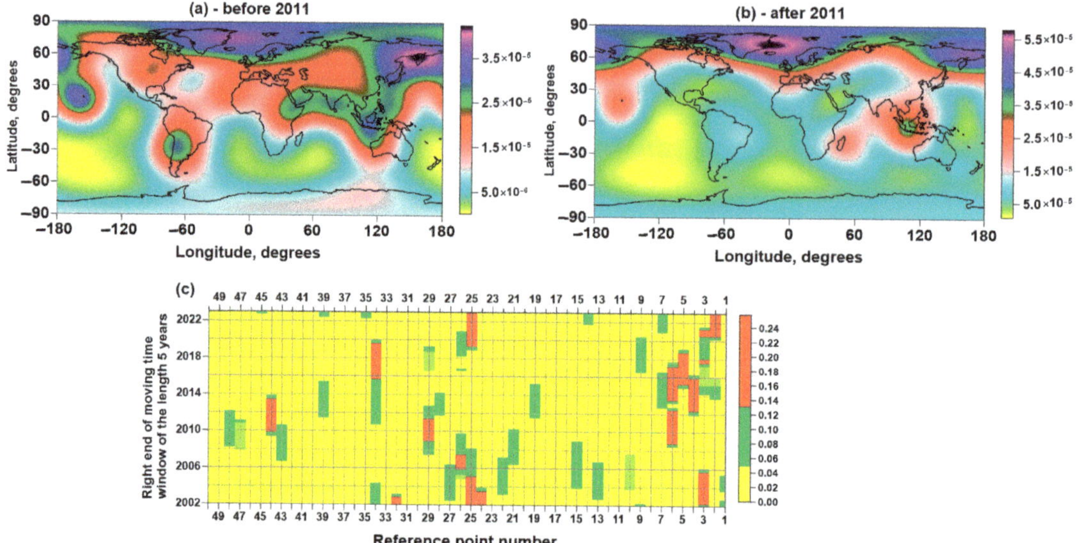

Figure 12. (**a,b**) Probability densities of the seismic-noise index γ response maxima to LOD before and after 2011; (**c**) histogram of numbers of reference points, in which the maximum values of the index γ response to LOD were realized in a sliding time window of 5 years.

The goal of such calculations for each reference point was the average correlations of responses over all other reference points. As a result of such averaging, 50 dependences of the same type as shown in Figure 5 were obtained. In order to obtain an integrated measure of the correlation of global noise responses to the LOD, the average dependences for all reference points needed to be averaged secondarily. As a result, we obtained the graph shown in Figure 13.

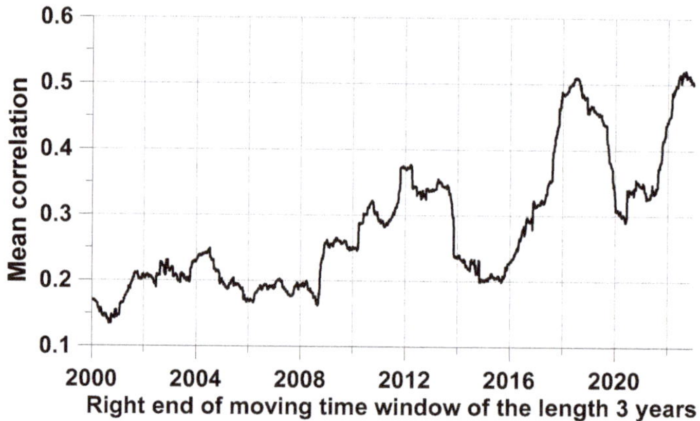

Figure 13. The average value of the correlation coefficients between the seismic-noise index γ responses per LOD for each of the 50 control points with responses at other points.

Figure 13 shows the change in time of the measure of synchronization of the response of seismic noise to the irregularity of the Earth's rotation. This graph also shows the growth of correlations, similar to the growth of the initial correlations in Figure 6.

7. Conclusions

The results of a detailed analysis of the properties of seismic noise based on the use of the Donoho–Johnston index, defined as the proportion of the maximum values of the moduli of the wavelet coefficients separating the "noise" coefficients, was presented. The main results of the conducted research were as follows:

(1)   Three characteristic time intervals for the behavior of seismic noise were identified: 1997–2003, 2004–2015 and 2016–2022.

(2)   For these time intervals, there was a parallel decrease in the average values of the DJ-index (interpreted as a simplification of the structure of seismic noise and an approximation of its properties to white noise) and an increase in average spatial correlations. According to general ideas about the behavior of complex systems, this means that the system is approaching a critical state, i.e., an increase in the probability of strong earthquakes.

(3)   After 2011 (after the Tohoku mega-earthquake in Japan on 11 March 2011), there was an explosive increase in the maximum distances at which strong correlations occur. Such behavior is known in the theory of critical phenomena as "an increase in the radius of fluctuation correlations" [26], which also heralds a sharp change in the state of complex systems.

(4)   The estimation of the correlation function between the logarithm of the released seismic energy and the average response of seismic noise to the irregularity of the Earth's rotation shifted towards negative time delays, which corresponds to the shift of the coherence maxima towards the peaks of the seismic-energy release. The maximum correlation fell on a time shift of 534 days. This estimate assumes that the burst of LOD response corresponding to the data analysis in 2020–2022 precedes a strong earthquake with an average delay of 1.5 years.

(5)   From the estimation of the change in time of the synchronization measure of the seismic-noise response to the irregularity of the Earth's rotation (Figure 13), one can also see a growth of correlations, similar to the growth of the initial correlations in Figure 6.

It should be noted that the noise level in real observations of the seismic background is inevitably extremely high. Transition to the low-frequency part of the spectrum allows the elimination of most of the man-made noise. Moreover, this transition makes it possible to study that part of the spectrum that is in the intermediate, little-studied region of periods between traditional seismology and gravimetry. The author's previous studies give hope that it is in this region of periods that the effects preceding strong earthquakes can be detected. This work is presented as a further development of these methods. The complexity of the data and processes underlying the formation of low-frequency seismic noise in the Earth's crust does not allow direct use of conventional statistical procedures for estimating confidence intervals, standard deviations and other measures of uncertainty. It is possible that further study of the statistical structure of low-frequency seismic noise will enable us to build such models that will make it possible to estimate the measures of uncertainty of the obtained statistical inferences, for example, using the bootstrap technique.

Funding: This research received no external funding.

Data Availability Statement: Open-access data were used from the sources: http://www.iris.edu/mda/_GSN, http://www.iris.edu/mda/G, http://www.iris.edu/mda/GE, https://hpiers.obspm.fr/iers/eop/eopc04/eopc04.1962-now and https://earthquake.usgs.gov/earthquakes/search/. All data accessed on 1 January 2023.

Appl. Sci. **2023**, *12*, 6958

Acknowledgments: This work was carried out within the framework of the state assi Institute of Physics of the Earth of the Russian Academy of Sciences (topic FMWU-2022-0018).

Conflicts of Interest: The authors declare no conflict of interest.

References

1. Lyubushin, A. Prognostic properties of low-frequency seismic noise. *Nat. Sci.* **2012**, *4*, 659–666. [CrossRef]
2. Lyubushin, A. Synchronization of Geophysical Fields Fluctuations. In *Complexity of Seismic Time Series: Measurement and Applications*; Chelidze, T., Telesca, L., Vallianatos, F., Eds.; Elsevier: Amsterdam, The Netherlands; Oxford, UK; Cambridge, MA, USA, 2018; Chapter 6; pp. 161–197. [CrossRef]
3. Lyubushin, A. Investigation of the Global Seismic Noise Properties in Connection to Strong Earthquakes. *Front. Earth Sci.* **2022**, *10*, 905663. [CrossRef]
4. Ardhuin, F.; Stutzmann, E.; Schimmel, M.; Mangeney, A. Ocean wave sources of seismic noise. *J. Geophys. Res.* **2011**, *116*, C09004. [CrossRef]
5. Aster, R.; McNamara, D.; Bromirski, P. Multidecadal climate induced variability in microseisms. *Seismol. Res. Lett.* **2008**, *79*, 194–202. [CrossRef]
6. Berger, J.; Davis, P.; Ekstrom, G. Ambient earth noise: A survey of the global seismographic network. *J. Geophys. Res.* **2004**, *109*, B11307. [CrossRef]
7. Friedrich, A.; Krüger, F.; Klinge, K. Ocean-generated microseismic noise located with the Gräfenberg array. *J. Seismol.* **1998**, *2*, 47–64. [CrossRef]
8. Kobayashi, N.; Nishida, K. Continuous excitation of planetary free oscillations by atmospheric disturbances. *Nature* **1998**, *395*, 357–360. [CrossRef]
9. Rhie, J.; Romanowicz, B. Excitation of Earth's continuous free oscillations by atmosphere-ocean-seafloor coupling. *Nature* **2004**, *431*, 552–554. [CrossRef] [PubMed]
10. Tanimoto, T. The oceanic excitation hypothesis for the continuous oscillations of the Earth. *Geophys. J. Int.* **2005**, *160*, 276–288. [CrossRef]
11. Fukao, Y.K.; Nishida, K.; Kobayashi, N. Seafloor topography, ocean infragravity waves, and background Love and Rayleigh waves. *J. Geophys. Res.* **2010**, *115*, B04302. [CrossRef]
12. Koper, K.D.; Seats, K.; Benz, H. On the composition of Earth's short-period seismic noise field. *Bull. Seismol. Soc. Am.* **2010**, *100*, 606–617. [CrossRef]
13. Nishida, K.; Montagner, J.; Kawakatsu, H. Global surface wave tomography using seismic hum. *Science* **2009**, *326*, 112. [CrossRef]
14. Donoho, D.L.; Johnstone, I.M. Adapting to unknown smoothness via wavelet shrinkage. *J. Am. Stat. Assoc.* **1995**, *90*, 1200–1224. [CrossRef]
15. Lyubushin, A. Trends of Global Seismic Noise Properties in Connection to Irregularity of Earth's Rotation. *Pure Appl. Geophys.* **2020**, *177*, 621–636. [CrossRef]
16. Lyubushin, A. Connection of Seismic Noise Properties in Japan and California with Irregularity of Earth's Rotation. *Pure Appl. Geophys.* **2020**, *177*, 4677–4689. [CrossRef]
17. Lyubushin, A. Low-Frequency Seismic Noise Properties in the Japanese Islands. *Entropy* **2021**, *23*, 474. [CrossRef] [PubMed]
18. Lyubushin, A. Global Seismic Noise Entropy. *Front. Earth Sci.* **2020**, *8*, 611663. [CrossRef]
19. Duda, R.O.; Hart, P.E.; Stork, D.G. *Pattern Classification*; Wiley-Interscience Publication: New York, NY, USA; Chichester, UK; Brisbane, Australia; Singapore; Toronto, ON, Canada, 2000.
20. Mallat, S.A. *Wavelet Tour of Signal Processing*, 2nd ed.; Academic Press: San Diego, CA, USA; London, UK; Boston, MA, USA; New York, NY, USA; Sydney, Australia; Tokyo, Japan; Toronto, ON, Canada, 1999.
21. Poli, P.; Shapiro, N.M. Rapid Characterization of Large Volcanic Eruptions: Measuring the Impulse of the Hunga Tonga Ha'apai Explosion from Teleseismic Waves. *Geophys. Res. Lett.* **2022**, *49*, e2022GL098123. [CrossRef]
22. Montelli, R.; Nolet, G.; Dahlen, F.A.; Masters, G. A catalogue of deep mantle plumes: New results from finite-frequency tomography. *Geochem. Geophys. Geosyst.* **2006**, *7*, Q11007. [CrossRef]
23. Shanker, D.; Kapur, N.; Singh, V. On the spatio temporal distribution of global seismicity and rotation of the Earth—A review. *Acta Geod. Geophys. Hung.* **2001**, *36*, 175–187. [CrossRef]
24. Bendick, R.; Bilham, R. Do weak global stresses synchronize earthquakes? *Geophys. Res. Lett.* **2017**, *44*, 8320–8327. [CrossRef]
25. Xu, C.; Sun, W. Co-seismic Earth's rotation change caused by the 2012 Sumatra earthquake. *Geod. Geodyn.* **2012**, *3*, 28–31. [CrossRef]
26. Gilmore, R. *Catastrophe Theory for Scientists and Engineers*; John Wiley and Sons, Inc.: New York, NY, USA, 1981.

Article

The 2022 Seismic Sequence in the Northern Adriatic Sea and Its Long-Term Simulation

Rodolfo Console [1,2,†], Paola Vannoli [1,*,†] and Roberto Carluccio [1,†]

1   Istituto Nazionale di Geofisica e Vulcanologia, 00143 Rome, Italy
2   Center of Integrated Geomorphology for the Mediterranean Area, 85100 Potenza, Italy
*   Correspondence: paola.vannoli@ingv.it
†   These authors contributed equally to this work.

Abstract: We studied the long-term features of earthquakes caused by a fault system in the northern Adriatic sea that experienced a series of quakes beginning with two main shocks of magnitude 5.5 and 5.2 on 9 November 2022 at 06:07 and 06:08 UTC, respectively. This offshore fault system, identified through seismic reflection profiles, has a low slip rate of 0.2–0.5 mm/yr. As the historical record spanning a millennium does not extend beyond the inter-event time for the largest expected earthquakes ($M \simeq 6.5$), we used an earthquake simulator to generate a 100,000-year catalogue with 121 events of $M_w \geq 5.5$. The simulation results showed a recurrence time (T_r) increasing from 800 yrs to 1700 yrs as the magnitude threshold increased from 5.5 to 6.5. However, the standard deviation σ of inter-event times remained at a stable value of 700 yrs regardless of the magnitude threshold. This means that the coefficient of variation ($C_v = \sigma / T_r$) decreased from 0.9 to 0.4 as the threshold magnitude increased from 5.5 to 6.5, making earthquakes more predictable over time for larger magnitudes. Our study supports the use of a renewal model for seismic hazard assessment in regions of moderate seismicity, especially when historical catalogues are not available.

Keywords: numerical modelling; earthquake simulator; statistical methods; earthquake clustering; northern Adriatic sea

Citation: Console, R.; Vannoli, P.; Carluccio, R. The 2022 Seismic Sequence in the Northern Adriatic Sea and Its Long-Term Simulation. *Appl. Sci.* **2023**, *13*, 3746. https://doi.org/10.3390/app13063746

Academic Editor: José A. Peláez

Received: 17 February 2023
Revised: 7 March 2023
Accepted: 14 March 2023
Published: 15 March 2023

1. Introduction

In regions of moderate seismicity, the recurrence times of damaging earthquakes are commonly longer than the time windows covered by reliable historical observations, even in Italy, which owns one of the longest records of past strong seismic events. These limitations of historical seismic data preclude the application of statistical analysis for seismic hazard assessment, even if they can be partly overcome by geological and paleoseismological investigations on specific faults. The information on past earthquakes is particularly poor for offshore seismic activity (where paleoseismological and geomorphological studies are not feasible), as is the case for the thrust fault systems located seaside of the northern Marche coast (central Italy), recently affected by a seismic sequence that has significantly concerned the population in a large area of central Italy (Istituto Nazionale di Geofisica e Vulcanologia website at http://terremoti.ingv.it; main event of the sequence at http://terremoti.ingv.it/en/event/33301831, accessed on 14 February 2023; ISIDe_Working_Group [1], Tertulliani et al. [2].

The use of earthquake simulators has become increasingly popular in generating synthetic earthquake catalogues with a vast number, even reaching millions, of events. This allows for statistical analyses to be conducted on simulated catalogues that are considerably more reliable than those conducted on real catalogues. However, criticism has been expressed regarding the practicality of simulated catalogues. Certain seismologists have pointed out that the algorithms used in earthquake simulators are based on oversimplified physical models and contain arbitrary assumptions, which pose significant challenges for creating a dependable representation of actual seismicity. Nevertheless, a wide consensus does exist on the potential of earthquake simulators to provide support for historical observations in the context of seismic hazard assessment, as long as a realistic physical model is available and the results are based on reliable source parameters. Some examples of research in this area include Schultz et al. [3], Christophersen et al. [4], Field [5]. In the frame of the ongoing debate on the usefulness of earthquake simulators, some valuable insights have also been gathered through the application of earthquake simulators to the seismicity of Italy, Greece, California, and Japan (Console et al. [6–9], Parsons et al. [10]). These insights have been gained by comparing the simulated catalogues with real catalogues from the respective regions, and the results have shown that the same algorithms can be applicable in different geographic regions and on different magnitude scales.

In a recent study by Console et al. [6], a new earthquake simulation code was developed, which included an enhanced Coulomb stress transfer among rupturing fault elements and increased production of multiple main shocks, frequently observed in earthquake sources of different mechanisms in Italy. To better understand the clustering behaviour of these events, a specific algorithm was created to detect and list inter-correlated earthquake clusters based on certain empirical rules.

The simulated catalogue produced by this code exhibits spatiotemporal features that imitate those frequently observed in reality. Using a stacking procedure, Console et al. [6] were able to represent the number of moderate-magnitude events that preceded and followed stronger earthquakes in a 100,000-year simulated catalogue. Their results for the central–northern Apennines region showed interesting long-term acceleration and quiescence patterns of seismic activity before and after mainshocks. They also analysed short-term patterns in periods of some weeks preceding and following every strong earthquake, confirming the simulator code's ability to reproduce typical foreshock–aftershock sequences.

In addition, the simulated catalogue for the central–northern Apennines region showed a decrease of b-values lasting some weeks before the occurrence of strong earthquakes, followed by a sudden increase at the time of these earthquakes. This pattern was previously observed by Gulia and Wiemer [11] in actual earthquake sequences. The fault model used in the simulations by Console et al. [6] consists of 43 fault systems divided into 198 quadrilaterals, which provide a suitable approximation of the seismic structures reported in the Database of Individual Seismogenic Sources (DISS) composite sources (DISS_Working_Group [12]; http://diss.rm.ingv.it/diss/, accessed on 14 February 2023).

Among these fault systems, the one labelled as ITCS106 (a NW–SE elongated fault system 70 km long in the Adriatic sea at about 25 km from the Italian coast) is believed to be responsible for the recent seismic sequence, which started with two relatively similar large mainshocks on 9 November 2022 at 06:07:25 UTC (M_w5.5) and 06:08:28 (M_l5.2), respectively. According to the definition adopted by Console et al. [6] this sequence, with two main earthquakes of similar magnitude, whose origin times differ by only one minute, can be retained a multiple shock.

No earthquakes from A.D. 1000 to 2020, with $M_w \geq 5.5$, are reported in the Parametric Catalogue of Italian Earthquakes (CPTI15; https://emidius.mi.ingv.it/CPTI15-DBMI15/, accessed on 14 February 2023) within a 5 km buffer from the ITCS106 fault system. The lack of historical and instrumental events pertinent to the ITCS106 earthquake source is consistent with the low slip rate of this source, as will be shown in more detail in the following sections. In this paper, we present a case study of application of an earthquake simulator as a contribution to seismic hazard assessment in an area of moderate seismicity, for which historical catalogues are not sufficient for this purpose.

2. Seismotectonic Settings

The northern Marche coastal belt and the Adriatic offshore area are characterised by a series of NW–SE trending, NE verging folds forming the easternmost edge of the Apennines thrust front (Figure 1; e.g., DISS_Working_Group [12], Vannoli et al. [13], Fantoni and

Franciosi [14], Kastelic et al. [15]). This Apennines thrust front has progressively migrated toward the east-northeast all through the Tertiary-Quaternary (e.g., Elter et al. [16]) and several geological evidence suggest that the folds are still growing and hence that the blind thrust fronts are active (e.g., Vannoli et al. [13]). Especially in the offshore area the interpretation of seismic reflection profiles is strategic to understanding the geometry of the active faults (e.g., Maesano et al. [17]).

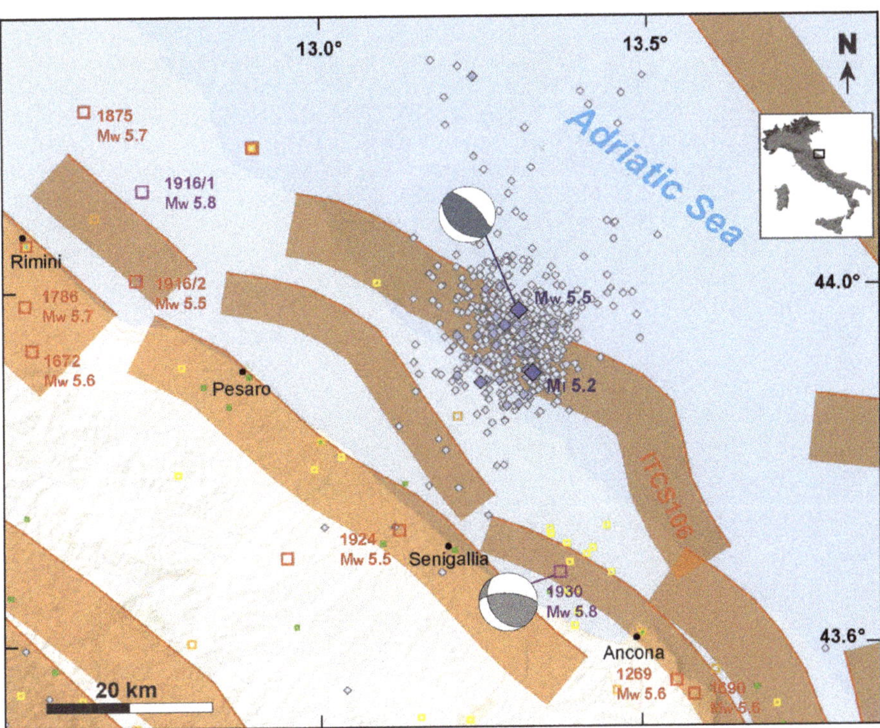

Figure 1. Seismotectonic framework of the northern Marche coastal and offshore area. The seismic sequence from 9 November to 14 February 2023 is shown in light blue (http://terremoti.ingv.it/, accessed on 14 February 2023); historical and instrumental earthquakes from CPTI15 are shown with coloured squares (Rovida et al. [18]), earthquakes with $M_w \geq 5.5$ are labelled (see Table 1); surface projection of Composite Seismogenic Sources from DISS 3.3.0 are shown with orange ribbons (https://diss.ingv.it, accessed on 14 February 2023; DISS_Working_Group [12]). Focal mechanisms of the 9 November 2022 earthquake and the 30 October 1930 event are from TDMT (http://webservices.ingv.it/webservices/ingv_ws_map/data/tdmt/19591/111604601_15 4_tdmt_reviewer_solution.pdf, accessed on 14 February 2023), and Vannoli et al. [19]), respectively.

The occurrence of historical and instrumental earthquakes (e.g., 1672, 1690, 1786, 1875, 1916, 1924, 1930, all with $Mw \geq 5.5$; Table 1; Figure 1) suggests that the thrust faults are also seismogenic. The earthquakes in the area have often generated a tsunami; subsequently, their causative faults have been able to produce significant vertical displacement of the ground surface or sea floor, and are very close to the coast or offshore (Table 1; Vannoli et al. [19]).

A relevant aspect of seismic activity in the study area, as in the rest of Italy (e.g., Vannoli et al. [20]), is the occurrence of seismic sequences with more than one main shock of similar magnitude (Table 1). These sequences can be distinguished from typical aftershock sequences because the difference between the largest and second-largest magnitudes is relatively small and does not follow the so-called "Bath Law" (see Console et al. [7]).

A seismic sequence with two events of similar magnitude also began on 9 November 2022. As a matter of fact, at 06:07:25 UTC a $M_w 5.5$ earthquake struck the northern Marche offshore area, and only one minute later, a $M_l 5.2$ earthquake struck about 7 km to the south (http://terremoti.ingv.it, accessed on 14 February 2023; Figure 1; Table 1). According to the hypocentral location of these two mainshocks, the ITCS106 fault system is retained the causative source of the sequence. More than 900 aftershocks occurred offshore Pesaro and Senigallia from 9 November 2022 to 14 February 2023 (Figure 1).

The occurrence of this sequence demonstrates that the buried offshore thrust front, included in the Database of Individual Seismogenic Sources since 2013 despite the absence of historical and recent seismicity (DISS_Working_Group [12]), is active and seismogenic. Specifically, the ITCS106 Pesaro mare-Cornelia source straddles the Adriatic sea just north of the city of Ancona, and it is part of the Umbro-Marche Apennines outer offshore thrust. ITCS106 is a blind fault system believed capable of infrequent moderate-size earthquakes. Its slip rate, a main ingredient of the earthquake simulators and seismic hazard model, is in the range of 0.20–0.52 mm/yr (DISS_Working_Group [12] and reference therein). Therefore, ITCS106 can produce only subtle displacement of the Adriatic sea floor, even when it is accumulated over several seismic cycles.

Table 1. Historical and instrumental earthquakes with $M_w \geq 5.5$ of the study area (Figure 1; CPTI15; Rovida et al. [18]; CFTI5Med; Guidoboni et al. [21]), and tsunami with their intensity (EMTC; Maramai et al. [22]). T: true; SA: Intensity Sieberg–Ambraseys scale; PI: Intensity Papadopoulos and Imamura scale; * unspecified day; ** the earthquake triggered a large landslide that fell into the sea on the eastern side of Monte Conero (CFTI5Med [21]); *** some historical sources describe the occurrence of a second large earthquake one hour after the first event (CFTI5Med [21]); **** EMS-98 scale; Tertulliani et al. [2].

ID	Date	$\mathbf{M_w}$	$\mathbf{I_o}$	Multiple	Associated Tsunami with Intensity
1	September 1269 *	5.6	VIII	n.a.	Generated by a landslide **
2	14 April 1672	5.6	VIII	n.a.	SA:3; PI:IV
3	23 December 1690	5.6	VIII	n.a.	SA:3; PI:IV
4	25 December 1786	5.7	VIII	T ***	No record
5	17 March 1875	5.7	VIII	n.a.	SA:3; PI:IV
6	17 May 1916	5.8	VIII	T	No record
7	16 August 1916	5.5	VI	T	SA:2; PI:IV
8	2 January 1924	5.5	VII-VIII	n.a.	No record
9	30 October 2022	5.5	V ****	T	No record

3. Earthquake Simulation Method

Our simulator code's algorithm was originally introduced by Console et al. [8], and subsequently modified in different versions with the inclusion of new features, as described by Console et al. [7,9]. The present study employs the following fundamental principles of the simulator algorithm:

1. The simulation of seismic sources involves the use of planar quadrilateral fault segments with specific spatial position, shape, and size. To accurately represent the various seismic events that may occur, each fault segment is discretized by square cells with sizes determined by the minimum magnitude of events expected in the output simulated catalogue.
2. At the start of the simulation, stress values are randomly assigned to each cell from a specified stress range. This randomization is essential to ensure that the simulation represents the natural variation in stress levels that exists in real-life seismic events.
3. Every fault segment is subjected to constant and uniform tectonic stress loading. This loading is based on geologically or geodetically inferred slip rate, providing the simulation with a more accurate representation of the tectonic stresses that occur in real seismic events.

4. The simulation involves the nucleation of a new event in a cell when its stress exceeds a given threshold strength. This threshold strength is essential to accurately model the conditions that trigger seismic events in reality.

5. Whenever a cell ruptures, a co-seismic stress-drop of 3.3 MPa is assigned to it. This stress-drop is a crucial aspect of the simulation as it represents the stress released by the rupturing cell, which is a significant factor in the generation of seismic events.

6. Following the rupture of a cell, the co-seismic Coulomb stress is transferred from the rupturing cell to all other neighbouring cells. This transfer occurs based on the theory of elasticity, taking into account the strike, dip, and rake of each cell.

7. The rupture expands to neighbouring cells based on heuristic rules that simulate a weakening mechanism. These rules are based on the behaviour of real-life seismic events and are essential to accurately represent the spread of a seismic event.

8. If the positive Coulomb stress transferred from neighbouring cells causes a cell's status to exceed the threshold strength, it is allowed to rupture more than once in the same earthquake. This is an important aspect of the simulation as it allows for the accurate representation of aftershocks and the complex nature of seismic events.

9. The simulation algorithm allows ruptures to jump from one fault segment to another (even if they belong to two different faults) if their distance is shorter than a limit assigned by the user. This feature is essential to accurately represent the interaction of seismic events and the way in which they can trigger each other.

10. The rupture stops when the stress of no neighbouring cell exceeds the computed strength threshold. This feature is important to ensure that the simulation accurately represents the natural behaviour of seismic events, which eventually reach a point where no further energy can be released.

The reason for choosing a trapezoidal shape for the seismic sources is to achieve a more precise representation of curved seismic structures. For instance, this helps to reduce gaps that may exist between adjacent rectangular segments when they do not have the same strike. It should be noted that using fault segments of rectangular or trapezoidal shapes to model seismic sources constitutes a simplification of the algorithm adopted in our physics-based simulation code. It is not a constraint that implies a rupture stops at the edges of every segment.

The simulator algorithm has two free parameters that control how a rupture nucleates, propagates, and stops. The first of these parameters is "Strength Reduction" (S-R), which reduces the strength at the edges of growing ruptures through a sort of weakening mechanism. Increasing this parameter favors the growth of ruptures and, consequently, it has the effect of decreasing the b-value of the frequency-magnitude distribution. The second parameter is "Aspect Ratio" (A-R), which is introduced to limit the effect of the S-R parameter when a rupture grows beyond certain specified limits. This parameter assumes a significant role in the frequency-magnitude distribution in the range of large magnitudes, as shown in Figure 5 of Console et al. [8].

It is important to note that neither of the free parameters mentioned above influences the long-term annual seismic moment rate of the simulated seismicity, which depends only on the slip rate assigned to the simulator within the kinematic fault model. In this study, we analysed the simulations computed by Console et al. [6] with the free parameters S-R = 0.2 and A-R = 5, obtaining a synthetic catalogue characterised by a b-value of the Gutenberg–Richter distribution equal to 0.95 and a realistic production of multiple events.

4. Simulation of the Seismicity in the Adriatic Thrust Fault Systems

In the application of the simulator, the quadrilateral segments representing the ITCS106 fault system were discretized in cells of 1.0 km × 1.0 km. As to the slip rate, we adopted the value of 0.5 mm/year corresponding to the top value of the range reported by the DISS 3.3.0 database for this source. We found in previous applications of the simulator to central and northern Apennines Console et al. [6,7,9] that the largest slip rate is the one that better matches the annual rate of seismic moment obtained from instrumental observations.

We chose for the synthetic catalogues a minimum magnitude of 4.2, which is produced approximately by the rupture of two cells. Magnitude 4.3 is not present in the output catalogue because the rupture of three cells produces an event of magnitude 4.4. The duration of the synthetic catalogue was 100,000 years. A warm up period, the events of which are not included in the output catalogue, is necessary to bring the system in a stable situation. This period should be longer than a few inter-event times of the strongest magnitude. In this regard, we chose a warm up period of 10,000 years.

In this study, still taking into account the interactions from other sources, we limited our analysis to the above-mentioned ITCS106 source in the Adriatic sea, where the November–February 2023 earthquake sequence took place. At first, for this source we considered the magnitude–frequency distribution. Figure 2 shows a clear deviation of the magnitude–frequency distribution obtained by our simulation algorithm from a plain linear Gutenberg–Richter distribution. In fact, the incremental magnitude–frequency distribution shows a fairly linear trend with a b-value slightly greater than 1.0 up to magnitude 5.5, but it exhibits a sharp "bump" around magnitude 6.5, This value could be defined as "characteristic" for the considered fault system, and is related to its dimensions. It can also be noted that the cumulative magnitude distribution of the synthetic catalogue is characterised by a very small b-value (i.e., a b-value slightly larger than 0) in the magnitude range 5.5 < M < 6.5.

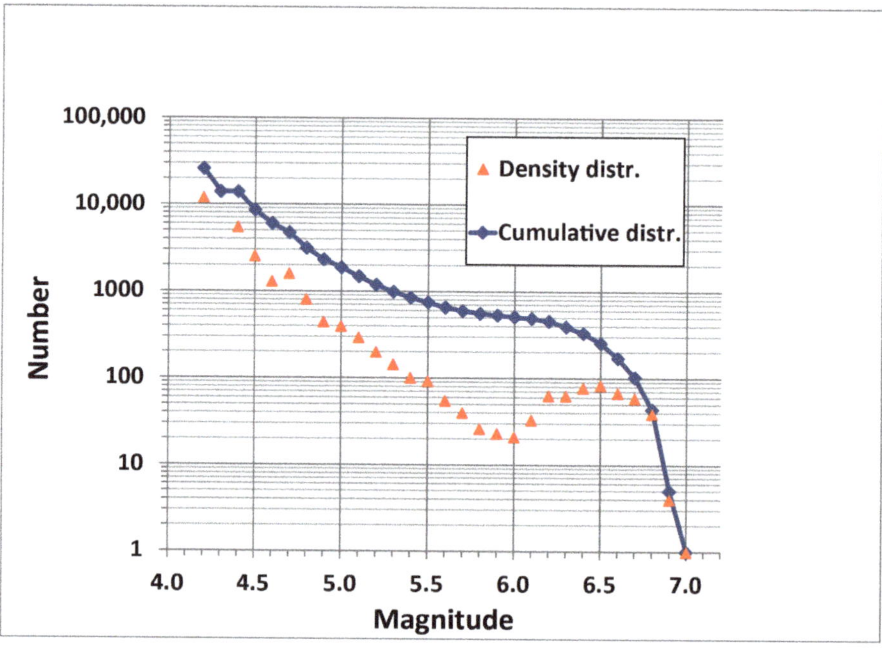

Figure 2. Incremental magnitude–frequency distribution of the earthquakes in the synthetic catalogue obtained from the simulation algorithm (triangles) and the corresponding cumulative distribution (diamonds).

Our results are consistent with those of other studies where the magnitude–frequency distribution is analysed on individual faults, ignoring the contribution of smaller surrounding sources. For instance, Parsons et al. [10], by means of various statistical methods, demonstrated strongly characteristic magnitude–frequency distributions on the San Andreas fault, with a nearly flat cumulative distribution in the 6.5–7.5 magnitude range, and higher rates of large earthquakes than what would be expected from a Gutenberg–Richter distribution. The difference of approximately 1 magnitude unit between the results obtained for the ITCS106 fault system (about 70 km long) and the Saint Andreas fault (several hundreds of km long) is easily justified by their different dimensions.

We focused our attention on the inter-event times between events for the simulated catalogue of 100,000 yrs for earthquakes of magnitude equal to or larger than magnitudes 5.5, 6.0 and 6.5, respectively. The smallest size of 5.5 was the magnitude threshold adopted for the pivot earthquake of a multiple event by Console et al. [6]. According to popular scaling relationships (Leonard [23]) magnitude 6.5 is expected for the largest size of an earthquake that may rupture the entire area of the ITCS106 fault system. Table 2 reports the results of the statistical analysis carried out on the simulated catalogue, and Figure 3 shows the number of inter-event times in bins of 250 yrs.

Table 2. Results of the statistical analysis of the simulated 100,000 yrs earthquake catalogue for the DISS composite source labelled ITCS106 for three different magnitudes thresholds. T_r = recurrence time, σ = standard deviation, C_v = coefficient of variation.

Magnitude Threshold	Number of Events	T_r(yrs)	σ(yrs)	C_v
5.5	121	814	733	0.90
6.0	84	1170	797	0.68
6.5	57	1713	690	0.40

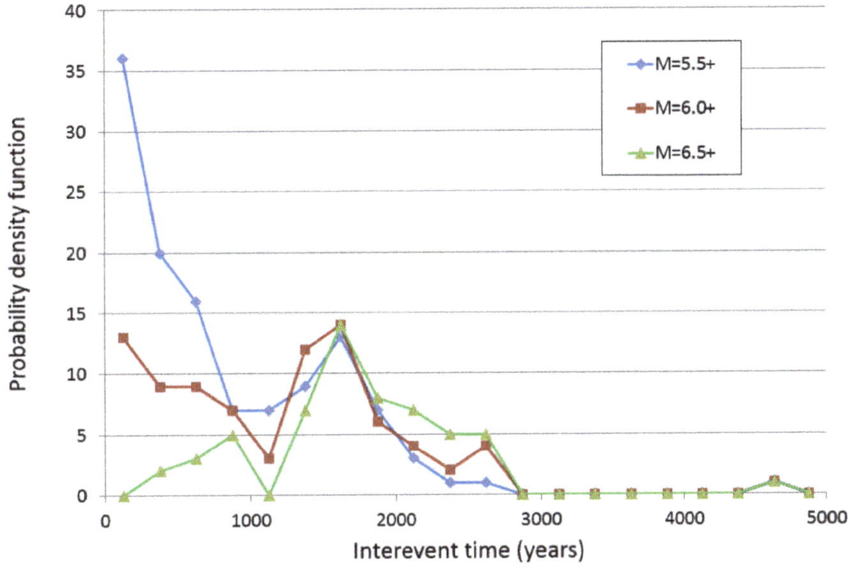

Figure 3. Density distributions of the number of inter-event times in bins of 250 yrs. The first bin corresponds to the range of 0–250 yrs and so on.

The plot of Figure 3 shows a bi-modal distribution of inter-event times with a local minimum at 1000–1250 yrs. Smaller inter-event times are pertinent to clustering (with particular regard to small magnitude earthquakes), while the maximum at 1500–1750 yrs denotes quasi-periodicity (typical of larger magnitude earthquakes).

As expected, the number of events decreases and the average inter-event time T_r increases when the magnitude threshold is increased (Table 2). However, the standard deviation σ maintains a value close to 700 yrs. In this way, the coefficient of variation (C_v), which is defined by the ratio between σ and T_r, decreases from 0.90 for $M \geq 5.5$ to 0.40 for $M \geq 6.5$. It is well known that C_v is equal to 1.0 for a sequence of events following a memoryless Poisson distribution and is equal to 0 for a perfectly periodical sequence.

Thus, the earthquake sequence of our simulated catalogue behaves more and more as a quasi-periodical process when, increasing the threshold magnitude of the analysis, the ruptured areas of the considered earthquakes approach the size of whole seismic structure.

Shortest inter-event times (smaller than 1000 yrs) are typical of small magnitude earthquakes (Table 2, and Figure 3), while the opposite situation happens for the inter-event times larger than 2000 years (Figure 3). An exceptionally long inter-event time of 4625 yrs is estimated between two earthquakes of magnitude larger than 6.5 (Figure 3).

The simulations analysed in this study were obtained adopting the largest slip rates reported for the earthquake source in the DISS database. If we adopted smaller slip rates, the time scale of our inter-event times would become inversely longer.

The synthetic 100,000-year catalogue highlights a spatiotemporal behaviour that can hardly be compared with analog features observed in reality, due to their brevity, but can be considered realistic in light of an earthquake cycle hypothesis.

Long-term patterns preceding and following a strong earthquake are shown in Figure 4. This figure is obtained by stacking the number of $M \geq 4.2$ earthquakes before and after an $M \geq 5.2$ earthquake within a distance of 20 km between their respective epicenters in bins of 5 years. Here, we can observe an increase in seismic activity starting a couple of centuries before a strong earthquake, and a sharp increase in occurrence rate in the first 5 years after that event. The latter feature is clearly related to aftershock sequences following strong events. After this aftershock phase, the plot shows a long-lasting quiescence with some modest fluctuations. This long-term pattern highlights the existence of a preparatory phase before every strong earthquake. Presumably, during this preparatory phase, stress accumulation causes an increase in the rate of moderate-magnitude events. This hypothesis is supported by a detailed study of the time history of the average stress and its standard deviation during several seismic cycles in the Corinth (Greece) and Nankai (Japan) fault systems (Console and Carluccio [24], Console et al. [25]).

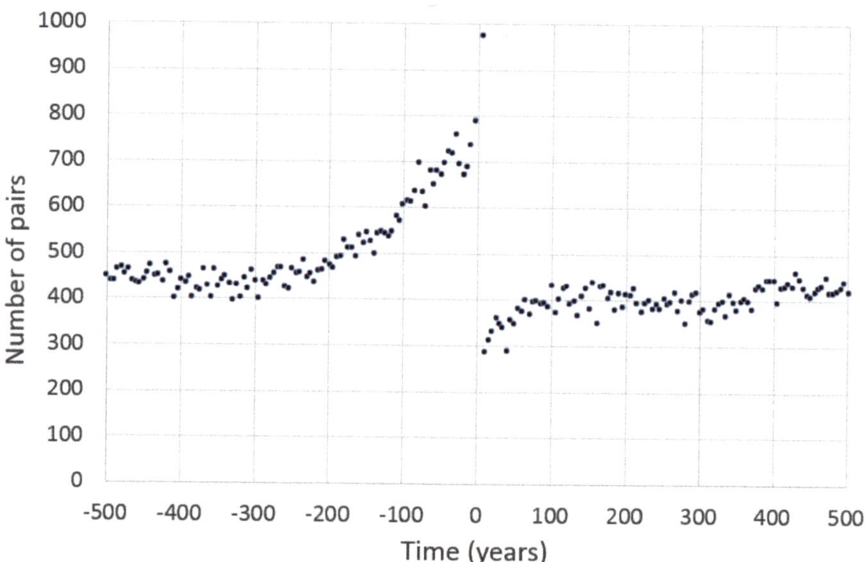

Figure 4. Stacked number of $M \geq 4.2$ earthquakes that preceded and followed an $M \geq 5.2$ earthquake within an epicentral distance of 20 km in the 100,000 years simulated catalogue.

5. Conclusions

A seismic sequence initiated by a pair of mainshocks of magnitude 5.5 and 5.2 took place in the Northern Adriatic sea on 9 November 2022 at 06:07 UTC. The two mainshocks were separated by only one minute in time and about 10 km in space (Figure 1). The sequence is believed to be generated by the ITCS106 seismogenic source of the DISS database, belonging to the series of NW–SE trending, NE verging folds forming the easternmost edge of the Apennines thrust front (Figure 1).

Due to its low slip rate, estimated between 0.2 and 0.5 mm/yr in the DISS database, the ITCS106 source is characterised by moderate seismic activity, so as no earthquakes with $M_w \geq 5.5$ were reported in the Parametric Catalogue of Italian Earthquakes from 1000 to 2020, before the two mainshocks of 9 November 2022. The occurrence of these earthquakes supports the hypothesis that even sources of low slip rate identified and characterised in the DISS database, for which little or no historical information is reported in the historical catalogues, may have seismic potential to be considered relevant in the context of seismic hazard assessment.

To quantify this hypothesis, we performed a statistical analysis of the results obtained by a previously developed simulator algorithm (Console et al. [6]) for the ITCS106 fault system. The results of our simulation consist of a catalogue of 100,000 yrs duration, containing 121, 84 and 57 earthquakes of magnitude equal to or exceeding a magnitude of 5.5, 6.0 and 6.5, respectively. The statistical analysis of the synthetic catalogue, shortly reported in Table 2, puts in evidence that the earthquakes belonging to the lowest magnitude class have an average inter-event time of 814 ± 733 yrs, while those belonging to the highest magnitude class have an average inter-event time of 1713 ± 690 yrs. Consequently, the simulated catalogue exhibits a nearly memoryless Poissonian behaviour for moderate-magnitude earthquakes (whose coefficient of variation is 0.9), and a time-dependent quasi-periodic behaviour for large-magnitude earthquakes (whose coefficient of variation is 0.4). The statistical features obtained by the application of our earthquake simulator appear consistent with the hypothesis of the earthquake cycle. In this respect, our work may be relevant for possible applications related to the recent method of earthquake nowcasting (Rundle et al. [26,27], Varotsos et al. [28], Christopoulos et al. [29]), in which interoccurrence natural time between strong earthquakes is used for the estimation of the current stage of the earthquake cycle (see e.g., Varotsos et al. [30,31]).

Another significant aspect of our results is the acceleration of seismic activity before strong earthquakes lasting a couple of centuries in the ITCS106 seismic source. We conclude by suggesting that statistical parameters derived by physics-based earthquake simulators, even recognizing the limitations connected with the extreme simplicity of simulators, could be of support for the comprehension and modelling of seismic processes in situations where there is a lack of suitable observations for earthquake hazard assessment.

Author Contributions: Conceptualization, R.C. (Rodolfo Console), P.V. and R.C. (Roberto Carluccio); Validation, R.C. (Rodolfo Console) and P.V.; Data curation, R.C. (Roberto Carluccio); Writing—review & editing, R.C. (Rodolfo Console), Paola Vannoli and R.C. (Roberto Carluccio). All authors have read and agreed to the published version of the manuscript.

Funding: This research received no external funding.

Data Availability Statement: The Italian Parametric Earthquake Catalogue (CPTI15) can be found at https://emidius.mi.ingv.it/CPTI15-DBMI15/ (accessed on 14 February 2023), the Database of Individual Seismogenic Sources (DISS) can be found at http://diss.ingv.it (accessed on 14 February 2023).

Acknowledgments: The authors are grateful to Jörg Renner, Eleftheria Papadimitriou and anonymous reviewers for their careful reading of the manuscript and useful suggestions.

Conflicts of Interest: The authors declare no conflict of interest.

Abbreviations

The following abbreviations are used in this manuscript:

CPTI	The Italian Parametric Earthquake Catalogue
DISS	The Database of Individual Seismogenic Sources
EMTC	Euro-Mediterranean Tsunami Catalogue
CFTI	Strong Italian Earthquakes Catalogue

References

1. ISIDe_Working_Group. *Italian Seismological Instrumental and Parametric Database (ISIDe)*; Istituto Nazionale di Geofisica e Vulcanologia: Roma, Italy, 2007. [CrossRef]
2. Tertulliani, A.; Antonucci, A.; Berardi, M.; Borghi, A.; Brunelli, G.; Caracciolo, C.H.; Castellano, C.; D'Amico, V.; Del Mese, S.; Ercolani, E.; et al. Gruppo Operativo Quest Rilievo Macrosismico MW 5.5 Costa Marchigiana del 9/11/2022 Rapporto Finale del 15/11/2022. Available online: https://www.earth-prints.org/handle/2122/15794 (accessed on 23 December 2022).
3. Schultz, K.W.; Sachs, M.K.; Yoder, M.R.; Rundle, J.B.; Turcotte, D.L.; Heien, E.M.; Donnellan, A. Virtual quake: Statistics, co-seismic deformations and gravity changes for driven earthquake fault systems. In Proceedings of the International Symposium on Geodesy for Earthquake and Natural Hazards (GENAH), Matsushima, Japan, 22–26 July 2014; Springer: Cham, Switzerland, 2017; pp. 29–37.
4. Christophersen, A.; Rhoades, D.; Colella, H. Precursory seismicity in regions of low strain rate: Insights from a physics-based earthquake simulator. *Geophys. J. Int.* **2017**, *209*, 1513–1525. [CrossRef]
5. Field, E.H. How physics-based earthquake simulators might help improve earthquake forecasts. *Seismol. Res. Lett.* **2019**, *90*, 467–472. [CrossRef]
6. Console, R.; Vannoli, P.; Carluccio, R. Physics-Based Simulation of Sequences with Foreshocks, Aftershocks and Multiple Main Shocks in Italy. *Appl. Sci.* **2022**, *12*, 2062. [CrossRef]
7. Console, R.; Murru, M.; Vannoli, P.; Carluccio, R.; Taroni, M.; Falcone, G. Physics-based simulation of sequences with multiple main shocks in Central Italy. *Geophys. J. Int.* **2020**, *223*, 526–542. [CrossRef]
8. Console, R.; Carluccio, R.; Papadimitriou, E.; Karakostas, V. Synthetic earthquake catalogs simulating seismic activity in the Corinth Gulf, Greece, fault system. *J. Geophys. Res. Solid Earth* **2015**, *120*, 326–343. [CrossRef]
9. Console, R.; Vannoli, P.; Carluccio, R. The seismicity of the Central Apennines (Italy) studied by means of a physics-based earthquake simulator. *Geophys. J. Int.* **2018**, *212*, 916–929. [CrossRef]
10. Parsons, T.; Geist, E.L.; Console, R.; Carluccio, R. Characteristic earthquake magnitude frequency distributions on faults calculated from consensus data in California. *J. Geophys. Res. Solid Earth* **2018**, *123*, 10–761. [CrossRef]
11. Gulia, L.; Wiemer, S. Real-time discrimination of earthquake foreshocks and aftershocks. *Nature* **2019**, *574*, 193–199. [CrossRef]
12. DISS_Working_Group. *Database of Individual Seismogenic Sources (DISS), Version 3.3.0: A Compilation of Potential Sources for Earthquakes Larger Than M 5.5 in Italy and Surrounding Areas*; Istituto Nazionale di Geofisica e Vulcanologia: Roma, Italy, 2021. Available online: http://diss.rm.ingv.it/diss/ (accessed on 14 February 2023). [CrossRef]
13. Vannoli, P.; Basili, R.; Valensise, G. New geomorphic evidence for anticlinal growth driven by blind-thrust faulting along the northern Marche coastal belt (central Italy). *J. Seismol.* **2004**, *8*, 297–312. [CrossRef]
14. Fantoni, R.; Franciosi, R. Tectono-sedimentary setting of the Po Plain and Adriatic foreland. *Rend. Lincei* **2010**, *21*, 197–209. [CrossRef]
15. Kastelic, V.; Vannoli, P.; Burrato, P.; Fracassi, U.; Tiberti, M.M.; Valensise, G. Seismogenic sources in the Adriatic Domain. *Mar. Pet. Geol.* **2013**, *42*, 191–213. [CrossRef]
16. Elter, P.; Giglia, G.; Tongiorgi, M.; Trevisan, L. Tensional and Compressional Areas in the Recent (Tortonian to Present) Evolution of the Northern Apennines. *Bollettino di Geofisica Teorica ed Applicata* **1975**, *17*, 3–18.
17. Maesano, F.E.; Toscani, G.; Burrato, P.; Mirabella, F.; D'Ambrogi, C.; Basili, R. Deriving thrust fault slip rates from geological modeling: Examples from the Marche coastal and offshore contraction belt, Northern Apennines, Italy. *Mar. Pet. Geol.* **2013**, *42*, 122–134. [CrossRef]
18. Rovida, A.N.; Locati, M.; Camassi, R.D.; Lolli, B.; Gasperini, P.; Antonucci, A. *Catalogo Parametrico dei Terremoti Italiani CPTI15, Versione 4.0*; Istituto Nazionale di Geofisica e Vulcanologia (INGV): Roma, Italy, 2022. [CrossRef]
19. Vannoli, P.; Vannucci, G.; Bernardi, F.; Palombo, B.; Ferrari, G. The source of the 30 October 1930 M w 5.8 Senigallia (Central Italy) earthquake: A convergent solution from instrumental, macroseismic, and geological data. *Bull. Seismol. Soc. Am.* **2015**, *105*, 1548–1561. [CrossRef]
20. Vannoli, P.; Bernardi, F.; Palombo, B.; Vannucci, G.; Console, R.; Ferrari, G. New constraints shed light on strike-slip faulting beneath the southern Apennines (Italy): The 21 August 1962 Irpinia multiple earthquake. *Tectonophysics* **2016**, *691*, 375–384. [CrossRef]
21. Guidoboni, E.; Ferrari, G.; Mariotti, D.; Comastri, A.; Tarabusi, G.; Sgattoni, G.; Valensise, G. *CFTI5Med, Catalogo dei Forti Terremoti in Italia (461 aC-1997) e nell'area Mediterranea (760 aC-1500)*; Istituto Nazionale di Geofisica e Vulcanologia (INGV): Rome, Italy, 2018. [CrossRef]
22. Maramai, A.; Graziani, L.; Brizuela, B. *Euro-Mediterranean Tsunami Catalogue (EMTC), Version 2.0*; Istituto Nazionale di Geofisica e Vulcanologia (INGV): Rome, Italy, 2019. [CrossRef]
23. Leonard, M. Self-consistent earthquake fault-scaling relations: Update and extension to stable continental strike-slip faults. *Bull. Seismol. Soc. Am.* **2014**, *104*, 2953–2965. [CrossRef]
24. Console, R.; Carluccio, R. Earthquake Simulators Development and Application. In *Statistical Methods and Modeling of Seismogenesis*; Wiley: New York, NY, USA, 2021; p. 27.
25. Console, R.; Carluccio, R.; Murru, M.; Papadimitriou, E.; Karakostas, V. Physics-Based Simulation of Spatiotemporal Patterns of Earthquakes in the Corinth Gulf, Greece, Fault System. *Bull. Seismol. Soc. Am.* **2022**, *112*, 98–117. [CrossRef]

26. Rundle, J.B.; Turcotte, D.; Donnellan, A.; Grant Ludwig, L.; Luginbuhl, M.; Gong, G. Nowcasting earthquakes. *Earth Space Sci.* **2016**, *3*, 480–486. [CrossRef]
27. Rundle, J.B.; Stein, S.; Donnellan, A.; Turcotte, D.L.; Klein, W.; Saylor, C. The complex dynamics of earthquake fault systems: New approaches to forecasting and nowcasting of earthquakes. *Rep. Prog. Phys.* **2021**, *84*, 076801. [CrossRef] [PubMed]
28. Varotsos, P.K.; Perez-Oregon, J.; Skordas, E.S.; Sarlis, N.V. Estimating the epicenter of an impending strong earthquake by combining the seismicity order parameter variability analysis with earthquake networks and nowcasting: Application in the Eastern Mediterranean. *Appl. Sci.* **2021**, *11*, 10093. [CrossRef]
29. Christopoulos, S.R.G.; Varotsos, P.K.; Perez-Oregon, J.; Papadopoulou, K.A.; Skordas, E.S.; Sarlis, N.V. Natural Time Analysis of Global Seismicity. *Appl. Sci.* **2022**, *12*, 7496. [CrossRef]
30. Varotsos, P.; Sarlis, N.; Skordas, E.; Uyeda, S.; Kamogawa, M. Natural-time analysis of critical phenomena: The case of seismicity. *Europhys. Lett.* **2010**, *92*, 29002. [CrossRef]
31. Varotsos, P.; Sarlis, N.V.; Skordas, E.S.; Uyeda, S.; Kamogawa, M. Natural time analysis of critical phenomena. *Proc. Natl. Acad. Sci. USA* **2011**, *108*, 11361–11364. [CrossRef] [PubMed]

MDPI
St. Alban-Anlage 66
4052 Basel
Switzerland
www.mdpi.com

Applied Sciences Editorial Office
E-mail: applsci@mdpi.com
www.mdpi.com/journal/applsci